Advances of Nanoparticles and Thin Films

Advances of Nanoparticles and Thin Films

Guest Editors

Bogdana Borca
Cristina Bartha

MDPI

Basel • Beijing • Wuhan • Barcelona • Belgrade • Novi Sad • Cluj • Manchester

Guest Editors

Bogdana Borca
Magnetism and
Superconductivity
National Institute of Materials
Physics
Magurele
Romania

Cristina Bartha
Magnetism and
Superconductivity
National Institute of Materials
Physics
Magurele
Romania

Editorial Office
MDPI AG
Grosspeteranlage 5
4052 Basel, Switzerland

This is a reprint of the Special Issue, published open access by the journal *Coatings* (ISSN 2079-6412), freely accessible at: https://www.mdpi.com/journal/coatings/special_issues/nanoparticles_films.

For citation purposes, cite each article independently as indicated on the article page online and as indicated below:

Lastname, A.A.; Lastname, B.B. Article Title. *Journal Name* **Year**, *Volume Number*, Page Range.

ISBN 978-3-7258-4301-5 (Hbk)
ISBN 978-3-7258-4302-2 (PDF)
https://doi.org/10.3390/books978-3-7258-4302-2

Cover image courtesy of Bogdana Borca and Cristina Bartha

Contents

About the Editors vii

Preface ix

Bogdana Borca and Cristina Bartha
Advances of Nanoparticles and Thin Films
Reprinted from: *Coatings* **2022**, *12*, 11, https://doi.org/10.3390/coatings12081138 1

Ovidiu Crisan, Alina Daniela Crisan, Gabriel Schinteie and Victor Kuncser
Highly Coercive $L1_0$-Phase Dots Obtained through Low Temperature Annealing for Nano-Logic Magnetic Structures
Reprinted from: *Coatings* **2023**, *13*, 2068, https://doi.org/10.3390/coatings13122068 4

Dana Gingasu, Daniela C. Culita, Jose Maria Calderon Moreno, Gabriela Marinescu, Cristina Bartha, Ovidiu Oprea, et al.
Synthesis of $CoFe_2O_4$ through Wet Ferritization Method Using an Aqueous Extract of Eucalyptus Leaves
Reprinted from: *Coatings* **2023**, *13*, 1250, https://doi.org/10.3390/coatings13071250 16

Viktor B. Ivanov, Elena V. Solina and Peter P. Levin
Photopolymerised Coatings with Nanopigments Based on Dye Mixtures
Reprinted from: *Coatings* **2023**, *13*, 300, https://doi.org/10.3390/coatings13020300 31

Cornel Staicu, Claudiu Locovei, Andrei Alexandru Dinu, Ion Burducea, Paul Dincă, Bogdan Butoi, et al.
Structural and Magnetic Specificities of Fe-B Thin Films Obtained by Thermionic Vacuum Arc and Magnetron Sputtering
Reprinted from: *Coatings* **2022**, *12*, 1592, https://doi.org/10.3390/coatings12101592 42

Catalin Palade, Adrian Slav, Ionel Stavarache, Valentin Adrian Maraloiu, Catalin Negrila and Magdalena Lidia Ciurea
Memory Properties of Zr-Doped ZrO_2 MOS-like Capacitor
Reprinted from: *Coatings* **2022**, *12*, 1369, https://doi.org/10.3390/coatings12091369 57

Anda Elena Stanciu, Gabriel Schinteie, Andrei Cristian Kuncser, Claudiu Locovei, Lucian Trupina, Nicusor Iacob, et al.
Magnetic Properties of Nanosized Fe and FeCo Systems on Trenched Mo Templates
Reprinted from: *Coatings* **2022**, *12*, 1366, https://doi.org/10.3390/coatings12091366 67

Petko Vitanov, Tatyana Ivanova, Hristosko Dikov, Penka Terziyska, Maxim Ganchev, Nikolay Petkov, et al.
Effect of a Discontinuous Ag Layer on Optical and Electrical Properties of ZnO/Ag/ZnOStructures
Reprinted from: *Coatings* **2022**, *12*, 1324, https://doi.org/10.3390/coatings12091324 82

Cezar Comanescu
Recent Advances in Surface Functionalization of Magnetic Nanoparticles
Reprinted from: *Coatings* **2023**, *13*, 1772, https://doi.org/10.3390/coatings13101772 97

Nicusor Iacob
Pitfalls and Challenges in Specific Absorption Rate Evaluation for Functionalized and Coated Magnetic Nanoparticles Used in Magnetic Fluid Hyperthermia
Reprinted from: *Coatings* **2025**, *15*, 345, https://doi.org/10.3390/coatings15030345 128

Petre Badica and Adam Lőrinczi
A Review on Preparation of Palladium Oxide Films
Reprinted from: *Coatings* **2024**, *14*, 1260, https://doi.org/10.3390/coatings14101260 . . . . **156**

Bogdana Borca
Advances in Organic Multiferroic Junctions
Reprinted from: *Coatings* **2024**, *14*, 682, https://doi.org/10.3390/coatings14060682 . . . . **188**

About the Editors

Bogdana Borca

Dr. Bogdana Borca is a senior scientist at the National Institute of Materials Physics, Romania. Her research is focused on characterizing and controlling the electronic and magnetic properties of nanoscale systems, including nanostructures, thin films, 2D materials, and molecular and metal–organic structures. She received her PhD degree in Physics in 2007 through a binational program between Joseph Fourier University (currently Grenoble Alpes), Grenoble, France, and Babeș-Bolyai University, Cluj-Napoca, Romania. Her research activity was carried out at Institut Néel, Grenoble, France, and was focused on the growth methods of nanostructures and on strategies to control their magnetic properties. From 2007 to 2011, Dr. Borca was a postdoctoral research fellow at Universidad Autónoma de Madrid, Spain, and from 2011 to 2015 at the Max Planck Institute for Solid State Research, Stuttgart, Germany. During this time, she investigated the electronic properties and functionalities of graphene on metals and of different organic molecules and nanostructures. Dr. Borca has also been a visiting scientist at the Institute for Applied Physics—Technische Universität Braunschweig, Germany; Physics of Interfaces and Nanomaterials Laboratory—University of Twente, Netherlands, and Peter Grünberg Institute—Forschungszentrum Jülich, Germany.

Cristina Bartha

Dr. Cristina Bartha is a research scientist at the National Institute of Materials Physics, Department of Magnetism and Superconductivity. She received her PhD in Physics from the Faculty of Physics of the University of Bucharest in 2011. Her work has consistently focused on understanding magnetic phenomena and magnetic field-induced effects in complex oxide materials such as perovskites, spinels, and garnets. These materials play a key role in emerging technologies including spintronics, biomedical applications, gas sensors, catalysts, and switching devices. Throughout her career, Dr. Bartha has collaborated with research institutions around the world. She was a DAAD Fellow at the Karlsruhe Institute of Technology (Germany) and has carried out research fellowships at several prestigious institutions, including Purdue University (USA), the Institute of Applied Chemistry Berlin, the UGC-DAE Consortium for Scientific Research (India), the University of Pavia (Italy), and Instituto Tecnológico e Nuclear (Portugal). Her innovative work has been recognized at international invention fairs, and she has received the Excellence Award and Gold Medal at PRO INVENT 2013 and the Silver Medal at EUROINVENT 2024. Beyond her research, she has been actively involved in organizing several national and international conferences.

Preface

Nanoparticles and thin films represent key areas in modern materials science, offering vast potential for technological applications due to their unique size-dependent properties. Originating from the visionary concept of nanotechnology, research in these fields has enabled the design and fabrication of advanced materials using both top-down and bottom-up approaches. This reprint brings together a series of studies and reviews exploring the latest developments in nanoparticle and thin-film systems. By synthesizing a wide range of research, we aim to provide a comprehensive overview of the field, focusing on both the theoretical principles and practical applications of these advanced materials. The reprint is designed as a valuable resource for researchers, engineers, and students interested in understanding the fundamental properties of nanoparticles and thin films, as well as the latest techniques used to fabricate and apply them in real-world contexts. A comprehensive article on these topics is *Advances of Nanoparticles and Thin Films.*

Recent progress in the development of nanostructured magnetic materials and functional thin films has opened new opportunities for applications in spintronics, electronics, and biomedicine. Innovative fabrication techniques, such as electron-beam lithography, have enabled the study *Highly Coercive L10-Phase Dots Obtained through Low Temperature Annealing for Nano-Logic Magnetic Structures* and the creation of precisely patterned magnetic FePt nanodots with enhanced structural and magnetic properties, for example, with a strong coercivity and saturation. These advancements demonstrate the potential for producing high-performance nanomagnetic components for nano-logic device integration using more efficient and cost-effective processes.

In the article *Synthesis of $CoFe_2O_4$ through Wet Ferritization Method Using an Aqueous Extract of Eucalyptus Leaves*, a green synthesis approach for magnetic nanoparticles was developed using plant-based extracts, offering an environmentally friendly alternative to conventional chemical methods. The resulting nanoparticles exhibit a well-defined crystalline structure and favorable magnetic properties, along with biofunctional characteristics such as antimicrobial and antibiofilm activity. This sustainable approach highlights the potential of natural resources in the fabrication of advanced materials for biomedical and environmental applications.

A new approach to pigment development was presented in the paper *Photopolymerized Coatings with Nanopigments Based on Dye Mixtures,* which uses dye mixtures combined with modified layered materials to produce nanopigments with improved color properties. These pigments exhibit improved stability under light exposure and better adhesion when integrated into coating or ink formulations. Structural and optical analyses confirm their nanoscale architecture and compatibility with various substrates, making them suitable for high-performance printing and coating applications.

The article *Structural and Magnetic Specificities of Fe-B Thin Films Obtained by Thermionic Vacuum Arc and Magnetron Sputtering* explores the structural and magnetic properties of Fe-B thin films, focusing on how different fabrication techniques and material compositions impact their performance. A comparative analysis of deposition methods revealed that film morphology, phase composition, and magnetic behavior are significantly affected by the synthesis route. Boron content plays a key role in promoting amorphization and enhancing functional properties such as wear resistance and magnetic anisotropy, making these films promising candidates for a wide range of advanced technological applications.

Advancements in memory technology have demonstrated that metal oxide semiconductor-like capacitors incorporating zirconia-based trilayer structures offer promising performance for non-volatile data storage. Through tailored fabrication techniques and thermal treatments presented in the article *Memory Properties of Zr-Doped ZrO_2 MOS-like Capacitor*, the structural and electrical properties of these devices can be significantly enhanced. Key factors such as crystallization behavior and defect engineering contribute to improved memory windows and long-term retention stability, highlighting their potential for integration into next-generation memory applications.

Nanostructured magnetic films of Fe and FeCo have shown highly tunable magnetic properties which are significantly influenced by their morphology and structural design, as demostrated in *Magnetic Properties of Nanosized Fe and FeCo Systems on Trenched Mo Templates*. By engineering features such as surface patterning and material composition, it is possible to control magnetic anisotropy and the behavior of magnetic domains across different temperature ranges. These systems exhibit complex interactions at low temperatures and more defined directional magnetization at ambient conditions, making them valuable for advanced magnetic technologies where precise control over magnetic behavior is essential.

Transparent conducting oxide structures based on multilayered ZnO/Ag/ZnO nanolaminates have been developed in *Effect of a Discontinuous Ag Layer on Optical and Electrical Properties of ZnO/Ag/ZnO Structures* using low-temperature deposition techniques, resulting in films with excellent optical transparency and electrical conductivity. The inclusion of conductive nanoclusters within the structure plays a crucial role in enhancing performance. These films demonstrate strong potential for use in optoelectronic and flexible electronic devices, offering a scalable, energy-efficient, and reliable alternative to traditional transparent conductors.

The article *Recent Advances in Surface Functionalization of Magnetic Nanoparticles* reviews recent progress in the surface functionalization of magnetic nanoparticles that have significantly expanded their capabilities across a range of applications, including imaging, catalysis, and targeted drug delivery. By employing techniques such as ligand modification, bioconjugation, and encapsulation, researchers have achieved greater control over particle behavior, improving their specificity, biocompatibility, and functional performance. These developments have enhanced the role of magnetic nanoparticles in both medical and technological fields, highlighting the importance of continued innovation in surface engineering to fully unlock their potential in interdisciplinary applications.

Furthermore, magnetic nanoparticles can be used in magnetic hyperthermia to generate localized heat within tumor tissues under radiofrequency fields and has gained recognition as a promising approach for cancer treatment. Beyond thermal effects, recent research has revealed that magnetic nanoparticles can also induce mechanical disruptions at the cellular level, particularly when exposed to low-frequency magnetic fields. Functionalizing these nanoparticles with specific molecules enhances targeting precision, allowing for more effective interaction with tumor cells. However, assessing treatment efficiency remains challenging due to variability in measuring techniques and the complexity of linking macroscopic heat generation metrics with nanoscale biological responses. These insights have been presented in *Pitfalls and Challenges in Specific Absorption Rate Evaluation for Functionalized and Coated Magnetic Nanoparticles Used in Magnetic Fluid Hyperthermia* and highlight the need for standardized evaluation methods and a deeper interdisciplinary understanding of both thermal and mechanical effects to fully exploit the potential of magnetic hyperthermia in clinical applications.

Recent developments in the fabrication of palladium oxide thin films have highlighted their growing potential in gas sensing applications across various industries. Distinct electrical properties of PdO offer advantages over traditional metal oxides, making it a promising material for detecting specific gases. In *A Review on Preparation of Palladium Oxide Films*, a wide range of synthesis techniques has been explored, leading to films with varied structural characteristics. However, challenges remain in achieving uniform and well-controlled film quality, as well as in understanding how different fabrication methods influence performance. While greener synthesis approaches are emerging, they often produce nanoparticles rather than continuous films. To fully unlock the potential of PdO in sensing technologies, further efforts are needed to refine deposition methods and develop efficient, scalable characterization tools.

The review *Advances in Organic Multiferroic Junctions* presents recent advances of organic multiferroic junctions as promising multifunctional components that combine magnetic and electric functionalities within a single flexible platform. By integrating ferroelectric and ferromagnetic materials, these structures exhibit tunable resistance behaviors in response to external stimuli, enabling applications in nonvolatile memory, sensing, and neuromorphic computing. Their flexibility, low power consumption, and potential for use in sustainable electronics make them attractive for next-generation devices. Ongoing research is focused on exploring new material combinations and improving interface design to enhance performance and ensure stability, thus facilitating broader implementation in advanced electronic systems.

Bogdana Borca and Cristina Bartha
Guest Editors

coatings

MDPI

Editorial

Advances of Nanoparticles and Thin Films

Bogdana Borca * and Cristina Bartha *

National Institute of Materials Physics, Atomistilor 405 A, 077125 Magurele, Romania
* Correspondence: bogdana.borca@infim.ro (B.B.); cristina.bartha@infim.ro (C.B.)

Nanoparticles and thin films are currently among the most active research fields in materials sciences for technological applications. "Nanotechnology", a concept first introduced in 1959 by Nobel laureate, Richard P. Feynman in his visionary lecture entitled "There's Plenty of Room at the Bottom" [1], is an essential pillar in technological development resulting in nanometers scale systems (nanoparticles and quantum systems, nanostructures and heterostructures, thin films and 2D materials, etc.). The interest in nanoscale systems arises not only from the purpose of miniaturization, but also from the fact that new properties are emerging at this length scale and that these properties change with their size, surface-to-volume ratio, and shape. The understanding and control of nanoscale properties has enabled scientists and engineers to design, theoretically model, produce, and characterize materials and advanced functional and multifunctional devices with current relevance. They will continue to contribute to many applications and an enormous body of research (e.g., information technology, electronics, spintronics, displays, memory units, sensors, biosensors, actuators, active surfaces with different characteristics, catalysis, energy harvesting, energy storage, environmental and safety concerns, healthcare, bioengineering, medicine, the drug industry, etc.) [2–5]. The development of these systems of organic or inorganic nanomaterials also induced the advancement of the technical equipment and methodologies for their production and characterization, even down to the atomic and single molecular scale, using, for example, tools such as Scanning Tunneling Microscopy [6,7]. The nanofabrication methods are classified in top-down and bottom-up approaches, which are based on physical and chemical and on dry and wet processes. The top-down methods are scaling-down techniques that imply the division of the bulk material into nanoscale structures following the physical routes, or lithography (including the standard: photolithography, phase-shift optical lithography, X-ray lithography, electron-beam lithography, focused-ion-beam lithography, and neutral-atomic-beam lithography, and softer ways: microcontact printing, nanoimprint, molding, and dip-pen lithography [8]) and chemical routes (including procedures such as: templated etching, selective dealloying, anisotropic dissolution, and thermal decomposition [9]). The bottom-up methods are scaling-up techniques that are based on self-assembly. They rely on the assemblage of atomic or molecular building-block units into larger structures, such as systems of nanowires [10,11] or 2D organic structures [12–14] driven by physical and chemical forces. All of these procedures involve physical deposition techniques (including physical vapor deposition techniques (PVD) such as vacuum thermal deposition, electron-beam deposition, laser-beam deposition, arc evaporation, molecular beam epitaxy, organic molecular beam epitaxy, ion plating evaporation, and sputtering methods) and chemical deposition techniques (including the sol-gel technique; chemical bath deposition, such as deep coating, spin-coating, and Langmuir–Blodgett deposition; the spray pyrolysis technique; plating, such as electroplating and electroless deposition; and chemical vapor deposition (CVD), including low-pressure CVD and plasma-enhanced CVD) [15].

The consistency of nanoparticle research is derived from the attempt to tailor new materials with desired properties via scrupulous consideration related to electronic phenomena induced by multiple valence ions, their location in the peculiar structure, the use of cutting-edge preparation procedures, and the ability to discern all appearing effects via a directional selection of the complementary investigation methods. It is known that synthesis methods are vital to the design of nanoparticles. The most used techniques for this type of material are chemical methods [2]. A disadvantage of these methods is the use of precursors that are most often toxic, thereby limiting the scope of the application of the material. Since many nanoparticles are used in medicine and/or the cosmetic industry, new synthesis methods (the so-called green chemistry) have been developed that use safe and environmentally friendly nontoxic reagents [16,17]. A modern trend in the design of nanomaterials, especially magnetic nanomaterials, is the formation of hybrid structures of nanoparticle–polymer types that allow the functionalization and control of properties through the structure and composition of the polymer [18]. Nanoparticle biosynthesis is an ecological, green, and non-toxic method of processing that involves microorganisms. It is particularly used in the synthesis of iron oxide, silver, nickel oxide, copper oxide nanomaterials, etc. [19–21]. The significant progress in recent years is linked on the one hand to the development and refinement of nanoparticle synthesis methods, and on the other hand, to the development of more complex characterization techniques and specific and efficient analysis methodologies. The experimental development coupled with the development of theoretical models allowed the elucidation of unexpected mechanisms and the proposal of new solutions that would lead to materials with high-performance properties in relation to the various applications. Finally, optimization and reproducibility are two fundamental concepts for the consistency of new phenomena and/or mechanisms.

Author Contributions: B.B. and C.B. contributed equally to the conceptualization, preparation, writing, and editing of the actual editorial article. All authors have read and agreed to the published version of the manuscript.

Funding: This work was developed within the scope of the projects: PN-III-P2-2.1-PED-2021-0378 (contract no. 575PED/2022) and PN-III-P2-2.1-PED-2021-2007 (contract no. 676PED/2022) that are funded by the Romanian Ministry of Research, Innovation, and Digitalization through UEFSCDI.

Institutional Review Board Statement: Not applicable.

Informed Consent Statement: Not applicable.

Data Availability Statement: Not applicable.

Conflicts of Interest: The authors declare no conflict of interest.

References

1. Feynman, R.P. There's Plenty of Room at the Bottom. *Eng. Sci.* **1960**, *23*, 22–36.
2. Nikam, A.V.; Prasad, B.L.V.; Kulkarni, A.A. Wet Chemical Synthesis of Metal Oxide Nanoparticles: A Review. *Cryst. Eng. Commun.* **2018**, *20*, 5091–5107. [CrossRef]
3. Dharmadasa, I.M. Advanced Thin Film Materials for Photovoltaic Applications. *Coatings* **2020**, *10*, 562. [CrossRef]
4. Botas, A.M.P. New Frontiers in Novel Optical Materials and Devices. *Coatings* **2022**, *12*, 856. [CrossRef]
5. Gandhi, A.C. Synthesis and Characterization of Functional Magnetic Nanomaterials. *Coatings* **2022**, *12*, 857. [CrossRef]
6. Schendel, V.; Barreteau, C.; Brandbyge, M.; Borca, B.; Pentegov, I.; Schlickum, U.; Ternes, M.; Wahl, P.; Kern, K. Strong Paramagnon Scattering in Single Atom Pd Contacts. *Phys. Rev. B* **2017**, *96*, 035155. [CrossRef]
7. Michnowicz, T.; Borca, B.; Pétuya, R.; Schendel, V.; Pristl, M.; Pentegov, I.; Kraft, U.; Klauk, H.; Wahl, P.; Mutombo, P.; et al. Controlling Single Molecule Conductance by a Locally Induced Chemical Reaction on Individual Thiophene Units. *Angew. Chem. Int. Ed.* **2020**, *59*, 6207. [CrossRef] [PubMed]
8. Cao, G. *Nanostructures and Nanomaterials. Synthesis, Properties & Applications*; ICP Imperial College Press: London, UK, 2004; ISBN 1-86094-480-9.
9. Yu, H.-D.; Regulacio, M.D.; Yea, E.; Han, M.-Y. Chemical Routes to Top-Down Nanofabrication. *Chem. Soc. Rev.* **2013**, *42*, 6006. [CrossRef] [PubMed]
10. Borca, B.; Frucharta, O.; David, P.; Rousseau, A.; Meyer, C. Kinetic Self-Organization of Trenched Templates for the Fabrication of Versatile Ferromagnetic Nanowires. *Appl. Phys. Lett.* **2007**, *90*, 142507. [CrossRef]
11. Borca, B.; Fruchart, O.; Kritsikis, E.; Cheynis, F.; Rousseau, A.; David, P.; Meyer, C.; Toussaint, J.-C. Tunable Magnetic Properties of Arrays of Fe(110) Nanowires Grown on Kinetically Grooved W(110) Self-Organized Templates. *J. Magn. Magn. Mater.* **2010**, *322*, 257. [CrossRef]

12. Barja, S.; Stradi, D.; Borca, B.; Garnica, M.; Díaz, C.; Rodriguez-García, J.M.; Alcamí, M.; de Parga, A.L.V.; Martín, F.; Miranda, R. Ordered Arrays of Metal-Organic Magnets at Surfaces. *J. Phys. Condens. Matter.* **2013**, *25*, 484007. [CrossRef] [PubMed]
13. Socol, M.; Trupina, L.; Galca, A.; Chirila, C.; Stan, G.E.; Vlaicu, A.; Stanciu, A.E.; Boni, A.G.; Botea, M. Anca Stanculescu Electro-Active Properties of Nanostructured Films of Cytosine and Guanine Nucleobases. *Nanotechnology* **2021**, *32*, 415702. [CrossRef] [PubMed]
14. Oyola-Reynoso, S.; Wang, Z.; Chen, J.; Çınar, S.; Chang, B.; Thuo, M. Revisiting the Challenges in Fabricating Uniform Coatings with Polyfunctional Molecules on High Surface Energy Materials. *Coatings* **2015**, *5*, 1002–1018. [CrossRef]
15. Jilani, A.; Abdel-wahab, M.S.; Hammad, A.H. Advance Deposition Techniques for Thin Film and Coating. In *Modern Technologies for Creating the Thin-Film Systems and Coatings*; Nikitenkov, N.N., Ed.; IntechOpen: Rijeka, Croatia, 2017.
16. Gingasu, D.; Mindru, I.; Culita, D.C.; Calderon-Moreno, J.M.; Bartha, C.; Greculeasa, S.; Iacob, N.; Preda, S.; Oprea, O. Structural, morphological and magnetic investigations on cobalt ferrite nanoparticles obtained through green synthesis routes. *Appl. Phys. A* **2021**, *127*, 892. [CrossRef]
17. Gingasu, D.; Mindru, I.; Ianculescu, A.-C.; Diamandescu, L.; Surdu, V.; Marinescu, G.; Bartha, C.; Preda, S.; Popa, M.; Chifiriuc, M.C. Soft Chemistry Synthesis and Characterization of CoFe1.8RE0.2O4 (RE3+ = Tb3+, Er3+) Ferrite. *Magnetochemistry* **2022**, *8*, 12. [CrossRef]
18. Li, Y.; Wang, N.; Huang, X.; Li, F.; Davis, T.P.; Qiao, R.; Ling, D. Polymer Assisted Magnetic Nanoparticles Assemblies for Biomedical Applications. *ACS Appl. Bio Mater.* **2020**, *3*, 121–142. [CrossRef] [PubMed]
19. Madubuonu, N.; Aisida, S.O.; Ali, A.; Ahmad, I.; Zhao, T.; Botha, S.; Maaza, M.; Ezema, F.I. Biosynthesis of Iron Oxide Nanoparticles via a Composite of Psidium Guavaja-Moringa Oleifera and Their Antibacterial and Photocatalytic Study. *J. Photochem. Photobiol. B Biol.* **2019**, *199*, 111601. [CrossRef] [PubMed]
20. Aisida, S.O.; Madubuonu, N.; Alnasir, M.H.; Ahmad, I.; Botha, S.; Maaza, M.; Ezema, F.I. Biogenic Synthesis of Iron Oxide Nanorods Using Moringa Oleifera Leaf Extract for Antibacterial Applications. *Appl. Nanosci.* **2020**, *10*, 305–315. [CrossRef]
21. Ugwoke, E.; Aisida, S.O.; Mirbahar, A.A.; Arshad, M.; Ahmad, I.; Zhao, T.K.; Ezema, F.I. Concentration Induced Properties of Silver Nanoparticles and Their Antibacterial Study. *Surf. Interfaces* **2020**, *18*, 100419. [CrossRef]

coatings

MDPI

Article

Highly Coercive L1$_0$-Phase Dots Obtained through Low Temperature Annealing for Nano-Logic Magnetic Structures

Ovidiu Crisan, Alina Daniela Crisan, Gabriel Schinteie and Victor Kuncser *

National Institute for Materials Physics, P.O. Box MG-7, Magurele, 077125 Bucharest, Romania; ocrisan@yahoo.com (O.C.); a_crisan2002@yahoo.com (A.D.C.); schinteie@infim.ro (G.S.)
* Correspondence: kuncser@infim.ro

Abstract: Nano-logic magnetic structures are of great interest for spintronic applications. While the methods used for developing arrays of magnetic L1$_0$-phase dots are, in most cases, based on deposition followed by annealing at high temperatures, usually around 700 °C, we demonstrate here a technique where a much lower annealing temperature (i.e., 400 °C) is needed in order to promote fully the disorder–order phase transformation and achievement of highly coercive L1$_0$-phase dots. In order to develop building blocks based on arrays of L1$_0$-phase FePt dots for further spintronic applications, an engraving technique using electron beam lithography is employed. This paper describes the fabrication, as well as the morphological and magnetic characterization, of regularly placed FePt dots of various shapes, as pre-requisites for integration into nano-logic devices. As a proof of concept, regular arrays of FePt circular dots were devised and their structural characterization, using X-ray diffraction (XRD) and transmission electron microscopy (TEM), was performed. It has been shown that annealing at only 400 °C for 30 min proved the occurrence of the tetragonal L1$_0$ phase. Moreover, structural characterization showed that the disorder–order phase transformation was complete with only the L1$_0$ phase detected in high resolution TEM. The magnetic characterization provided more insight into the potential of such arrays of magnetic devices with convenient values of magnetic coercivity, remanent and saturation magnetization. These findings show good potential for developing regular arrays of uniformly shaped magnetic entities with encouraging magnetic performances in view of various applications.

Keywords: regular arrays; FePt; L1$_0$ phase; magnetic nanostructures

1. Introduction

There has been a surge of interest in the investigation of size-dependent phenomena as a response to the industry needs for highly performant nanodevices. The recent award of the Nobel Prize in Chemistry 2023 to the pioneers of the development of quantum dots fabrication [1] shows how important and necessary is the study of size-dependent phenomena at nanoscale. A special case is of the nanoscale-structured thin films and coatings. Of particular interest is the magnetic dots or nanostructures that are suitable for spintronic applications as well as for heated-dot magnetic recording (HDMR). The formation of hybrid nanostructures in spintronic devices has been recently investigated [2] as a response to specific needs, especially in nanodiagnostics, nanoelectronics and adjacent fields, where combined optical and magnetic responses to various excitations is required for various types of sensing. With the advent of high-accuracy and high-resolution fabrication technologies, such as lithography, coupling phenomena at the nanoscale in magnetic dots and in arrays of dots may become accessible. The integration of magnetic and semiconductor components adds new capabilities to electronic devices [3]. While spin phenomena have long been investigated within the context of conventional ferromagnetic materials, the study of spin generation, relaxation and spin-orbit coupling in non-magnetic materials has taken off only recently with the advent of hybrid spintronics. Therefore, many novel materials and associated nanoarchitectures can find their greatest potential in both science and technology. On the other hand, the heated-dot magnetic recording (HDMR) principle, derived from the heat-assisted magnetic recording (HAMR), was shown to be a targeted application where a key system is represented by FePt nanoclusters [4,5]. In this view, Mohan et al. [4] and Medwal et al. [5] have proposed the fabrication of self-stabilized FePt non-stoichiometric clusters by a chemical route, i.e., reduction using oleic acid and oleylamine. An annealing at 600 °C transformed these FePt colloidal nanoparticles into carbon-coated $L1_0$ FePt-phase nanoclusters. As proposed by Seagate a few years ago [6], the heated dot magnetic recording (HDMR) is a promising technology suitable for increasing the density of media used for magnetic recording to values around 4 Tb/in^2. It is indeed true that the heated-dot magnetic recording technology needs arrays of magnetic dots, periodically dispersed, non-interacting and well within the nanometer-size range, with stable magnetic response and high uniaxial perpendicular magnetic anisotropy. While the chemical route described in [4,5] provides non-interacting, small nanoparticles, their carbonaceous coating can impede the recording processes by diminishing the magnetic response. The lithographical route proposed in [6] has the advantage of providing regular disposal of these small nanoclusters. Here, the media is made of a continuously deposited FePt film and a lithographical path is employed to provide and define regular arrays of periodically disposed magnetic dot, which are subsequently heated to achieve the $L1_0$ FePt phase.

A systematic study on the size dependence of the coercivity in FePt magnetic dots has been developed in [7,8]. In these works, FePt nanodot arrays were fabricated by film pattering via electron beam lithography followed by ion etching. Their magnetic properties were investigated using the technique of X-ray magnetic circular dichroism combined with magneto-optic Kerr effect (MOKE) magnetometry. It was shown that circular magnetic dots with thickness of 6 nm and sizes between 30 and 100 nm exhibit a coercivity increase at decreasing sizes upon annealing at quite high temperatures [8]. However, it was found that not all of the dots were achieving the $L1_0$ FePt phase and, in addition, defects at the damaged edges may drastically influence the magnetic properties.

At variance to the above-mentioned fabrication route, the initial pre-pattering of a photoresist with the subsequent deposition and processing of the magnetic material is also a reliable solution, developed especially for biosensing applications. In this respect, integrated platforms and arrays of sensors where the detection is guided by magneto-transport phenomena are used, without the need for perpendicular anisotropy in the ferromagnetic layers. For instance, pre-patterned media was used for proving the efficiency of giant magnetoresistance (GMR) spin valve sensors designed for the detection of superparamagnetic nanoparticles as potential biomolecular labels in magnetic biodetection technology [9]. In this example, a patterned self-assembly method, involving a polymer-mediated process via e-beam lithography, was used. Other applications of integrated magnetic/spintronic device microarrays in biomedical research are reported in [10–12].

Due to the possibilities for the manipulation of spin configuration in magnetic dots fabricated by nanolithography, there is a high potential for making nanodevices an alternative to the presently employed sensors, but ones that are more sensitive and with lower power and lower costs [11].

A lot of the recent studies of FePt nanodots report on other various issues and the applications of such systems. For instance, in [13], the FePt magnetic dots were considered in the case of heated-dot magnetic recording (HDMR) applications. Moreover, some coercivity enhancement was recently reported in the P-doped FePt prepared by electrodeposition and subsequent annealing [14]. In another simulation study, the $L1_0$ FePt dots were investigated for their potential applicability in bit patterned media [15].

The current paper describes the fabrication and morpho-structural and magnetic characterization of arrays of magnetic dots of various shapes, made of non-stoichiometric $L1_0$ FePt, as building blocks to be further used as integration platforms for nano-logic devices. Such devices with various shapes and geometries and with convenient magnetic responses are suitable for both bio-sensing and potentially for HDMR applications. The pre-patterning route of a photoresist followed by the deposition of the metallic layers was found to be very suitable for a subsequent thermal treatment at a relatively low temperature. Given the expected influence of the damaged edges of the dots on the magnetic properties, the specific parameters of the engraving process were also investigated.

2. Materials and Methods

2.1. Method of Pre-Patterning and Magnetic Elements Nanofabrication

The building block needed for the integration platform was made by using electron beam lithography. The specific steps involved in the engraving technique are the standard ones. They are based on an e-beam exposure of a resist layer deposited onto the desired substrate. The e-beam lithography system uses hardware typical for the scanning electron microscopy for precise guiding of an electron beam, of nanometer size, onto the substrate, so as to form a latent image in the resist layer. The consequence of the exposure is to render the resist either more soluble (positive tone resist) or less soluble (negative tone resist) in the developer solution. The resulting pattern is furthermore transferred via etching or by depositing other materials on top of the remaining pattern. By iterating a number of steps of this type, complex structures of very short length scales can be built up. The envisaged nanostructures were fabricated in the dedicated class 1000 cleanroom. The fabrication tool used was a Raith150-Two high resolution e-beam lithography machine, dotted with Wet Etch cabinets for acids and bases/solvents, photolithography cabinet with PC-controlled spinner and hot plates, a mask aligner EVG 620 NT and a scanning electron microscopy SEM Hitachi S3400 with laser interferometer positioning stage from Raith (Raith GmbH, Dortmund, Germany). In order to obtain magnetic nanostructures with specific shapes and geometries, we considered the electron beam lithography technique. For this purpose, two photoresist layers were applied to a Si(111) substrate with a native SiO_2 layer on top, i.e., methyl methacrylate (MMA EL10) and poly-methyl methacrylate (PMMA A3). Each layer was deposited via spin-coating for 60 s, at a speed of 4000 rpm and acceleration of 500 rpm/s. After the deposition of each layer, the sample was heated at 175 °C for 10 s in order to evaporate the solvent. Nanofabrication of the magnetic element was considered, in the first instance, in engraving high aspect ratio wires and, secondly, in engraving the desired shapes in the photoresist, followed by the deposition of metallic layers. The engraving was realized via electron beam lithography using a current of 15 pA. The size of the electron beam was 2 nm, whereas the field magnification was 800×. It is known that in electron beam lithography, the resist performance in terms of exposure dose latitude, sensitivity, contrast, roughness and resolution are influenced by the nature of the resist, the developer type and composition and the development technique. In our case, the dose was incremented with 0.1 $\mu C/cm^2$ starting from 0.6 $\mu C/cm^2$ and ending at 0.79 $\mu C/cm^2$. The development was realized keeping the sample in a solution of methyl isobutyl ketone (MiBK) and isopropyl alcohol (IPA) (1:3) for 8 s and then in IPA for 1 min. Finally, the sample was dried with nitrogen gas flow. After the development, the reactive ion etching technique was implemented in order to produce a better definition of the patterned shapes. A 10 s reactive ion etching process was realized in the dedicated UHV chamber at a pressure of 15 mbar.

A layer of Pt with 30 nm thickness was deposited over the engraved desired shapes, seconded by a layer of Fe with 40 nm thickness. The deposition rate in the case of Pt was 0.25 nm/s, whereas the value corresponding to Fe deposition was 0.19 nm/s. The working pressure inside the deposition chamber was 10^{-7} mbar. The lift-off process was realized by placing the sample in acetone for 2 min at 50 °C and, afterwards, in acetone at room temperature for 30 min. After the removal of the photoresist, the nanostructure was furthermore annealed at 400 °C for 30 min in order to promote the atomic inter-diffusion and the formation of the $L1_0$ FePt phase.

2.2. Characterization Methods

Several investigation techniques were then used to study the structural and morphological aspects of dots as well as the phase evolution with the annealing temperature. Also, a thorough study of magnetic properties in various stages of annealing as a function of applied field or as a function of temperature were performed. The morpho-structural studies were performed via X-ray diffraction (XRD), scanning electron microscopy (SEM) and high-resolution transmission electron microscopy (HRTEM) coupled with selected area electron diffraction (SAED). For XRD, a Bruker D8 Advance (Bruker AXS GmbH, Karlsruhe, Germany) with Cu Kα radiation wavelength of 1.54 Å was used. For the SEM imaging, Lyra 3XMU with Focused Ion Beam SEM-FIB sample preparation stage, from Tescan (Tescan France, Fuveau, France), was used. The facility is equipped with secondary electron detectors with a resolution of 5 nm at 30 kV. Transmission electron microscopy in high resolution mode was recorded using a JEM ARM 200 F electron microscope operated at 200 kV acceleration voltage, from JEOL (JEOL Ltd., Hertfordshire, UK). The magnetic characterization was performed via a SQUID (Superconducting QUantum Interference Device) unit of a MPMS (Magnetic Properties Measurement System) from Quantum Design (Quantum Design Europe GmbH, Darmstadt, Germany), under magnetic applied fields of up to 12 Tesla. The measurements were taken both in parallel and perpendicular geometry and at temperatures ranging from 5 K to 300 K. The parallel and perpendicular refers to the direction of the applied magnetic field with respect to the array plane.

3. Results and Discussion

3.1. Complex Morphology Analysis

Scanning electron microscopy (SEM) images of the obtained engraved hallow nanostructures have been collected. The first set of engraved wires, with a depth of approximately 200 nm, have been obtained in a sequence of exposure doses ranging from 0.60 $\mu C/cm^2$ up to 0.79 $\mu C/cm^2$. The SEM image presented in Figure 1a exhibits these traits in which the doses are increasing from the top of the image to the bottom. The best formed trait, shaped as rectangle with an optimal aspect ratio, was found to occur for an exposure dose of about 0.67 $\mu C/cm^2$. A second set of engraved wires, with a depth of 100 nm obtained under exposed doses ranging from 0.76 to 0.79 $\mu C/cm^2$, are shown in Figure 1b.

In order to obtain regular shapes of various types of magnetic structures (rectangles, triangles and circular) using electron beam lithography, reactive ion etching was implemented to improve the definition of the patterned hallow shapes. The considered hallow structures engraved on photoresist films of 200 nm thickness consist of circular structures with diameters of 200 nm with different spacing (Figure 1c), equilateral triangles with characteristic lengths of 200 nm and 400 nm, respectively (Figure 1d), and simple circular structures with diameters of 335 nm as well as rectangular structures formed by overlapping the circular structures on the y direction over 200 nm (Figure 1e). The photoresist application, the metal deposition on the hallow structures and the lift-off procedures were similar across all the deposited samples. For each hallow shape, different dose values were considered from 0.6 $\mu C/cm^2$ to 0.79 $\mu C/cm^2$, with 0.1 $\mu C/cm^2$ discretization.

Finally, Figure 1e shows a regular array of FePt circular dots obtained after the lift-off process following the growth of the successive Pt and Fe layers according to the previous section, on the array of engraved circular structures with diameters of 335 nm.

We have further chosen these regular arrays of FePt circular magnetic dots, for structural and magnetic characterization, considering the potential of such nanostructures to be developed into an array of nanodevices, individually addressable. The same annealing treatment was applied for the considered magnetic nanostructures in order to induce the $L1_0$ FePt phase of high coercivity. The influence of the dose value during the engraving process on the phase formation was also investigated.

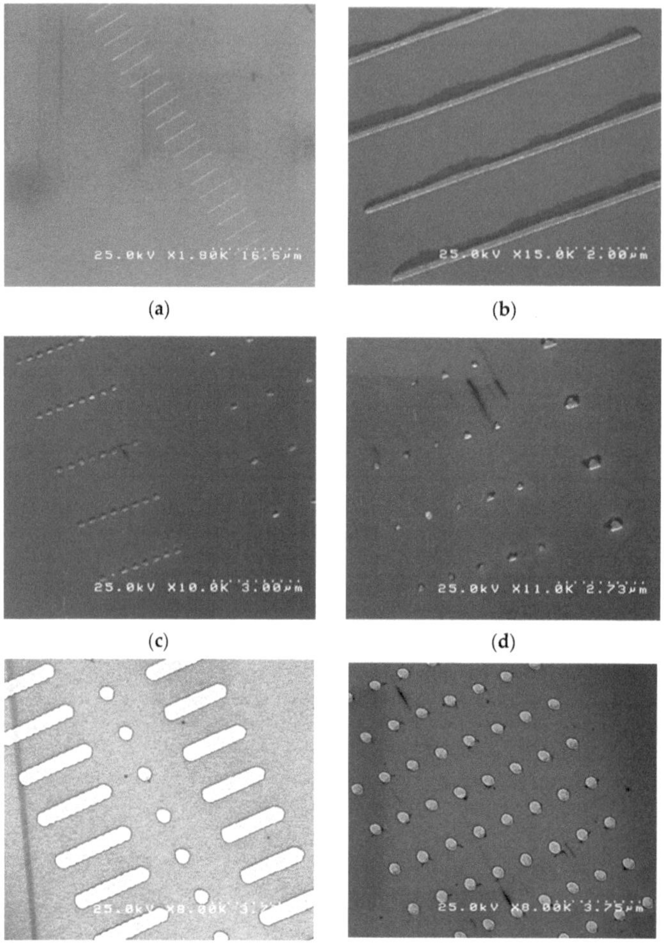

Figure 1. (**a**) SEM image of engraved wires with a depth of 200 nm, obtained at different doses ranging from 0.6 μC/cm^2 (**up**) to 0.79 μC/cm^2 (**down**). The ruler on bottom right represents the overall scale of the image. (**b**) Engraved wires with a depth of 100 nm, each one corresponding to a dose in the interval 0.76–0.79 μC/cm^2 in descending order from top to bottom. (**c**) Engraved circular structures obtained at various doses in the interval 0.76–0.79 μC/cm^2 in descending order from top to bottom. On the left side are circular structures with an exposed diameter of 200 nm spaced at a distance of 200 nm, while on the right side are circular structures with an exposed diameter of 200 nm spaced at a distance of 1200 nm. (**d**) Equilateral FePt triangles with exposed characteristic dimension of 200 nm spaced at a distance of 1200 nm corresponding to a dose in the interval 0.76–0.79 μC/cm^2 in descending order from top to bottom (on the **left side**) and equilateral triangles with exposed characteristic dimensions of 400 nm spaced at a distance of 1200 nm corresponding to the same doses (**right side**). (**e**) Rectangular hallow structures formed via circular structures with the exposed diameter of 335 nm, overlapped over 200 nm (column of structures on the left and on the right of the image); circular structures with exposed diameter of 335 nm (central column). (**f**) Regular array of FePt circular magnetic dots, 335 nm in size, spaced at about 1200 nm in both x and y directions. Exposure dose was about 0.79 μC/cm^2.

The creation of a wide area of magnetic dots or nanostructures, regularly dispersed and integrated onto functionalized substrates with logic capabilities, would allow for the fabrication of complex architecture structures where the magnitude and direction of magnetic moments may be individually addressed and exploited for applications in the fields of magnetic nanosensors and molecular detection.

Systems exhibiting giant magnetoresistive effect (GMR) materials are generally made of conductive regions, alternated with or separated by magnetic regions [16,17]. In a trilayer-like GMR system, a conductive layer is sandwiched between two ferromagnetic layers of different magnetization reversal mechanisms. A spin reconfiguration in the ferromagnetic layers due to the application of a magnetic field causes the spin-dependent scattering of conduction electrons moving through the nm thick conductive layer. That means the resistance of the conductive film in the trilayer spin valve system switches from a low value in cases of parallel spin configuration in the two neighboring ferromagnetic layers to a higher value in cases of an antiparallel spin configuration in the two ferromagnetic layers. Usually, the different magnetic reversal in the two ferromagnetic layers can be induced by pinning one of the layers to an antiferromagnetic one, with the generation of a unidirectional anisotropy and an exchange bias effect at the level of the pinned layer. Alternatively, one of the layers could have a much higher coercive field than of the second layer with soft magnetic properties (low coercive field) [18]. The presence of a small external magnetic field will easily change the magnetic configuration of the soft magnetic layer and, so, it can be easily detected by the changes in the resistivity of the spin valve system. This change in resistivity will be induced either in the case of in plane or out of plane anisotropy of the low coercive free and the high coercive pinned layers, the only condition being to have the driving field coplanar with the anisotropy axis in the layers. In this respect, the $L1_0$-phase of FePt films of high coercivity can be successfully used in spin valve systems, with either perpendicular anisotropy, usually requested by the magnetic recording elements, or in-plane anisotropy, requested for other types of applications (bio-medical, displacement or angular sensors). However, magnetic anisotropy can be changed from in plane to perpendicular to the plane by tuning the thickness of the FePt layers or the specific preparation conditions. In addition, it was shown in [19] that in the case of in-plane anisotropy of the free soft magnetic layer of an exchange bias spin valve structure, its hysteretic opening can be drastically diminished if the anisotropy direction of the free layer can be oriented perpendicularly to the pinning direction. This adjustment can be achieved by a so-called biasing process through a patterned hard magnetic layer next to the spin valve structure. Although Nd-Fe-B biasing layers are actually used in this respect, FePt biasing layers can be proposed in the future due to their much higher corrosion resistance, within the condition of competitive production costs.

3.2. *Structural Analysis of the Regular Arrays of Magnetic Structures*

The regular arrays of magnetic dots, periodically dispersed, that are imaged in Figure 1f have been structurally investigated by means of X-ray diffraction, in a grazing incidence geometry. Such grazing geometry of the X-ray beam incidence has been carefully chosen, with an angle of 0.8 up to 1.5 degrees, in such a manner as to minimize the effect and influence of the substrate contribution into the overall patterns. The diffractograms have been recorded between 20 and 95 degrees in 2 theta. For better clarity of the structure of the regular arrays, the diffractogram of the pristine substrate in the absence of the engraved regular arrays was also recorded and subsequently subtracted from the diffractogram of the investigated samples, the difference being normalized afterwards for a better clarity. We have thus obtained a clear X-ray spectrum, consisting of diffracted beams originating only from the regular array of FePt circular dots. It is known that, usually, the bulk FePt binary alloy, either equiatomic or off-stoichiometric, is formed in a disordered A1-type cubic structure and, upon annealing, it undergoes a structural disorder–order phase transformation. Following such a transformation, the FePt would exhibit the ordered, tetragonal

$L1_0$ FePt phase, which is the phase responsible for the hard magnetic properties of the system.

In Figure 2 are depicted the XRD patterns corresponding to the circular FePt dots obtained via exposure doses of 0.6 μC/cm^2 and at 0.79 μC/cm^2, respectively, the growth of Pt (30 nm thickness) and Fe (40 nm thickness) films, the lift-off process and subsequent annealing at 400 °C for 30 min. All the main Bragg lines of tetragonal $L1_0$ FePt are observed in both samples.

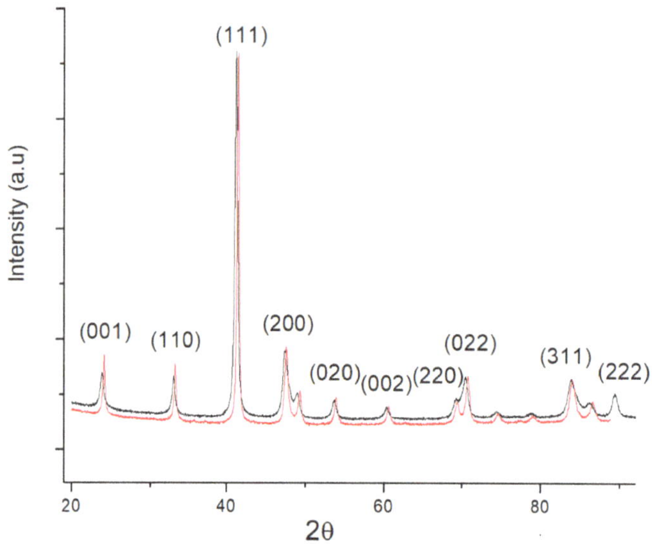

Figure 2. X-ray diffractograms for two of the regular arrays of FePt circular structures, engraved with an exposure dose of 0.6 μC/cm^2 (black line) and, respectively, 0.79 μC/cm^2 (red line).

There is a strong presence of the superlattice peaks: (001) and (110) diffraction lines, as well as the presence of the main (111) Bragg lines followed by other superlattice peaks: distinctly split (200) fundamental reflection, recorded at around 48–49 degrees in 2 theta. It has to be mentioned that the two main superlattice peaks of tetragonal $L1_0$, the (001) and (110) reflections, occurs for the Cu Kα radiation we used at about 23° and 32°, respectively.

It also has to be mentioned that the main six Bragg lines of the tetragonal FePt, from lower to higher angles, are, in order, attributed to (001), (110), (111), (200), (020), (002), (220), (022), (311) and (222) *hkl* planes reflections. The peak assignation has been performed in agreement with the ICSD card 03-065-9121. The L10 phase lattice parameters, calculated from the full-profile analysis were found to be a = 0.384(3) nm and c = 0.317(5) nm, with an ordering parameter c/a = 0.825. In addition to the occurrence of the superlattice peaks (001) and (110) another distinct signature of the occurrence of the tetragonal $L1_0$ phase is the fact that the (200), (220) and (311) peaks are split into two. Such a feature has been encountered in other works on FePt [20–22] and has been unambiguously regarded as further proof of the occurrence of the $L1_0$ phase, this being due to the tetragonally distorted planes, giving rise to the spectral splitting observed in these peaks. Upon the fitting of the XRD spectra, we have also derived the average crystallite size of the $L1_0$ FePt. It has been thus observed that the average grain size of the tetragonal $L1_0$ phase increases with the exposure dose from 20 nm (for engraving with a dose of 0.6 μC/cm^2) to 24 nm (for engraving with a dose of 0.79 μC/cm^2). We conclude that in these XRD results, only the $L1_0$ phase has been identified. In relation to previous studies [7,8], where it has been shown that not all the FePt dots were transformed from a cubic A1 to tetragonal $L1_0$ structure, which was detrimental to the magnetic anisotropy, in our case, all the observed dots seemed to be in their ordered $L1_0$ tetragonal crystal structure.

3.3. High-Resolution Transmission Electron Microscopy and Electron Diffraction Results

The microstructure of the granular regions on the magnetic dots' surfaces, in the sample presented in Figure 1f, has been imaged with the help of the transmission electron microscopy in the high-resolution mode, coupled with the observation of the electron diffraction patterns, within the imaged areas on the dot surface. In the imaging area, the sample has been thinned down by ion etching of the substrate with the focused ion beam module of the microscope. A typical magnified image obtained in HRTEM is shown in Figure 3, on the left side. The high-resolution image, recorded in phase contrast imaging mode, reveals small nanocrystals of ordered tetragonal $L1_0$ FePt symmetry. The observed microstructure is apparently made of ordered FePt crystallites having a size between 4 and 8 nm, approximately, which are distributed within a wide area between 13 and 18 nm. The reticular interplanar distances, clearly observed and measured on the image, are proving the occurrence of the (110) and (1-10) superlattice reflections, which are indeed typical for the tetragonal $L1_0$ FePt phase. This confirms well, in a more illustrative way, the findings obtained from the XRD analysis, detailed above. Another main Bragg reflection of the $L1_0$ phase, the (200) reflection, is also observed and accurately measured on the image through the measuring of the interplanar distance. Taking into account that the limiting critical size for single-magnetic domains is 55 nm for $L1_0$ FePt nanograins [23], it can safely be concluded that the FePt crystallites are magnetically single domains, irrespective of their crystal symmetry or of their degree of crystallographic ordering. On the right side of the image of Figure 3, the selected area electron diffraction (SAED) image is depicted, corresponding to the area that is imaged in real space in Figure 3, on the left side. Being essentially a map of the reciprocal space, the SAED image illustrates via the brightness of the observed spots, the intensity of the corresponding planar reflection, indicating thus quantitatively the abundance of the ordered FePt grains. It is seen that SAED analysis also confirms the occurrence of the ordered $L1_0$ FePt phase in the imaged magnetic dot. The presence of the main superlattice spots, the (001) and (110) Bragg reflections of the $L1_0$ phase, in the SAED patterns, proves the occurrence of both crystallographic orientations, with the tetragonal c-axis perpendicular and parallel to the surface normal. Here, the atomic planes of a (110) nature are seen, which again confirms the occurrence of the superlattice reflections.

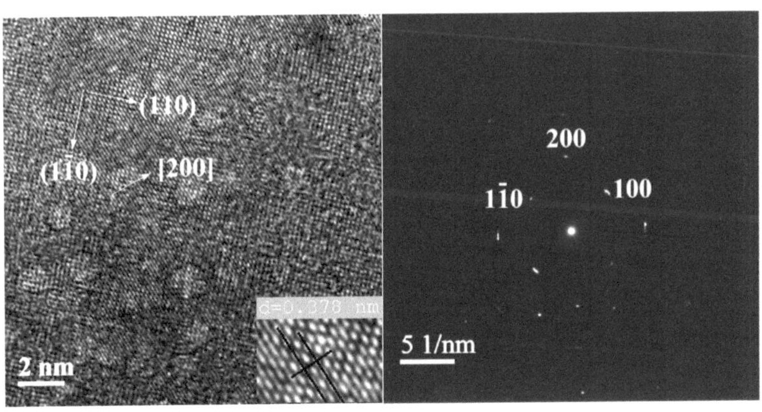

Figure 3. HRTEM image of an FePt magnetic dot (**left side**). Atomic planes are identified, and reticular distances allow for the identification of several superlattice peaks of $L1_0$ FePt. The electron diffraction pattern associated with the FePt magnetic dot (**right side**). In the reciprocal space, superlattice reflections are also observed.

It is to be noted though that while Figure 3 is only an illustrative example, several other high-resolution images have been recorded, and nanocrystals of a tetragonal symmetry with detectable (001) planes, with a reticular distance of d_{001} = 3.78 Å, are observed in many other imaged areas, thus confirming the extended $L1_0$ ordering alongside the magnetic dot surface. It has been explained before that two crystallograhic variants are retrieved, one having the c-axis in plane and the other one having the c-axis in an orientation normal to the observing plane. Another (001) superlattice reflection, with a reticular distance of d_{001} = 3.67 Å, is also identified in various other imaged areas. All the *c* lattice parameter values measured in the observed nanocrystals with tetragonal symmetry are in good agreement with the XRD results. In the same image, a second superlattice spot attributed to the (110) Bragg reflection is observed, with a reticular distance d_{110} = 1.91 Å.

All the structural data are thus consistent with each other and are unambiguously proving the formation of the highly ordered $L1_0$ FePt, of tetragonal symmetry in the analyzed samples.

3.4. *Magnetic Investigations of the Regular Arrays of Magnetic Nanostructures*

In order to assess the magnetic performances of the arrays, an investigation using the Physical Property Measurement System (PPMS) was undertaken. Magnetic hysteresis loops were measured both at 5K and at 300K in parallel and perpendicular orientation, in an applied field of up to 12 Tesla. Figure 4 depicts the magnetization reversal behavior of the regular array of FePt circular dots, namely the sample imaged in Figure 1f.

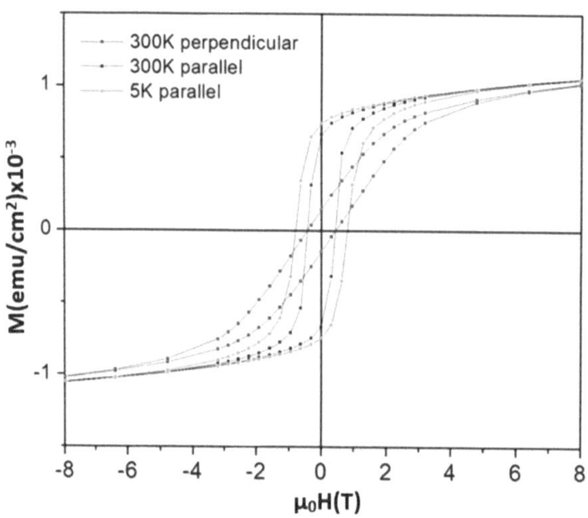

Figure 4. In this figure, 300 and 5K hysteresis loops of the regular arrays of FePt circular dots presented in Figure 1f. The loops were recorded in perpendicular and parallel applied magnetic fields, with respect to the array plane. High coercive fields correspond to the sample engraved with the maximum exposure rate of 0.79 $\mu C/cm^2$.

As mentioned before, the diffraction measurements did not provide any significant structural differences between samples with various exposure rates of the engraving. Only a slightly larger size of the $L1_0$ FePt crystallites at increasing doses could be mentioned, with no direct influence on the magnetic reversal. In this respect, we have chosen to show only the sample engraved with the maximum exposure rate of 0.79 $\mu C/cm^2$. Notably, the sample engraved at 0.6 $\mu C/cm^2$ had shown quite similar magnetic reversal, proving that the involved change in the engraving rate does not change the damaging of the dot edges.

The shape of the hysteresis loops is compatible with the presence of the $L1_0$ FePt phase, exhibiting increased saturation magnetization, remanent magnetization and magnetic coercivity. Due to the distance between the dots, their magnetic interaction should be neglected and the hysteresis loop should reflect an average of the magnetization reversal of each individual dot.

The magnetic hysteresis loops, recorded both in perpendicular and parallel geometry, do not show shoulders on the demagnetization curve. This indicates, on one side, the monophased magnetic character of the sample and, on the other side, it assures a full exchange coupling between FePt nanograins across all the investigated arrays. This is in agreement with the observed microstructure, composed of hard $L1_0$ binary FePt.

Another important observation is the fact that, at 300K, there are strong differences between the shape of the hysteresis loops recorded in parallel and in perpendicular geometry. The elongated shape of the loop in perpendicular geometry, as opposed to the rectangular shape in the case of the loop in parallel geometry, witness a pronounced in-plane magnetocrystalline anisotropy. Indeed, it was shown that such nano-islands of FePt can exhibit a strong magnetocrystalline anisotropy of about 7×10^6 J/m^3 [24] which should be correlated to a high coercive field along the in-plane magnetic easy axes. It was shown in [18,25] that in the case of no superposition between the coercive or switching field of the magnetization reversal along an easy axis and the saturation field of the magnetization reversal perpendicular to the easy axis, the magnetization reversal mechanism occurs through nucleation and movements of magnetic domains. In the present case, the coercive field measured in parallel geometry at 300 and 5K is of 4.5 and 8.0 kOe (i.e., μ_0H is 0.45 T and 0.8 T), respectively. Moreover, the saturation field at 300 K in perpendicular geometry is about five times higher than the coercive field in parallel geometry, providing evidence for the formation of magnetic domains in the dots. Indeed, taking into account the limiting critical size of 55 nm for single-magnetic domains for $L1_0$ FePt nanograins as reported in [23], the present magnetic dots with a lateral size of 335 nm and 70 nm in thickness are susceptible to the formation of magnetic domains. However, the polycrystalline configuration of the dots with well-formed $L1_0$ FePt nanograins, of 5 to 20 nm in size, of enhanced magnetocrystalline anisotropy suggests the consideration of the magnetic dot as an Imry and Ma type random anisotropy ferromagnet [26] consisting of a superposition of interacting magnetic domains of Stoner–Wohlfarth type associated to the $L1_0$ FePt nanograins. This behavior is in agreement with the structural observation of different orientations of the c axis of the FePt nanograins, also offering complementary information for the mainly in-plane random distribution of the c axis.

For the calculation of the specific magnetization values, we assumed the sample geometry as being the one observed in the SEM images, such as Figure 1f, with a total assumed thickness of the dots close to the nominal one of 70 nm. For the calculation of the filling factor, we assumed a regular array, as in the one in Figure 1f, where the dots of 335 nm diameter are spaced by 1200 nm in each direction. Taking into account the size of the image and the number of dots per image, we have calculated a filling factor of about 12.7%. The values of magnetization reported on the ordinate of Figure 4 refer to the magnetic moment on the surface unit of the measured sample. This means that the magnetic moment at saturation per unit surface of magnetic material is about 1×10^{-3} emu/cm^2 times 100/12.7, i.e., about 7.9×10^{-3} emu/cm^2. For a film with a nominal thickness of 70 nm, this results in a saturation magnetization of about 1100 emu/cm^3, which is in close agreement with the spontaneous magnetization of the $L1_0$ FePt bulk alloys. Also, the remanent magnetization values are high, namely 750 emu/cm^3 (690 emu/cm^3 at 300K), comparable with those of the $L1_0$ FePt bulk alloys. It is important to notice that in the investigated array of FePt circular dots, the microstructure consists of only the tetragonal $L1_0$ phase and this microstructure is well reflected in the overall magnetic behavior. The saturation magnetization is also proven to be high, a good prerequisite for obtaining an elevated maximum energy product (BH_{max}) which is the Figure of Merit for any nanomagnetic structure, albeit nanoparticle, thin film or structured dot. These findings are encouraging, as they show the potential for developing regular arrays of uniformly shaped magnetic entities with good magnetic performances in view of potential uses in various applications, from biomagnetic sensing through magnetoresistance effects to heated-dot magnetic recording.

4. Conclusions

A versatile method of engraving magnetic materials with various shapes, sizes and orientations using e-beam lithography has been described as potentially suitable for achieving building blocks for the further development of nanomagnet-logic structures for several types of applications. The method has been shown to be reliable in providing various regularly spaced shapes for magnetic material, with characteristics that are possible to be controlled by varying the exposure dose during the pre-patterning. We have focused then on the structural and magnetic characterization of a nanodevice consisting of regular arrays of FePt circular dots. Structural characterization using X-ray diffraction and HRTEM had shown that the proposed low time and low temperature annealing procedure provided monodomain FePt magnetic dots, where the tetragonal $L1_0$ phase is already formed, as proven by the occurrence of the superlattice peaks. Moreover, the magnetic characterization provided more insight into the potential of such arrays of magnetic devices. Hysteresis loops recorded for the array of FePt dots were consistent with the findings from XRD and provided good values of magnetic coercivity (8 kOe), remanent magnetisation (750 emu/cm^3) and saturation magnetization (1100 emu/cm^3), values that are comparable to those obtained for bulk $L1_0$ alloys. It is, in fact, for the first time that the formation of the $L1_0$ phase in FePt dots, with high coercivity, has been reported for such a low temperature of annealing (400 °C) and short annealing time (30 min). These findings are encouraging, as they show the potential for developing, at much lower production costs, regular arrays of uniformly shaped magnetic entities with strong magnetic performances, in view of various applications.

Author Contributions: Conceptualization, V.K.; methodology, O.C.; formal analysis and investigation, A.D.C. and G.S.; writing—original draft preparation, A.D.C.; writing—review and editing, O.C. and V.K. All authors have read and agreed to the published version of the manuscript.

Funding: This research was funded by Romanian Ministry of Research, Innovation and Digitalization, through IDEI grant PCE 79/2022, PN-III-P4-PCE-2021-0573 as well as through the Core Program at NIMP, Project PN23080102.

Data Availability Statement: The data presented in this study are available on request from the corresponding author. The data are not publicly available due to foreseen patenting potential.

Conflicts of Interest: The authors declare no conflict of interest. The funders had no role in the design of the study; in the collection, analyses or interpretation of data; in the writing of the manuscript or in the decision to publish the results.

References

1. Murray, C.B.; Norris, D.J.; Bawendi, M.G. Synthesis and characterization of nearly monodisperse CdE (E = sulfur, selenium, tellurium) semiconductor nanocrystallites. *J. Am. Chem. Soc.* **1993**, *115*, 8706–8715. [CrossRef]
2. Liu, W.Q.; Wong, P.K.L.; Xu, Y.B. Hybrid spintronic materials: Growth, structure and properties. *Prog. Mater. Sci.* **2019**, *99*, 27–105. [CrossRef]
3. Vedmedenko, E.Y.; Kawakami, R.K.; Sheka, D.D.; Gambardella, P.; Kirilyuk, A.; Hirohata, A.; Binek, C.; Chubykalo-Fesenko, O.; Sanvito, S.; Kirby, B.J.; et al. The 2020 magnetism roadmap. *J. Phys. D Appl. Phys.* **2020**, *53*, 453001. [CrossRef]
4. Mohan, J.R.; Medwal, R.; Gupta, S.; Gogia, K.; Vas, J.V.; Gupta, R.; Deka, A.; Rawat, R.S.; Subramanian, A.; Fukuma, Y. Nonstoichiometric FePt Nanoclusters for Heated Dot Magnetic Recording Media. *ACS Appl. Nano Mater.* **2021**, *4*, 7079–7085. [CrossRef]
5. Medwal, R.; Gautam, S.; Gupta, S.; Chae, K.H.; Asokan, K.; Deen, G.R.; Rawat, R.S.; Katiyar, R.S.; Annapoorni, S. Self-Stabilized Carbon-$L1_0$ FePt Nanoparticles for Heated Dot Recording Media. *IEEE Magn. Lett.* **2018**, *9*, 5504105. [CrossRef]
6. Kuo, D.S.; Lee, K.Y.; Yang, X.M.; Xiao, S.G.; Hsu, Y.; Yu, Z.; Feldbaum, M.; Klemmer, T.; Kubota, Y.; Thiele, J.-U.; et al. Heated Dot Magnetic Recording Media—Path to 10 TDOTS/in^2. In Proceedings of the 2016 International Conference of Asian Union of Magnetics Societies (ICAUMS), Tainan, Taiwan, 1–5 August 2016; p. 1.
7. Yan, Z.J.; Takahashi, S.; Hasegawa, T.; Ishio, S.; Kondo, Y.; Ariake, J. Towards an understanding of microstructure of patterned FePt dots by magnetometry using pulse fields. *J. Magn. Magn. Mater.* **2014**, *349*, 5–8. [CrossRef]

8. Yan, Z.J.; Takahashi, S.; Hasegawa, T.; Ishio, S.; Kondo, Y.; Ariake, J.; Xue, D.S. Understanding magnetic properties of arrays of small FePt dots with perpendicular anisotropy. *J. Magn. Magn. Mater.* **2012**, *324*, 3737–3740. [CrossRef]
9. Li, G.X.; Sun, S.H.; Wilson, R.J.; White, R.L.; Pourmand, N.; Wang, S.X. Spin valve sensors for ultrasensitive detection of superparamagnetic nanoparticles for biological applications. *Sens. Actuators A Phys.* **2006**, *126*, 98–106. [CrossRef]
10. Chen, Y.; Wang, X.; Sun, Z.; Li, H. The application of spintronic devices in magnetic bio-sensing. In Proceedings of the 2nd Asia Symposium on Quality Electronic Design (ASQED), Penang, Malaysia, 3–4 August 2010; pp. 230–234. [CrossRef]
11. Su, D.Q.; Wu, K.; Saha, R.; Peng, C.Y.; Wang, J.P. Advances in Magnetoresistive Biosensors. *Micromachines* **2020**, *11*, 34. [CrossRef]
12. Dey, C.; Yari, P.; Wu, K. Recent advances in magnetoresistance biosensors: A short review. *Nano Futures* **2023**, *7*, 012002. [CrossRef]
13. Strungaru, M.; Nguyen, B.T.; Yuanmae, K.; Evans, R.F.L.; Chantrell, R.W.; Chureemart, P.; Chureemart, J. HAMR switching dynamics and the magnetic recording quadrilemma. *J. Magn. Magn. Mater.* **2022**, *564*, 17004. [CrossRef]
14. Ying, Y.; Xie, L.; Zheng, J.W.; Yu, J.; Li, W.C.; Qiao, L.; Cai, W.; Li, J.; Che, S.L. Enhancement of Coercivity in the P-Doped FePt Nanocrystalline Film Prepared by Electrodeposition. *J. Supercond. Novel Magn.* **2022**, *35*, 1533–1538. [CrossRef]
15. Khunkitti, P.; Wannawong, N.; Jongjaihan, C.; Siritaratiwat, A.; Kruesubthaworn, A.; Kaewrawang, A. Micromagnetic Simulation of L10-FePt-Based Exchange-Coupled-Composite-Bit-Patterned Media with Microwave-Assisted Magnetic Recording at Ultrahigh Areal Density. *Micromachines* **2021**, *12*, 1264. [CrossRef] [PubMed]
16. Crisan, A.D.; Vasiliu, F.; Nicula, R.; Bartha, C.; Mercioniu, I.; Crisan, O. Thermodynamic, structural and magnetic studies of phase transformations in MnAl nanocomposite alloys. *Mater. Charact.* **2018**, *140*, 1–8. [CrossRef]
17. Crisan, A.D.; Crisan, O.; Randrianantoandro, N.; Valeanu, M.; Morariu, M.; Burkel, E. Crystallization processes in Fe-Pt-Nb-B melt spun ribbons. *Mater. Sci. Eng. C-Biomim. Supramolec. Syst.* **2007**, *27*, 1283–1285. [CrossRef]
18. Kuncser, V.; Palade, P.; Kuncser, A.; Greculeasa, S.; Schinteie, G. Engineering Magnetic Properties of Nanostructures via Size Effects and Interphase Interactions. In *Size Effects in Nanostructures: Basics and Applications*; Kuncser, V., Miu, L., Eds.; Springer Series in Materials Science; Springer: Berlin/Heidelberg, Germany, 2014; Volume 205, pp. 169–237.
19. Rijks, T.G.S.M.; Jonge, W.J.M.; Folkerts, W.; Kools, J.C.S.; Coehorn, R. Magnetoresistance in NiFe/Cu/NiFe/FeMn spin valves with low coercivity and ultrahigh sensitivity. *Appl. Phys. Lett.* **1994**, *65*, 916–918. [CrossRef]
20. Hong, Y.; de Moraes, I.; Eslava, G.G.; Grenier, S.; Bellet-Amalric, E.; Dias, A.; Bonfim, M.; Ranno, L.; Devillers, T.; Dempsey, N.M. A high throughput study of both compositionally graded and homogeneous Fe-Pt thin films. *J. Mater. Res. Technol.* **2022**, *18*, 1245–1255. [CrossRef]
21. Yu, J.; Xiao, T.; Wang, X.; Zhou, X.; Wang, X.; Peng, L.; Zhao, Y.; Wang, J.; Chen, J.; Yin, H.; et al. A Controllability Investigation of Magnetic Properties for FePt Alloy Nanocomposite Thin Films. *Nanomaterials* **2019**, *9*, 53. [CrossRef]
22. Xie, Z.; Cai, Y.; Tang, M.; Zhou, J.; Liu, J.; Peng, J.; Jiang, T.; Shi, Z.; Chen, Z. Fluence and Temperature Dependences of Laser-Induced Ultrafast Demagnetization and Recovery Dynamics in $L1_0$-FePt Thin Film. *Materials* **2023**, *16*, 5086. [CrossRef]
23. Kikuchi, N.; Okamoto, S.; Kitakami, O. The critical size between single domain and multidomain in $L1_0$-FePt. *J. Appl. Phys.* **2008**, *103*, 07D511. [CrossRef]
24. Sun, S. Recent Advances in Chemical Synthesis, Self-Assembly, and Applications of FePt Nanoparticles. *Adv. Mater.* **2006**, *18*, 393–403. [CrossRef]
25. Stanciu, A.E.; Schinteie, G.; Kuncser, A.C.; Locovei, C.; Trupina, L.; Iacob, N.; Leca, A.; Borca, B.; Kuncser, V. Magnetic properties of nanosized Fe and FeCo systems on trenched Mo templates. *Coatings* **2022**, *12*, 1366. [CrossRef]
26. Imry, Y.; Ma, S.-K. Random-Field Instability of the Ordered State of Continuous Symmetry. *Phys. Rev. Lett.* **1975**, *35*, 1399. [CrossRef]

coatings

MDPI

Article

Synthesis of $CoFe_2O_4$ through Wet Ferritization Method Using an Aqueous Extract of Eucalyptus Leaves

Dana Gingasu [1,*], Daniela C. Culita [1], Jose Maria Calderon Moreno [1], Gabriela Marinescu [1], Cristina Bartha [2,*], Ovidiu Oprea [3], Silviu Preda [1], Mariana Carmen Chifiriuc [4,5,6] and Marcela Popa [5]

[1] Ilie Murgulescu Institute of Physical Chemistry, Romanian Academy, 202 Splaiul Independentei, 060021 Bucharest, Romania; dculita@icf.ro (D.C.C.); calderon@icf.ro (J.M.C.M.); gmarinescu@icf.ro (G.M.); predas01@yahoo.co.uk (S.P.)
[2] National Institute of Materials Physics, 405A Atomistilor Street, P.O. Box Mg-7, 077125 Magurele, Romania
[3] Faculty of Chemical Engineering and Biotechnologies, University Politehnica of Bucharest, 1-7 Polizu Street, 011061 Bucharest, Romania; ovidiu73@yahoo.com
[4] Microbiology Department, Faculty of Biology, University of Bucharest, 91-95 Splaiul Independentei, 050095 Bucharest, Romania; carmen_balotescu@yahoo.com or carmen.chifiriuc@unibuc.ro
[5] Research Institute of the University of Bucharest—ICUB, University of Bucharest, 91-95 Splaiul Independentei, 050095 Bucharest, Romania; bmarcelica@yahoo.com or marcela.popa@bio.unibuc.ro
[6] Romanian Academy, 125 Calea Victoriei, 010071 Bucharest, Romania
* Correspondence: d_gingasu@yahoo.com or dgingasu@icf.ro (D.G.); cristina.bartha@infim.ro (C.B.)

Abstract: This study explored a new green approach of the wet ferritization method to obtain magnetic cobalt ferrite ($CoFe_2O_4$) by using eucalyptus leaves aqueous extract as a reducing/chelating/capping agent. The spinel single cubic phases of prepared samples were proved by powder X-ray diffraction (XRD), Fourier-Transform Infrared (FTIR) and Raman spectroscopy. The average crystallite size is in the range between 3 and 20 nm. The presence of the functional groups coating the obtained material is confirmed from FTIR and thermal analysis. The scanning electron microscopy (SEM) analysis showed a morphology consisting of nanoparticle aggregates. Raman spectroscopy detects the characteristic bands of spinel-type $CoFe_2O_4$. Magnetic investigations reveal the formation of ferromagnetic compounds with cubic magnetic anisotropy and a blocking temperature around 140 K, specific for this type of material. The biosynthesized $CoFe_2O_4$ could be an attractive candidate for biomedical applications, exhibiting promising antimicrobial and antibiofilm activity, particularly against Gram-negative bacteria and fungal strains.

Keywords: spinel ferrite; green synthesis; plant extract; magnetic properties; biomedical applications

1. Introduction

In recent years, there has been an increasing interest in the synthesis of nano-sized magnetic ferrites due to their diverse potential applications as permanent magnets, storage devices, magnetic recording devices, ferrofluids, magnetic drug delivery, catalysts, etc. [1–3].

Among them, cobalt ferrite ($CoFe_2O_4$) has a special place, being a hard magnetic material with good chemical stability, high coercivity (Hc), and moderate magnetization (Ms) [4]. Cobalt ferrite has received a renewed research focus due to its multiple biomedical applications in magnetic refrigeration, magnetic resonance imaging, hyperthermia treatment, and in the field of nano-biosensors [5–7].

Nano-sized cobalt ferrite can be obtained by using a wide range of chemical and physical methods [3,8,9]. Some of these methods present a series of impediments such as the use of toxic solvents, prolonged high temperature heat treatments, and high energy consumption with risks for the environment. To reduce the negative impact on the environment, cleaner/greener synthesis methods are required. Green synthesis methods are guided by the twelve principles of green chemistry [10]. A green approach of chemistry synthesis methods assume, among other things, the use of natural raw materials, simple equipment, and reduced synthesis times. All of these things lead to a lower cost of obtaining nanomaterials compared to other methods [11,12].

Among the biological agents, plants seem to be the best candidates, being easily accessible, widely distributed, and providing a vast reserve [13]. Moreover, every part of a plant could be useful in the synthesis of nanomaterials. Plant extracts contain phytochemicals (polysaccharides, phenolic acids, flavonoids, etc.) that have multiple functions (reducing, capping, chelating, and even stabilizing) in the synthesis of nanomaterials. The variation of these active biocomponents concentrations, their compositions, as well as their interactions with metal ions are among the main factors that feature the diversity of sizes and shapes of the obtained nanoparticles [14]. The biomolecules from the plant extracts are immobilized on the surface of nanoparticles during the synthesis process, leading to modified/functionalized/coating of nanoparticle surfaces that results in an increased stability of the biosynthesized nanoparticles. These processes also improve material properties and thus expand their ranges of applications.

In the last decade, a special focus was given to cobalt ferrite synthesized by green chemistry methods using different types of plant extracts. Routray et al. [15] reported the use of Aloe vera in the sol–gel auto-combustion process, to obtain $CoFe_2O_4$ for high frequency application, magneto recording devices. Mahajan et al. [16] revealed tulsi seed and garlic extract-based synthesis of $CoFe_2O_4$ with good antibacterial activity. A green synthesis method of cobalt ferrite using extracts of *Artemisia annua* L. "hairy" roots was demonstrated by Kobylinska et al. [17] and the effect and role of green synthesis parameters were evaluated. In our previous papers, we described the synthesis of $CoFe_2O_4$ nanoparticles employing aqueous extracts of sesame seed extract, ginger root, cardamom seeds, tamarind fruit as fuel or reducing/chelating/capping agents [18–20]. The obtained $CoFe_2O_4$ nanoparticles exhibited very good antimicrobial properties, demonstrated by the low minimal inhibitory concentration values against a wide range of microbial strains, suggesting their potential for the development of novel antimicrobial agents [18].

The *Eucalyptus* genus from the *Myrtaceae* family contains about 600 species native to Australia. The eucalyptus plant leaves extracts have various effects such as antibacterial, antihyperglycemic, antioxidant etc. [21].

The phytochemical studies of eucalyptus leaves highlighted the presence of alkaloids, aminoacids, carbohydrates, flavonoids, glycosides, tannins, phenols, terpenoids, steroids, and saponins [22,23]. The leaves extracts contain phenolcarboxylic acids (gallic and ellagic acids), hydroxycinnamic acids (caffeic, ferulic, chlorogenic, p-coumaric acids), flavonoids (lutelin, myricetin, quercetin, kaempferol), coumarin (coumarin, skinnin, scopolin), etc. [22], one of the major constituents being 1,8-cineole (between 73 and 85%) [21]. These components have an important role in the green synthesis process. The hydroxyl and carboxyl groups from the active biomolecules facilitate the interaction with the metal ions and have a high involvement in the metal ion reduction [24]. A high content of polyphenols leads to the formation of a protective coating around the synthesized nanoparticle, thus harboring functional groups (charged groups) that create repulsive forces preventing nanoparticle agglomeration and instability [25]. Generally, the green synthesis of nanoparticles has three phases [26]: (i) the activation phase consisting in the reduction of metal ions to zero-valent metal and nucleation; (ii) the growth phase, in which the combination of several nanoparticles leads to the formation of larger ones, with increased thermodynamic stability; (iii) the termination phase is essential for defining the size and shape of the resulted nanoparticle.

Previous studies revealed the use of eucalyptus extract, which is easy to obtain and sustainable, in the synthesis of metal nanoparticles such as Ag, Zn, Fe [25,27,28] and metal oxide TiO_2, ZnO, La_2O_3 [29–31]. These nanoparticles could be good candidates for biomedical, ecological, and agricultural applications. However, the use of this extract in the synthesis of $CoFe_2O_4$ nanoparticles has not yet been reported.

Considering the above aspects, in the present study we show for the first time in the literature the synthesis of cobalt ferrite ($CoFe_2O_4$) through the wet ferritization method using an aqueous extract of eucalyptus leaves. The physico-chemical characterization of the obtained cobalt ferrite nanoparticles (XRD, SEM, thermal analysis, FTIR, Raman, and magnetic measurements) has been carried out. The antimicrobial activity of the obtained cobalt ferrite nanoparticles has also been investigated to formulate possible practical applications of the obtained materials.

2. Materials and Methods

2.1. Materials

The iron ($Fe(NO_3)_3·9H_2O$) and the cobalt ($Co(NO_3)_2·6H_2O$) nitrates were of reagent quality (Merck, Darmstadt, Germany). Eucalyptus leaves (Fares brand, Eucalyptus globulus, contain volatile oil in which the main component is 1,8-cineole) were purchased from the local health food store and the ammonia solution 25% from Chimreactiv, Bucharest, Romania.

2.2. Preparation of the Plant Extract

The eucalyptus aqueous extract was prepared according to the following protocol: 2.5 g eucalyptus dry leaves were mixed with 100 mL of distilled water and heated up to 80 °C for 30 min under magnetic stirring. After cooling and filtering, a dark orange extract (pH = 5) was obtained.

2.3. Synthesis of Cobalt Ferrites

The metal nitrates ($2Fe^{3+}:1Co^{2+}$) were added slowly into the 100 mL eucalyptus aqueous extract and the pH was raised to 10 by adding NH_4OH 25%. During this time, under stirring, a dark brown precipitate was formed. The obtained suspension was maintained under continuously stirring at ~80 °C for 3 h. After this time, the weak magnetic dark precipitate was separated and washed until the washing water had a pH of 7. It was dried on P_4O_{10} until a black powder was obtained (N sample). To improve the crystallinity of the N sample, an additional thermal treatment at 800 °C/1 h was applied (N-800).

2.4. Characterization Techniques

The X-ray diffraction patterns were recorded using a Rigaku Ultima IV diffractometer (Tokyo, Japan). The equipment was set up in parallel beam geometry, using Cu Kα radiation (λ = 1.5406), CBO optics, a graphite monochromator, and 40 kV and 30 mA operating outputs. The measurements were carried out in the θ-2θ mode, with a step size of 0.02° and a scan speed of 1°/min. The phase identification was done using the Search/Match method, coupled with the ICDD PDF-2. The lattice constant was refined using the diffraction line position, adjusted by the external standard method. The crystallite size (d) was calculated from the diffraction line width (integral width) using the Scherrer equation:

$$d = 0.9\lambda/\beta\cos\theta,$$

where d is the crystallite size, λ is the wavelength of the X-ray used, θ is the angle of reflection, and β is the full width at half maximum or the broadening of the diffraction line in radians. X-ray fluorescence (XRF) was used for elemental analysis. The measurements were performed using a Rigaku ZSX Primus II spectrometer (Rigaku Corp., Tokyo, Japan), equipped with 4.0 kW X-ray Rh tube. EZ−scan combined with Rigaku SQX fundamental parameters module of ZSX v5.18 software (standard less) was used for data analysis. The microstructure of the obtained powders was investigated by scanning electron microscopy (SEM) in a FEI Quanta 3D FEG apparatus (Hillsboro, OR, USA) operating at accelerating voltages in 2–5 kV range, using secondary electron images. IR spectra were recorded on JASCO FTIR 4100 spectrophotometer (Tokyo, Japan) in the 4000–400 cm^{-1} range, with solid samples prepared as KBr pellet. Unpolarized Raman spectra were achieved in a Horiba Jobin–Yvon LabRam HR spectrometer (Kyoto, Japan) with a He-Cd laser, using excitation of 325 nm wavelength. In order to avoid sample damage by irradiation, the power on the sample was kept below 0.5 mW. The UV Raman spectra were collected on a 40× objective, in a UV-enhanced CCD camera and corrected by subtracting the background spectrum. The Raman shift was recorded from 300 to 900 cm^{-1} and the integration time of the measurements 60 s. The thermal analysis TG-DSC was performed with a STA 449C F3 system, TG-DSC (thermogravimetry—differential scanning calorimetry) from Netzsch (NETZSCH-Gerätebau GmbH, Selb, Germany), from room temperature up to 900 °C, in a constant air flow of 50 mL/min. A Bruker Tensor 27 (Bruker Co., Ettlingen, Germany) with thermostatic gas cell was used to record the FTIR spectra of the evolved gases. Magnetic properties were investigated with a MMPS SQUID magnetometer from Quantum Design (San Diego, CA, USA). Hysteresis curves have been acquired at four different temperatures (10 K, 100 K, 200 K, and 300 K, respectively) under an applied magnetic field of −50,000 Oe and +50,000 Oe. The evolution of magnetization under field cooling (FC) and zero field cooling (ZFC) conditions as a function of temperature was analyzed under a magnetic field of 500 Oe.

2.5. Antimicrobial Activity Assays

The antimicrobial activity of the obtained cobalt ferrite nanoparticles has been assessed against four bacterial and one fungal standard strain, i.e., *Staphylococcus aureus* ATCC 25923, *Enterococcus faecalis* ATCC 29212, *Escherichia coli* ATCC 25922, *Pseudomonas aeruginosa* ATCC 27853, and *Candida albicans* ATCC 10231, at a density of 0.5 McFarland. The selected strains have been tested both in planktonic and biofilm growth states, by using qualitative (adapted disk diffusion method allowing the quantification of growth inhibition zones diameters on solid agar) and quantitative (broth microdilution method and violet crystal microtiter allowing to establish the minimal inhibitory concentration—MIC and minimal biofilm inhibitory concentration-MBIC) assays [32,33].

The tested concentration in the qualitative assay was 10 mg/mL, while in the quantitative assays, two-fold serial dilutions ranging from 10 to 0.0195 mg/mL have been obtained in the liquid culture medium. In the quantitative assays, the MIC and MBIC have been established by inspecting the turbidity and respectively the color intensity of the well contents, in comparison with the aspect of the negative and positive controls. The lowest concentrations inhibiting the microbial growth revealed by the absence of turbidity for the planktonic growth and the absence of the blue color for the biofilm growth have been noted as MIC and respectively, MBIC. The assays were performed in duplicate [32,33].

3. Results and Discussion

The cobalt ferrite nanoparticles were obtained through the wet ferritization green approach, using eucalyptus leaves aqueous extract. The phytochemical basis of the plant extract is adequate, containing the bioactive molecules involved in the reduction (enol groups) and capping (methoxy/phenyl groups) action [34]. The biomolecules present in the eucalyptus leaves extract form coordinated metal complexes with the Co^{2+} and Fe^{3+} ions. The ferritization process takes place through the degradation of complex precursors (metals/active biomolecules) directly in the reaction medium, by varying the pH and the temperature of the reaction medium [18]. A possible mechanism of $CoFe_2O_4$ formation using the eucalyptus extract is presented in Scheme 1.

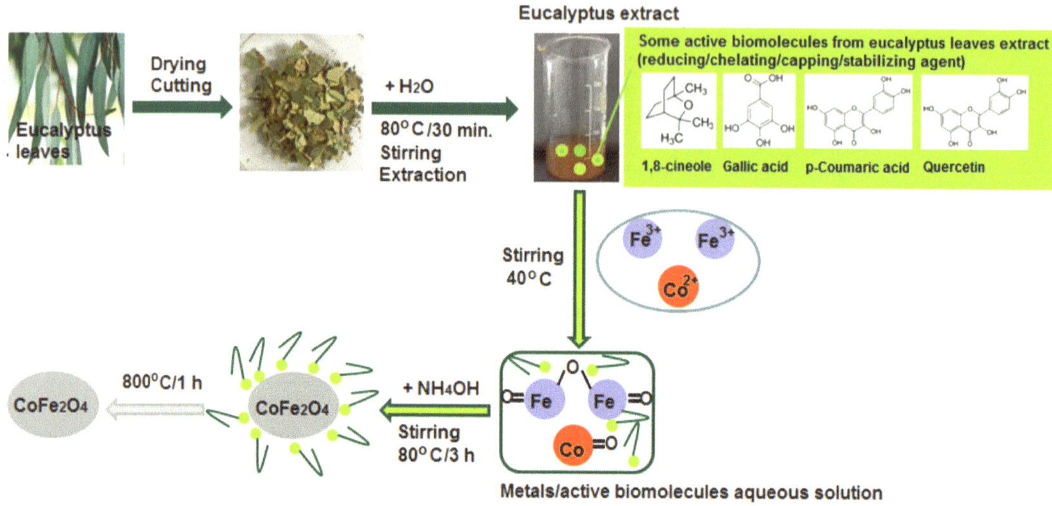

Scheme 1. A possible $CoFe_2O_4$ formation mechanism by a green approach of the wet ferritization method, using eucalyptus leaves aqueous extract.

3.1. X-ray Fluorescence and X-ray Diffraction

X-ray fluorescence (XRF) spectrometry was used to determine the elemental composition of both samples, N and N-800. The ratio between Co and Fe was 32.51:67.49 by mass percentage for the N sample and 33.21:66.79 for the N-800 sample, respectively. These ratios are close to the theoretical composition of $CoFe_2O_4$.

The crystal structure, phases, and purity of the as-synthesized sample (N sample) and the thermally-treated sample (N-800 sample) were investigated by X-ray diffraction (XRD). Figure 1 shows the XRD patterns of both powder samples. It was found that all diffraction lines in Figure 1b, corresponding to the crystal planes (111), (220), (311), (222), (400), (331), (422), (511), (440), (531), (620), (533), (622), and (444) could be indexed to a cubic inverse spinel structure. No additional diffraction lines corresponding to other phases or an amorphous halo were observed. Based on the XRD pattern of the sample N-800, coupled with XRF results, the spinel phase could be indexed to $CoFe_2O_4$, which matches well against ICDD file no. 22-1086. Figure 1a shows the XRD patterns of the sample N, the as-prepared sample. The diffraction lines are broad, but the positions of the lines are specific to a spinel structure. As for the thermally treated sample, the lack of an amorphous phase (amorphous halo), combined with the mass ratio between Co and Fe, indicates the presence of a spinel structure, with a composition of $CoFe_2O_4$.

The sharp and narrow diffraction lines of the cobalt ferrite calcined at 800 °C for 1 h (N-800) indicate the better crystallinity of the sample. The lattice parameters for the N and N-800 samples were 8.3114(10) Å and 8.38364(2) Å, respectively, being in good agreement with the literature data [18]. The crystallite sizes, calculated using the Scherrer equation, were 3 nm (N sample) and 20 nm (N-800 sample). The calculated values are lower than the values of $CoFe_2O_4$ samples (~26 nm) obtained by the sol–gel method using olive leaves aqueous extracts [35].

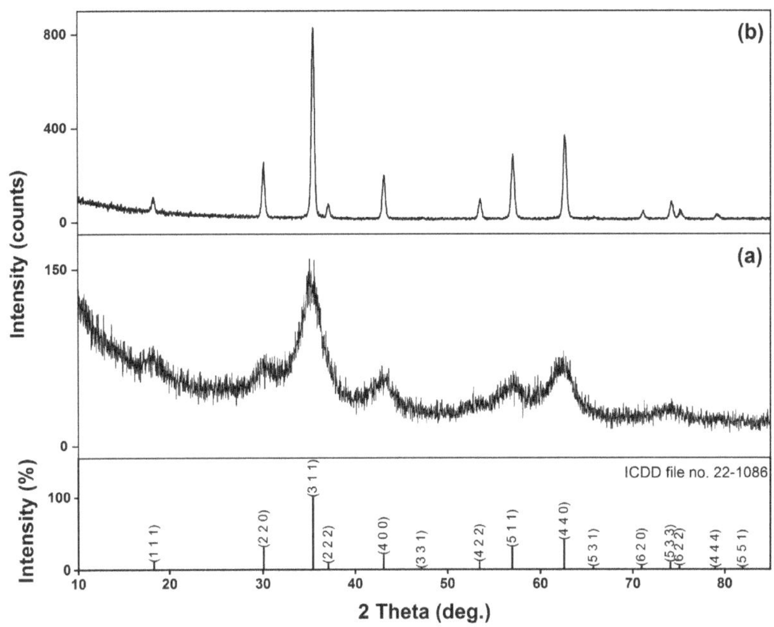

Figure 1. XRD patterns of spinel ferrites: (**a**) N; (**b**) N-800 sample.

3.2. Scanning Electron Microscopy

The morphology of the $CoFe_2O_4$ powders obtained by wet ferritization using eucalyptus extract was revealed by SEM images (Figure 2). Micro-sized aggregates are observed in the 10,000× micrograph in Figure 2a with aggregates size up to ten microns. The aggregates have a fluffy appearance, observed in the 200,000× micrograph (Figure 2b) and consist of loosely aggregated nanoparticles embedded in a disordered matrix. The nanoparticles have typical sizes below 10 nm, distinguishable in the 1,000,000× image in the inset in Figure 2b. The particle size distribution histogram is shown in Figure 2c. After thermal treatment at 800 °C for 1 h there is a densification of the aggregates into micro-sized solid polycrystalline particles (Figure 2d) consisting of primary $CoFe_2O_4$ nanocystallites sized about 20 nm (Figure 2e), as shown in the particle size distribution histogram (Figure 2f).

3.3. FTIR and Raman Spectroscopy

The IR spectra of the $CoFe_2O_4$ samples were recorded in the 4000–400 cm^{-1} range and were compared with the IR spectrum of eucalyptus leaf dry extract (Table 1 and Figure S1, see Supplementary Materials).

Literature data mention that the IR spectra recorded for different types of eucalyptus extract revealed the functional groups which could be the responsible candidates for the obtained nanoparticles [28–31].

The IR spectrum of sample N contains the intense bands characteristic of spinel ferrites, but also weak vibration bands characteristic of the carboxylate groups from the eucalyptus extract and traces of nitrate anion (Table 1). After the treatment at 800 °C/1 h (N-800 sample), these bands disappear and only the bands of cobalt ferrite in the 600–400 cm^{-1} region, assigned to the stretching vibrations of the metal–oxygen bond [4,5], are observed (Table 1).

Table 1. The characteristic bands of eucalyptus leaves dry extract and synthesized $CoFe_2O_4$ samples.

Eucalyptus Leaves Dry Extract	N	N-800	Assignement
3382 vs, br	3419 vs, br	-	$\nu OH/H_2O$
2915 m, sh	-	-	$\nu CH_{2\ asym}$
2850 w, sh	-	-	$\nu CH_{2\ sym}$
-	1635 s	-	νOH in water
1616 vs	-	-	$\nu COO^{-}{}_{asym}$
1396 m	1384 m	-	$\nu COO^{-}{}_{sym}/\nu NO_3^{-}$
1072 s	-	-	$\nu C\text{-}O\text{-}C_{sym}$
-	1031 w	987 w	-
-	578 s	573 vs	νM-O spinel
	486 w	490 m	νM-O spinel

w = weak, m = medium, s = strong, vs = very strong, sh = shoulder, br = broad.

Figure 2. SEM micrographs at different magnifications: 10,000× (**a**,**d**); 200,000× (**b**,**e**); 1,000,000× (inset in **b**); and particle size distributions (**c**,**f**) of the $CoFe_2O_4$ powders obtained by wet ferritization using eucalyptus extract, as prepared (**a**–**c**), and calcined at 800 °C/1 h (**d**–**f**).

Additionally, the cobalt ferrites obtained by wet ferritization were studied by Raman vibrational spectroscopy. The Raman spectra (Figure 3) of the N and N-800 powders present six vibrational modes at 683, 615, 562, 494, 460, and 325 cm^{-1}. After calcination at 800 °C, N-800 shows very similar Raman spectra, including the same vibrational modes and relative intensities, indicating the same crystalline structures and types of bonds in N and N-800, so our discussion is valid for the Raman spectra of both materials. The measured spectra correspond to spinel-type structures. Broad Raman peaks indicate the presence of disorder and defects within the crystal structure. Group theory predicts five active Raman modes for spinel-type structures (A_{1g} + E_g + $3F_{2g}$) which are typically observed at Raman shifts from 200 to 800 cm^{-1} [36–38]. The highest-frequency A_{1g} mode at 683 cm^{-1} is assigned to the symmetric stretching vibrations of the oxygen atoms with respect to the metal ion in the tetrahedral sites, a mode that can be also described as tetrahedral breath mode [36,39]. Spinel inversion is usually considered the cause for the appearance of an additional A_{1g}

peak in the Raman spectra, observed as a weak feature at 615 cm^{-1}, according to [39]. The F_{2g} mode at 562 cm^{-1} is very weak. The strong F_{2g} band at 450–500 cm^{-1}, that corresponds to the bending vibrations of oxygen atoms in the tetrahedral sites, shows two distinct peaks at 494 and 460 cm^{-1}, indicating different local environments experienced by the Fe^{3+} ions in the tetrahedral sites, leading to distinct vibrational frequencies. The E_g mode, associated with the stretching vibration of the octahedral units, is observed at 325 cm^{-1}.

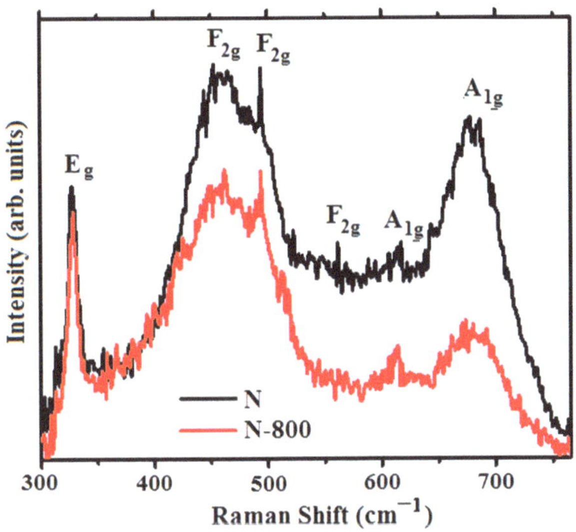

Figure 3. Raman spectra of $CoFe_2O_4$ powders obtained by wet ferritization using eucalyptus extract, as prepared (N sample), and calcined at 800 °C/1 h (N-800 sample).

3.4. Thermal Analysis

The thermal behavior of the $CoFe_2O_4$ obtained by wet ferritization (N sample) was followed to confirm the presence of the functional groups of eucalyptus extract on the particles' surface (Figure 4).

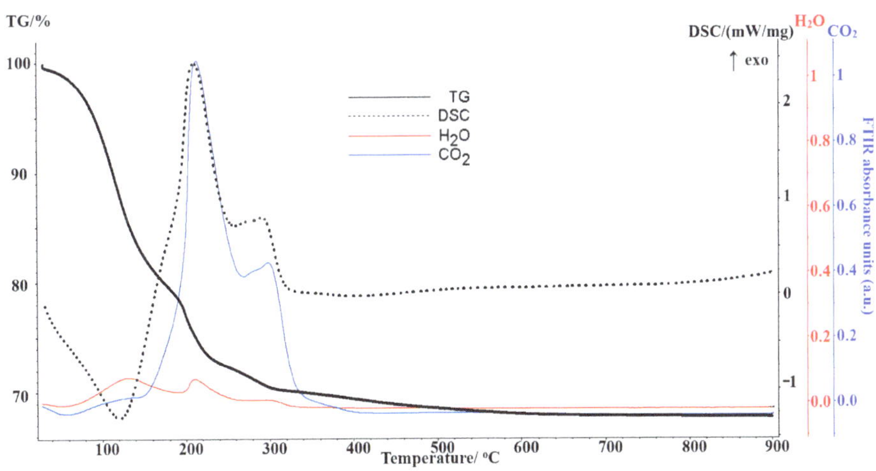

Figure 4. TG-DSC analysis (N sample) with extracted traces for CO_2 and H_2O evolved gases identified by FTIR wavenumbers 2354 and 3735 cm^{-1}.

The degradative-oxidative process can be grouped in three temperature intervals. The first step between room temperature and 180 °C is associated with an endothermic effect on the DSC curve (weight loss ~20.5%). The gases evolved in this step indicate water as the main product together with traces of CO_2, and, also, the less stable/volatile components (Figure 5) from eucalyptus extract [40]. The second mass loss step (~9%), between 180 and 320 °C, is accompanied by a strong exothermic effect on the DSC curve. This step corresponds to the total oxidation of the organic part to CO_2 and H_2O. The carbonaceous residue from eucalyptus extract is eliminated in the third step between 320 and 700 °C (weight loss ~2.5%). The thermal decomposition is complete below 700 °C with the formation of $CoFe_2O_4$ (ox%~68%). Corroboration of these results with the IR spectrum of the N sample suggests the presence of organic components from the plant extract on the prepared cobalt ferrite.

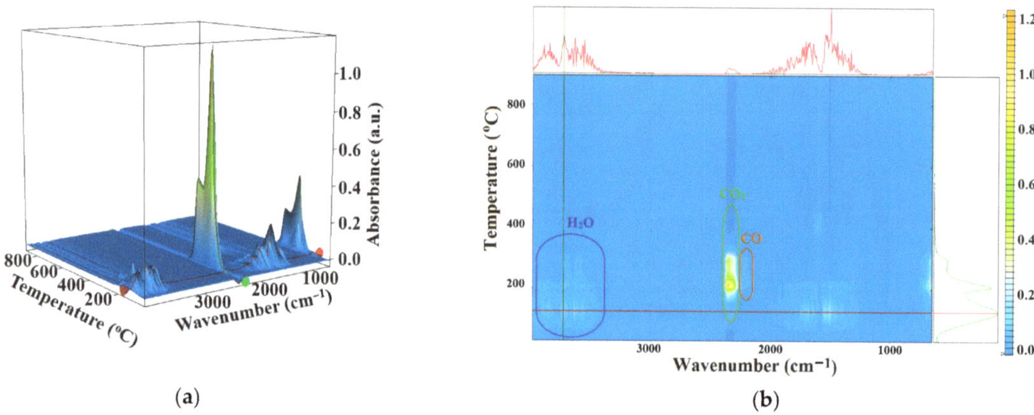

Figure 5. FTIR 3D diagram for the evolved gases from thermal analysis of N sample (**a**) and its projection in the wavenumber/temperature plane, indication of temperature intervals in which H_2O, CO_2, and CO are identified (**b**); the absorbance increases from light blue to orange-yellow.

The participation of the eucalyptus extract to the cobalt ferrite synthesis is indicated by the difference observed in the thermal analysis of the N sample vs. eucalyptus extract powder (Figures S2 and S3). While in the case of the N sample the oxidation reactions of organic traces take place up to 350 °C, for the eucalyptus extract the thermal analysis indicates a strong oxidation process between 435–680 °C, when high quantities of CO_2, H_2O, and hydrocarbon fragments are eliminated (Figure S3). This different thermal behavior of the organic part can indicate a dramatic change in composition of eucalyptus extract due to its participation in the process of cobalt ferrite synthesis.

3.5. Magnetic Measurements

The hysteresis curves corresponding to the two samples, N and N-800, respectively, are shown in Figure 6.

Both samples are magnetically ordered at both low and room temperatures and exhibit ferromagnetic behavior. The samples were measured starting from the demagnetized state (virgin curve visible especially for sample N-800, Figure 7b) at B = 0. At a low applied magnetic field, the magnetization process is reversible, magnetic moments being aligned along an axis of low magnetization. When the magnetic field increases, magnetization becomes irreversible, the magnetic moments aligning with the direction of the applied field. Magnetizations increase rapidly at low fields (0–10,000 Oe), tending to saturate as the magnetic field strength increases. The saturation magnetization values obtained at 10 K are 45.85 emu/g for sample N and 73.16 emu/g for the calcined sample N-800. These values are lower than the saturation magnetization of bulk cobalt ferrite, 93.9 emu/g when measured at 5 K, and 80.8 emu/g at ambient temperature [41], but comparable to the values obtained on nanoparticles synthesized by wet chemistry [42,43]. The coercive magnetic fields (Hc) obtained at 10 K were 6566 Oe for the N sample and 13,372 Oe for the N-800 sample and much lower at room temperature, respectively 107 Oe for the N sample and 2223 Oe for the N-800 sample. These values can be correlated with the particle sizes, i.e., for the N sample we obtained the smallest crystallite size in XRD (~3 nm) and the smallest coercive field, while for the heat-treated sample, N-800, the highest value for both crystallite size (~20 nm) and coercive field was obtained. Another important aspect is the coercivity value at 300 K of the N sample; it is very small, but not zero. This means that a small fraction of the nanoparticles are not superparamagnetically relaxed even at this temperature. This behavior is similar to that reported by Palade et al. [42]. The ratio (R = Mr/Ms) between remanent magnetization (Mr) and magnetization at saturation (Ms) was also calculated. At 10 K, the obtained values for R were 0.49 for the N sample and 0.89 for the N-800 sample. According to the Stoner–Wohlfarth model [44], for an ensemble of non-interacting single magnetic domain nanoparticles with uniaxial symmetry, under the condition of a random distribution of easy magnetization axes, this factor should be 0.5. It has also been reported that, for spherical nanoparticles with cubic magneto-crystalline anisotropy, the value of R is 0.83 [45]. In our case, the corresponding R value for N-800 sample clearly shows the existence of a cubic magnetic anisotropy provided by the magnetocrystalline anisotropy of $CoFe_2O_4$ that exceeds the shape anisotropy effect. The N sample, with R of 0.49, is characterized by a single non-interacting magnetic domain.

Figure 7 shows the magnetizations as a function of temperature of the two samples measured in two different modes: zero field cooling (ZFC) and field cooling (FC) under an applied magnetic field of 0.05 T.

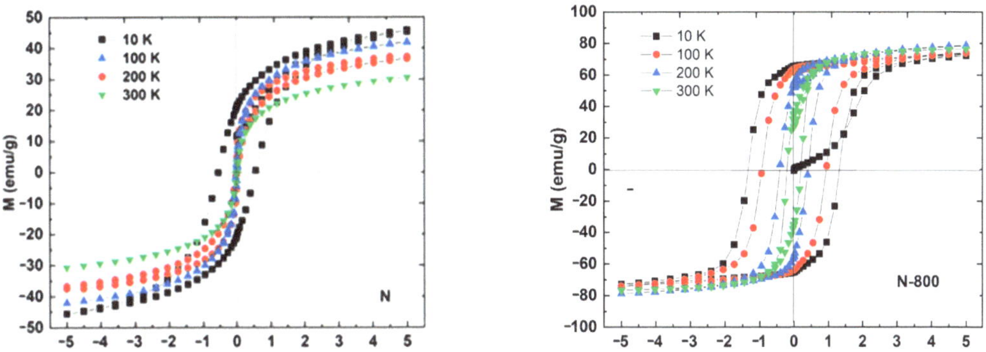

(a) (b)

Figure 6. Hysteresis curves at selected temperatures in the range of 10–300 K for samples: N (**a**) and N-800 (**b**).

In ZFC mode, the two samples were cooled in a zero magnetic field to 5 K. At this temperature, a magnetic field of 500 Oe was applied and the magnetic moments were measured as a function of temperature in the process of sample heating. In FC mode, samples were cooled in an applied external magnetic field (B = 0.05 T), measuring the evolution of magnetic moments as a function of temperature. It can be seen that the two curves coincide with each other at room temperature, but as the temperature decreases, they begin to separate. Both ZFC curves show a maximum at 140 K representing the blocking temperature (T_b). It is known that T_b depends on several factors, such as: particle size, effective anisotropy constant, applied magnetic field, experimental measurement time, etc. If the measured sample has almost equal particle size (small particle distribution), then the blocking temperature will be almost the same for all particles. This is also the case of our samples that have almost the same particle distribution (≥5 nm), hence the same T_b. Differences may occur at large particle distributions (≥100 nm). Over this temperature, FC curves increase as the temperature decreases. At this temperature, the nanoparticles have enough thermal energy to overcome the magnetic energy barrier separating the two possible magnetic orientations of the magnetic moments.

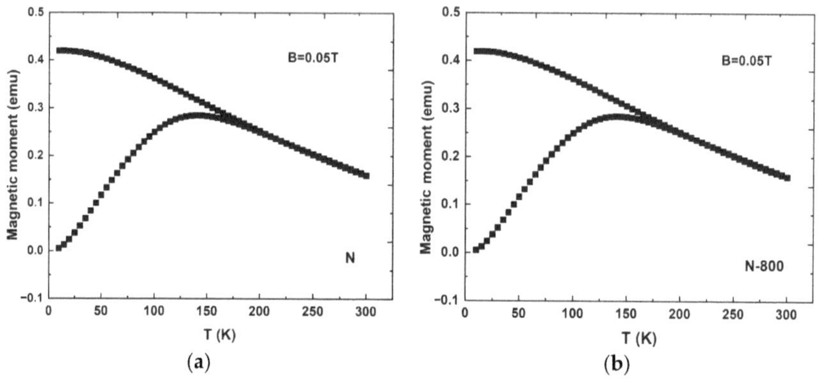

Figure 7. Temperature dependences of magnetic moments (ZFC and FC measurements) for samples: N (**a**) and N-800 (**b**). Measurements were carried out in magnetic field B = 0.05 T.

3.6. Antimicrobial Activity

The results of the qualitative assay indicated that the two $CoFe_2O_4$ nanoparticles exhibited relatively similar antimicrobial activity against four of the tested strains, including two Gram-negative bacterial strains (*P. aeruginosa* and *E. coli*), one Gram-positive bacterial strain (*S. aureus*) and the fungal strain (*C. albicans*), the average diameters of the growth inhibition zones ranging between 9.5 and 7.25 mm, with the highest and respectively lowest efficiency recorded for N against *P. aeruginosa* and *S. aureus*, respectively. None of the tested nanoparticles inhibited the growth of *E. faecalis* on the solid medium (Figure 8).

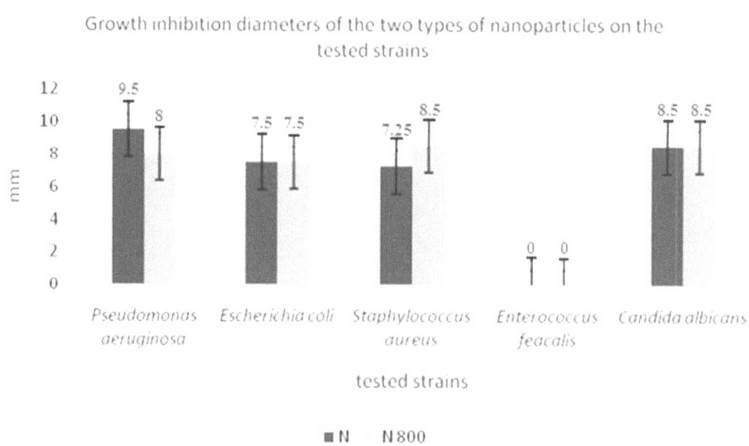

Figure 8. The results of the qualitative assay of the antimicrobial activity quantified by the growth inhibition zone diameters (mm).

Taking into account the limitations of the qualitative assay consisting in the unknown diffusion patterns of novel compounds in the solid medium, making it impossible to have a glimpse into the concentration–activity relationship, we have further tested different binary concentrations of the obtained nanoparticles (from 5 to 0.0195 mg/mL) in liquid medium assays, in order to establish quantitative parameters of their antimicrobial and anti-biofilm activity, respectively the MIC and MBIC values.

The results of the quantitative assay revealed that in all cases the MIC and MBIC values ranged from 5 to 10 mg/mL (Figure 9), suggesting a good efficiency of these nanoparticles both against planktonic and sessile microorganisms, due probably to the good penetration of the obtained materials through the protective biofilm matrix. The *E. faecalis* strains were indeed the most resistant, both in planktonic and biofilm growth states, followed by *S. aureus*, particularly in the biofilm growth state. These results suggest that the Gram-positive strains seem to be more resistant to the obtained nanoparticles, compared to the Gram-negative bacteria and fungal strain, probably due to differences in the structure of the bacterial wall. The obtained nanoparticles are probably better interacting with the outer membrane of the Gram-negative bacteria, than with the thick peptidoglycan cellular wall of the Gram-positive ones through the carboxylate groups. Once adhered to the Gram-negative bacteria outer membrane, they will destabilize it and favor the microbial cell lysis. The qualitative assay shows that N nanoparticles are more active than N-800 against planktonic and biofilm embedded *E. coli* as well as against planktonic *S. aureus* strain.

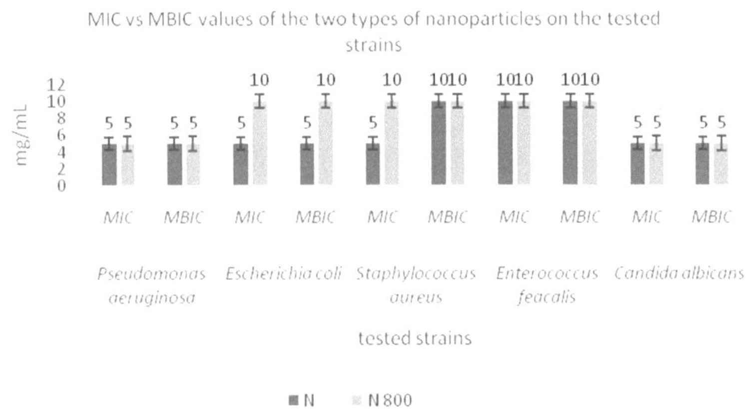

Figure 9. The results of the quantitative assay of the antimicrobial activity against planktonic and biofilm growing strains quantified by the MIC and MBIC (mg/mL) values.

4. Conclusions

In this study, a facile and eco-friendly method of plant-based synthesis of $CoFe_2O_4$ using eucalyptus leaves extract is reported for the first time in the literature. XRD patterns revealed $CoFe_2O_4$ with a face-centered cubic spinel structure. SEM analysis shows that wet ferritization using eucalyptus extract results in disordered aggregates consisting of noncrystallites sized below 10 nm, in agreement with the average crystalline size measured from diffraction patterns. Vibrational Raman spectroscopy confirms the formation of a $CoFe_2O_4$ spinel phase and evidences a high degree of spinel inversion. Thermogravimetric analysis also indicated three steps weight loss for the functional groups-coated $CoFe_2O_4$, confirming the presence of the organic components from the plant extract. The magnetic behavior of the two samples is strongly influenced by particle size and shows a rapid increase in magnetization at low fields with saturation at 5 T. The existence of a cubic magnetic anisotropy provided by the magnetocrystalline anisotropy of this material has been demonstrated. The blocking temperature is similar for both samples and coincides with that reported on nanoparticles of this type processed by other chemical methods. The results of the qualitative and quantitative bioassays confirm their potential for the development of novel antimicrobial agents, active both against planktonic and biofilm embedded microbial cells, being particularly efficient on the Gram-negative and fungal strains. Eucalyptus extract can be a valuable natural resource in the synthesis of cobalt ferrite. Plant-based synthesis methods are opening new ways to obtain various materials by cleaner pathways.

Supplementary Materials: The following supporting information can be downloaded at: https://www.mdpi.com/article/10.3390/coatings13071250/s1, Figure S1: FTIR spectra of the samples; Figure S2: Thermal analysis of dry eucalyptus extract; Figure S3: The FTIR 3D diagram for the evolved gases from thermal analysis of eucalyptus extract and its projection in the wavenumber/temperature plane.

Author Contributions: Conceptualization, D.G.; Investigation, D.C.C., J.M.C.M., G.M., C.B., O.O., S.P., M.C.C. and M.P.; Supervision, D.G. and C.B.; Writing—original draft, D.G., D.C.C., J.M.C.M., G.M., C.B., O.O., S.P. and M.C.C.; Writing—review & editing, D.G. and C.B. All authors have read and agreed to the published version of the manuscript.

Funding: This research was funded by the Romanian Ministry of Research and Innovation (MCI), UEFISCDI, through the grant PN-III-P2-2.1-PED-2021-2007 (contract no. 676 PED/2022).

Institutional Review Board Statement: Not applicable.

Informed Consent Statement: Not applicable.

Data Availability Statement: Not applicable.

Conflicts of Interest: The authors declare no conflict of interest.

References

1. Kefeni, K.K.; Msagati, T.A.M.; Mamba, B.B. Ferrite nanoparticles: Synthesis, characterisation and applications in electronic device. *Mater. Sci. Eng. B* **2017**, *215*, 37–55. [CrossRef]
2. Ahmad, S.I. Nano cobalt ferrites: Doping, structural, low-temperature, and room temperature magnetic and dielectric properties—A comprehensive review. *J. Magn. Magn. Mater.* **2022**, *562*, 169840. [CrossRef]
3. Asadi, R.; Abdollahi, H.; Gharabaghi, M.; Boroumand, Z. Effective removal of Zn (II) ions from aqueous solution by the magnetic $MnFe_2O_4$ and $CoFe_2O_4$ spinel ferrite nanoparticles with focuses on synthesis, characterization, adsorption, and desorption. *Adv. Powder Technol.* **2020**, *31*, 1480–1489. [CrossRef]
4. Aghrich, K.; Abdellaoui, M.; Mamouni, N.; Bellaouchou, A.; Fekhaoui, M.; Hlil, E.K.; Mounkachi, O. Experimental and first-principles study of the origin of the magnetic properties of $CoFe_2O_4$ spinel ferrite. *Appl. Phys. A* **2020**, *126*, 940. [CrossRef]
5. Manohar, A.; Geleta, D.D.; Krishnamoorthi, C.; Lee, J. Synthesis, characterization and magnetic hyperthermia properties of nearly monodisperse $CoFe_2O_4$ nanoparticles. *Ceram. Int.* **2020**, *46*, 28035–28041. [CrossRef]
6. Velho-Pereira, S.; Noronha, A.; Mathias, A.; Zakane, R.; Naik, V.; Naik, P.; Salker, A.V.; Naik, S.R. Antibacterial action of doped $CoFe_2O_4$ nanocrystals on multidrug resistant bacterial strains. *Mater. Sci. Eng. C* **2015**, *52*, 282–287. [CrossRef]
7. Shakil, M.S.; Hasan, M.A.; Uddin, M.F.; Islam, A.; Nahar, A.; Das, H.; Khan, M.N.I.; Dey, B.P.; Rokeya, B.; Hoque, S.M. In vivo toxicity studies of chitosan-coated cobalt ferrite nanocomplex for its application as MRI contrast dye. *ACS Appl. Bio Mater.* **2020**, *3*, 7952–7964. [CrossRef]
8. Mindru, I.; Gingasu, D.; Diamandescu, L.; Patron, L.; Marinescu, G.; Culita, D.C.; Calderon Moreno, J.M.; Preda, S.; Oprea, O.; Parvulescu, V. $CoFe_{2-x}Cr_xO_4$ ferrites: Synthesis, characterization and their catalytic activity. *Chem. Pap.* **2018**, *72*, 3203–3213. [CrossRef]
9. Imanipour, P.; Hasani, S.; Seifoddini, A.; Nabiałek, M. Synthesis and characterization of zinc and vanadium Co-substituted $CoFe_2O_4$ nanoparticles synthesized by using the sol-gel auto-combustion method. *Nanomaterials* **2022**, *12*, 752. [CrossRef]
10. Anastas, P.T.; Warner, J.C. *Green Chemistry: Theory and Practice*; Oxford University Press: New York, NY, USA, 1998.
11. Nguyen, N.T.T.; Nguyen, L.M.; Nguyen, T.T.T.; Nguyen, T.T.; Nguyen, D.T.C.; Van Tran, T. Formation, antimicrobial activity, and biomedical performance of plant-based nanoparticles: A review. *Environ. Chem. Lett.* **2022**, *20*, 2531–2571. [CrossRef]
12. Bukhari, A.; Ijaz, I.; Gilani, E.; Nazir, A.; Zain, H.; Saeed, R.; Alarfaji, S.S.; Hussain, S.; Aftab, R.; Naseer, Y. Green synthesis of metal and metal oxide nanoparticles using different plants' parts for antimicrobial activity and anticancer activity: A review article. *Coatings* **2021**, *11*, 1374. [CrossRef]
13. Sila, J.M.; Kiio, I.; Mwaura, F.B.; Michira, I.; Abongo, D.; Iwuoha, E.; Kamau, G.N. Green synthesis of silver nanoparticles using *Eucalyptus corymbia* leaves extract and antimicrobial applications. *Nano Hybrids Compos.* **2019**, *25*, 32–45. [CrossRef]

14. Shah, M.; Fawcett, D.; Sharma, S.; Tripathy, S.K.; Poinern, G.E.J. Green synthesis of metallic nanoparticles via biological entities. *Materials* **2015**, *8*, 7278–7308. [CrossRef] [PubMed]
15. Routray, K.L.; Saha, S.; Behera, D. Green synthesis approach for nano sized $CoFe_2O_4$ through aloe vera mediated sol-gel auto combustion method for high frequency devices. *Mater. Chem. Phys.* **2019**, *224*, 29–35. [CrossRef]
16. Mahajan, P.; Sharma, A.; Kaur, B.; Goyal, N.; Gautam, S. Green synthesized (*Ocimum sanctum* and *Allium sativum*) Ag-doped cobalt ferrite nanoparticles for antibacterial application. *Vacuum* **2019**, *161*, 389–397. [CrossRef]
17. Kobylinska, N.; Klymchuk, D.; Shakhovsky, A.; Khainakova, O.; Ratushnyak, Y.; Duplijc, V.; Matvieieva, N. Biosynthesis of magnetite and cobalt ferrite nanoparticles using extracts of "hairy" roots: Preparation, characterization, estimation for environmental remediation and biological applications. *RSC Adv.* **2021**, *11*, 26974. [CrossRef]
18. Gingasu, D.; Mindru, I.; Mocioiu, O.C.; Preda, S.; Stanica, N.; Patron, L.; Ianculescu, A.; Oprea, O.; Nita, S.; Paraschiv, I.; et al. Synthesis of nanocrystalline cobalt ferrite through soft chemistry methods: A green chemistry approach using sesame seed extract. *Mater. Chem. Phys.* **2016**, *182*, 219–230. [CrossRef]
19. Gingasu, D.; Mindru, I.; Preda, S.; Calderon Moreno, J.M.; Culita, D.C.; Patron, L.; Diamandescu, L. Green synthesis of cobalt ferrite nanoparticles using plant extract. *Rev. Roum. Chim.* **2017**, *62*, 647–655.
20. Gingasu, D.; Mindru, I.; Culita, D.C.; Calderon Moreno, J.M.; Bartha, C.; Greculeasa, S.; Iacob, N.; Preda, S.; Oprea, O. Structural, morphological and magnetic investigations on cobalt ferrite nanoparticles obtained through green synthesis routes. *Appl. Phys. A* **2021**, *127*, 892. [CrossRef]
21. Chandorkar, N.; Tambe, S.; Amin, P.; Madankar, C. A systematic and comprehensive review on current understanding of the pharmacological actions, molecular mechanisms, and clinical implications of the genus *Eucalyptus*. *Phytomed. Plus* **2021**, *1*, 100089. [CrossRef]
22. Avidzba, Y.N.; Koshovyi, O.M.; Komisarenko, A.M. Research of eucalyptus leaves dry extract obtained after essential oil extraction. *Pharma Inov. J.* **2014**, *3*, 18–22.
23. Shala, A.Y.; Gururan, M.A. Phytochemical properties and diverse beneficial roles of *Eucalyptus globulus* Labill.: A review. *Horticulturae* **2021**, *7*, 450. [CrossRef]
24. Santos, S.A.O.; Pinto, R.J.B.; Rocha, S.M.; Marques, P.A.A.P.; Neto, C.P.; Silvestre, A.J.D.; Freire, C.S.R. Unveiling the chemistry behind the green synthesis of metal nanoparticles. *Chem. Sus. Chem.* **2014**, *7*, 2704–2711. [CrossRef] [PubMed]
25. El-Seedi, H.R.; El-Shabasy, R.M.; Khalifa, S.A.M.; Saeed, A.; Shah, A.; Shah, R.; Iftikhar, F.J.; Daim, M.M.A.; Omri, A.; Hajrahand, N.H.; et al. Metal nanoparticles fabricated by green chemistry using natural extracts: Biosynthesis, mechanisms, and applications. *RSC Adv.* **2019**, *9*, 24539. [CrossRef] [PubMed]
26. Balciunaitiene, A.; Liaudanskas, M.; Puzeryte, V.; Viškelis, J.; Janulis, V.; Viškelis, P.; Griškonis, E.; Jankauskaite, V. *Eucalyptus globulus* and *Salvia officinalis* extracts mediated green synthesis of silver nanoparticles and their application as an antioxidant and antimicrobial agent. *Plants* **2022**, *11*, 1085. [CrossRef]
27. Munir, H.; Mumtaz, A.; Rashid, R.; Najeeb, J.; Zubair, M.T.; Munir, S.; Bilal, M.; Cheng, H. Eucalyptus camaldulensis gum as a green matrix to fabrication of zinc and silver nanoparticles: Characterization and novel prospects as antimicrobial and dye-degrading agents. *J. Mater. Res. Technol.* **2020**, *9*, 15513–15524. [CrossRef]
28. Alves Puiatti, G.; Passos de Carvalho, J.; Teixeira de Matos, A.; Pereira Lopes, R. Green synthesis of Fe0 nanoparticles using Eucalyptus grandis leaf extract: Characterization and application for dye degradation by a (Photo) Fenton-like process. *J. Environ. Manag.* **2022**, *311*, 114828. [CrossRef]
29. Torres-Limiñana, J.; Feregrino-Pérez, A.A.; Vega-González, M.; Escobar-Alarcón, L.; Cervantes-Chávez, J.A.; Esquivel, K. Green synthesis via Eucalyptus globulus L. extract of Ag-TiO_2 catalyst: Antimicrobial activity evaluation toward water disinfection process. *Nanomaterials* **2022**, *12*, 1944. [CrossRef]
30. Barzinjy, A.A.; Azeez, H.H. Green synthesis and characterization of zinc oxide nanoparticles using *Eucalyptus globulus* Labill. leaf extract and zinc nitrate hexahydrate salt. *SN Appl. Sci.* **2020**, *2*, 991. [CrossRef]
31. Maheshwaran, G.; Muneeswari, R.S.; Bharathi, A.N.; Kumar, M.K.; Sudhahar, S. Eco-friendly synthesis of lanthanum oxide nanoparticles by Eucalyptus globulus leaf extracts for effective biomedical applications. *Mater. Lett.* **2021**, *283*, 128799. [CrossRef]
32. Iconaru, S.L.; Ciobanu, C.S.; Predoi, D.; Motelica-Heino, M.; Negrilă, C.C.; Badea, M.L.; Predoi, M.V.; Chifiriuc, C.M.; Popa, M. Nitrogen and bromide Co-doped hydroxyapatite thin films with antimicrobial properties. *Coatings* **2021**, *11*, 1505. [CrossRef]
33. Marinescu, G.; Culita, D.C.; Mocanu, T.; Mitran, R.-A.; Petrescu, S.; Stan, M.S.; Chifiriuc, M.C.; Popa, M. New nanostructured materials based on mesoporous silica loaded with Ru(II)/Ru(III) complexes with anticancer and antimicrobial properties. *Pharmaceutics* **2023**, *15*, 1458. [CrossRef] [PubMed]
34. Pradeep, M.; Kruszka, D.; Kachlicki, P.; Mondal, D.; Franklin, G. Uncovering the phytochemical basis and the mechanism of plant extract-mediated eco-friendly synthesis of silver nanoparticles using ultra-performance liquid chromatography coupled with a photodiode array and high-resolution mass spectrometry. *ACS Sustain. Chem. Eng.* **2022**, *10*, 562–571. [CrossRef]
35. Banifatemi, S.S.; Davar, F.; Aghabarari, B.; Segura, J.A.; Alonso, F.J.; Ghoreishi, S.M. Green synthesis of $CoFe_2O_4$ nanoparticles using olive leaf extract and characterization of their magnetic properties. *Ceram. Int.* **2021**, *47*, 19198–19204. [CrossRef]
36. Hosterman, B.D. Raman Spectroscopic Study of Solid Solution Spinel Oxides. Ph.D. Dissertation, University of Nevada, Las Vegas, NV, USA, August 2011. [CrossRef]
37. Slotznick, S.P.; Shim, S.-H. In situ Raman spectroscopy measurements of $MgAl_2O_4$ spinel up to 1400 °C. *Am. Mineral.* **2008**, *93*, 470–476. [CrossRef]

38. D'Ippolito, V.; Andreozzi, G.B.; Bersani, D.; Lottici, P.P. Raman fingerprint of chromate, aluminate and ferrite spinels. *J. Raman Spectrosc.* **2015**, *46*, 1255–1264. [CrossRef]
39. Chandramohan, P.; Srinivasan, M.P.; Velmurugan, S.; Narasimhan, S.V. Cation distribution and particle size effect on Raman spectrum of $CoFe_2O_4$. *J. Solid State Chem.* **2011**, *184*, 89–96. [CrossRef]
40. Motelica, L.; Ficai, D.; Oprea, O.; Ficai, A.; Trusca, R.-D.; Andronescu, E.; Holban, A.M. Biodegradable alginate films with ZnO nanoparticles and Citronella essential oil—A novel antimicrobial structure. *Pharmaceutics* **2021**, *13*, 1020. [CrossRef]
41. Grigorova, M.; Blythe, H.J.; Blaskov, V.; Rusanov, V.; Petkov, V.; Masheva, V.; Nihtianova, D.; Martinez, L.; Munoz, J.S.; Mikhov, M. Magnetic properties and Mössbauer spectra of nanosized $CoFe_2O_4$ powders. *J. Magn. Magn. Mater.* **1998**, *183*, 163–172. [CrossRef]
42. Palade, P.; Comanescu, C.; Kuncser, A.; Berger, D.; Matei, C.; Iacob, N.; Kuncser, V. Mesoporous cobalt ferrite nanosystems obtained by surfactant-assisted hydrothermal method: Tuning morpho-structural and magnetic properties via pH-variation. *Nanomaterials* **2020**, *10*, 476. [CrossRef]
43. Hunpratub, S.; Phokha, S.; Kidkhunthod, P.; Chanlek, N.; Chindaprasirt, P. The effect of cation distribution on the magnetic properties of $CoFe_2O_4$ nanoparticles. *Results Phys.* **2021**, *24*, 104112. [CrossRef]
44. Stoner, E.C.; Wohlfarth, E.P. A mechanism of magnetic hysteresis in heterogeneous alloys. *Philos. Trans. R. Soc. A* **1948**, *240*, 599. [CrossRef]
45. Chikazumi, S. *Physics of Ferromagnetism*; Oxford University Press: Oxford, UK, 1997.

coatings

Article

Photopolymerised Coatings with Nanopigments Based on Dye Mixtures

Viktor B. Ivanov [1], Elena V. Solina [1] and Peter P. Levin [1,2,*]

[1] Department of Dynamics of Chemical and Biological Processes, Semenov Federal Research Center of Chemical Physics, Russian Academy of Sciences, Moscow 119991, Russia
[2] Laboratory of Photo- and Chemiluminescent Processes, Emanuel Institute of Biochemical Physics, Russian Academy of Sciences, Moscow 119334, Russia
* Correspondence: levinp@mail.ru or levinp@sky.chph.ras.ru; Tel.: +7-4959397314

Abstract: The properties of new coatings with nanopigments (NPs) prepared by the dye mixture treatment of montmorillonite modified with cationic surfactants were investigated. The colorimetric parameters of polymer films with nanopigments were determined. The mutual interference of the dyes in their adsorption on nanoparticles and their colourfastness (photostability) to light was analysed. The properties of oligomer/monomer compositions with synthesised nanopigments designed for the printing of non-metallic surfaces were discussed. The structure of nanopigments and polymerised oligomer/monomer compositions was revealed by small-angle X-ray diffraction, UV–Vis spectra, and atomic force microscopy. NPs with a wide colour range were prepared using dye mixtures. One other advantage of these NPs over NPs based on individual dyes is the possibility to increase the total concentration of the colouring components in the NPs and to increase the colourfastness.

Keywords: coating; nanopigments; dye; montmorillonite; surfactant; surfaces printing

1. Introduction

Photopolymerised compositions are widely used in various fields of technology, but their use as materials for 3D printing [1], in holography [2,3], and in flexographic printing [4] seems to be particularly promising. The use of nanomaterials for coatings is also encouraging [5]. It seems obvious that the combination of these technologies can lead to the production of new materials and products with new, unique properties.

Nanopigments (NPs) produced by different methods represent a rather interesting class of nanomaterials. Therefore, their preparation and the analysis of their properties can be quite interesting to investigators in various branches of chemistry and chemical technology [6–14]. The main efforts are currently concentrated on the synthesis and study of inorganic NPs [6–14]. Such nanoparticles, in particular, can be used in order to obtain magnetically sensitive holographic images [3].

However, the development of a method for the preparation of organic pigments with a nano-sized structure is also of interest [12–14]. Intermediate to these research topics is the synthesis of hybrid organic–inorganic NPs [15–21], which can be achieved in particular by applying organic dyes on inorganic nanoparticles, such as natural montmorillonite [16] or montmorillonite modified by cationic surfactants (MMCS) [19–21]. The advantage of MMCS is the good compatibility of NPs with monomers, oligomers, and polymers, which significantly contributes to high colour quality. The essential limitations of such NPs are the relatively low concentrations of the colouring component (a dye)—which usually does not exceed 30 wt%—and restricted colour range. The latter circumstance is associated less with the "discreteness" of colours of individual dyes and more with the incapability of many of them to withstand the severe conditions of use in NPs, particularly during photopolymerisation in oligomer/monomer compositions. As shown previously, the colour range can be significantly improved using NPs based on dye mixtures [19,20].

The goals of this investigation were to extend the class of NPs based on MMCS and dye mixtures, to analyse their composition and properties, and to analyse the possibility of using NPs in oligomer/monomer compositions.

The scientific novelty of this work was in analysing the possibility of obtaining nanopigments based on dye mixtures, evaluating the adsorption of each of the dyes from their mixture, and identifying the specific effects of increasing the total amount of adsorbed dyes compared to individual dyes. The dyes' nanopigment colourfastness to light in a polymer matrix, along with the effects of the mutual influence of dyes on colourfastness, was studied. The effects of the nanopigments on the photopolymerisation of oligomer/monomer compositions in comparison with that of the conventional pigments were evaluated. The structure and the resistance of the coatings to physical and chemical influences were analysed.

Our approach to solving these problems is based on three hypotheses: (1) dyes can be adsorbed independently on the surface of montmorillonite, and this can lead to an increase in the total content of the colouring component in the NPs; (2) dyes can mutually affect their resistance to light due to shielding action or intermolecular energy transfer; (3) NPs based on mixtures of dyes, as well as NPs based on individual dyes, can be used in photopolymerising compositions intended, in particular, for printing on textile materials.

Experimental verification of these hypotheses constitutes the novelty of this work and its potential practical significance.

The choice of Cloisite 20A and Cloisite 10A products as nanoparticles for the production of NPs was due to the fact that they contain the same modifying surfactants, but at significantly different contents. This allows modification of the NPs' properties. An important additional argument was the availability of data obtained earlier for similar NPs based on individual dyes [21], as well as the high quality and commercial availability of the products.

The choice of dyes for the production of NPs was determined by the expediency of testing the possibility of using dyes of various colours and different classes (i.e., cationic, anionic, and dispersed) used in the textile industry. In particular, this guarantees the availability of acceptable hygienic characteristics. An additional factor, as before, was the availability of data on the properties of NPs based on the individual use of some of these dyes.

2. Experimental

2.1. Materials

The samples of montmorillonite modified by the cationic surfactants $(HT)_2N^+(CH_3)_2Cl^-$ (MMCS-1) and $HTN^+(CH_3)_2CH_2C_6H_5Cl^-$ (MMCS-2)—manufactured by Southern Clay Products, Inc. (Austin, TX, USA) under the brand names Cloisite 20A and Cloisite 10A, respectively—were used as nanoparticles for NP preparation, where HT is alkyl C_{18} (~65%), C_{16} (~30%), and C_{14} (~5%).

The dyes Basic Red 13 (CI 48015), Basic Red 18 (CI 11085), Acid Red 18 (CI 16255), Acid Red 73 (CI 27290), Acid Red 441, Mordant Red 7 (CI 18760), Direct Red 239 (CI 29160), Acid Yellow 76 (CI 18850), Basic Yellow 13 (CI 48056), Direct Yellow 1 (CI 22250), Disperse Yellow 13 (CI 58900), Acid Violet 49 (CI 42640), Direct Violet 9 (CI 27885), Acid Green 25 (CI 61570), Direct Blue 151 (CI 24175), Direct Blue 199 (CI 74190), Disperse Blue 7 (CI 62500), Disperse Blue 56 (CI 63285), Direct Brown 95 (CI 30145), Disperse Orange 11 (CI 60700), Disperse Orange 30 (CI 11119), Disperse Red 50 (CI 11226), Reactive Red 247 (produced by "Novocheboksarsky Chemical Plant" OJSC (Novocheboksarsk, Russia)), and Reactive Blue 38 (SCPL, India) were used as colouring components of the NPs.

For NP preparation and film-casting, analytical-grade ethanol, acetone, benzene, and methylene chloride were used.

The main binder components for the oligomer/monomer compositions were a laboratory sample of oligourethane methacrylates with an average molecular weight of 1400, based on oligotetrahydrofuran with tolylene isocyanate end groups (OUMA), which were synthesised by the Polymer Department of Semenov Federal Research Center for Chemical Physics, Russian Academy of Sciences, Moscow Russia. The industrial products of the interaction of 2,4-toluylene diisocyanate with Laprol 2000 and hydroxypropyl methacrylate (OUMA-2002T) or hydroxypropyl methacrylate (OUMA MTM, "Khimtranzit" LLC, (Dzerzhinsk, Russia)) were used as well.

In order to increase the colour composition fixing rate, trimethylolpropane triacrylate ("Yarsintez" JSC RI, Yaroslavl, Russia) was used as a crosslinking agent.

The photoinitiator was 2,6-di-tert-butyl anthraquinone (NIOPiK, Moscow, Russia), and the non-ionogenic surfactants were polypropylene glycol (Schuchardt, Munchen, Germany), polyoxyethylene sorbitol monostearate (Tween 20, Merk KGaA, Darmstadt, Germany), and polyethylene glycol (Ferak Berlin GmbH, Berlin, Germany).

Poly(methyl methacrylate) (PMMA) of the Plexiglas 8N brand with $M \approx 160{,}000$ and polystyrene (PS) of the PSM-115 brand were the model media for the study of the NPs' colorimetric parameters and the phototransformations of the dyes in terms of composition.

The choice of dyes was determined by the need to study dyes of different classes (i.e., cationic, anionic, dispersed, and active), by their colour characteristics, and by the ability to compare the obtained results with similar data for nanopigments based on individual dyes.

The purity of the nanopigments was determined by the purity of the modified montmorillonite and dyes used. These products were directly obtained from manufacturers who guarantee their quality in accordance with specifications. Chemically pure solvents and distilled water used during the application of dyes cannot lead to contamination of the resulting nanopigments.

2.2. Methods

Dyed films with NPs were formed by the slow evaporation of the solvent from PMMA or PS solution in a benzene–methylene chloride mixture. To improve the distribution of NPs in the polymer solution, samples were preliminarily processed using an ultrasonic disperser for 30–60 s.

The oligomer/monomer compositions were produced by mixing oligourethane methacrylate (67.5 wt%), trimethylolpropane triacrylate (22.5 wt%), 2,6-di-tert-butyl anthraquinone (2.6 wt%), NPs (3.7–7.4 wt%) and, in individual cases, non-ionogenic surfactants (0–7 wt%). The mixtures were then processed by an ultrasonic disperser for 30–60 s.

Photochemical aging was applied using a Q-SUN (Q-Lab corporation, Cleveland, OH, USA) device utilising filter systems matching the spectral composition of sunlight under natural conditions (i.e., radiation wavelength greater than 290 nm), with a light intensity of 1000 W/m^2.

The colorimetric parameters of dyed films in the CIELAB-76 system were determined using a ColorFlex spectrocolorimeter (HunterLab, Reston, VA, USA). The values of L^*, a^*, and b^* were determined directly from the measurements of the reflection spectra and their processing by the ColorFlex instrument software (version 1.32B).

Colourfastness was tested in accordance with the following standards: ISO 105-B04 (to light), ISO 105-C06 (to washing), ISO 105-E01 (to distilled water), ISO 105-E04 (to perspiration), and ISO 105-X12 (to dry rubbing).

The NP sizes and the surface parameters of the dyed compositions were determined using an atomic force microscope (NT-MDT, Ltd, Moscow, Russia). Small-angle X-ray measurements were performed using a Bruker NANOSTAR device (Bruker Inc., Billerica, MA, USA).

The UV–Vis absorption spectra of dyes in solutions and films were recorded using a MultiSpec-1501 spectrophotometer (Shimadzu, Kyoto, Japan).

3. Results and Discussion

3.1. NP Synthesis Based on Dye Mixtures

NPs based on dye mixtures were prepared in accordance with the synthesis techniques for NPs based on individual dyes [21].

NP synthesis was carried out by simply adding a mixture of dyes in water or a mixture of water and acetone (for cationic and anionic dyes), or in acetone (for dispersed dyes), to modified montmorillonite while stirring the solution, after which they were isolated by filtration. The dye content in the NPs was determined based on the differences in their contents in the solution before and after treatment of the montmorillonite, estimated based on the changes in the absorption spectra of the solutions.

It was shown that, after the addition of the dye, the filtrate can be reused for the synthesis of NPs; therefore, this method of obtaining NPs is largely waste-free.

Numerous experimental data indicated that the total dye concentration in NPs depends on a number of factors, including the following:

- The structure, ratio, and total concentration of the applied dyes in the solution;
- The structure of the surfactants used for MMCS production and their contents of montmorillonite;
- The ratio of the quantity of MMCS to the volume of the dye solution.

The data obtained provide qualitative proof that, during the preparation of NPs based on dye mixtures, the dye features and surfactant structural effects are the same as those for the individual dyes [21].

3.2. Some Features and Advantages of NPs

Colorimetric analysis of PMMA with NPs based on dye mixtures (Table 1) proves that their use can yield paints of various colours and hues.

Table 1. Conditions of nanopigments preparation, dyes' loading in nanopigments, and colour coordinates of the coloured PMMA films.

MMCS	Dyes (wt%)	Solvent	Dye Loading in NP [a] (wt%)	NP Content in PMMA (wt%)	Colour Coordinates		
					L*	a*	b*
MMCS-2	Disperse Blue 56 (10)/Disperse Yellow 13 (10)	EtOH	17.0	8	18.50	−41.11	0.10
MMCS-2	Disperse Blue 56 (5)/Disperse Yellow 13 (15)	EtOH	16.1	10	23.28	−46.16	19.57
MMCS-1	Direct Blue 199 (5)/Direct Yellow 1(15)	EtOH/H_2O	13.4	10	27.97	−52.43	36.56
MMCS-1	Direct Blue 199 (10)/Direct Yellow 1 (10)	EtOH/H_2O	13.3	10	17.52	−43.47	18.55
MMCS-2	Disperse Orange 30 (15)/Disperse Blue 7 (15)	Me_2CO/H_2O	15.0/15.0	30	31.87	−32.91	25.44
MMCS-2	Direct Red 239 (30)/Direct Violet 9 (10)	EtOH/H_2O	16.9/7.8	4	8.50	34.55	11.92
MMCS-2	Direct Red 239 (20)/Direct Violet 9 (20)	EtOH/H_2O	10.5/16.4	4	4.97	23.32	4.56
MMCS-1	Basic Yellow 24 (20.4)/Direct Yellow 1 (10)	EtOH/H_2O	20.3/9.8	8	58.67	32.10	93.80
MMCS-2	Acid Violet 49 (18)/Disperse Blue 56 (18)	Me_2CO/H_2O	2.4/18.0	8	1.96	3.63	−6.62
MMCS-2	Basic Red 18 (30)/Acid Violet 49 (3.3)	EtOH/H_2O	19.6/3.3	8	4.71	15.78	4.70
MMCS-2	Acid Violet 49 (20)/Acid Yellow 76 (30)	EtOH/H_2O	6.7/28.8	4	0.73	0.38	0.13
MMCS-2	Direct Yellow 1 (20)/Direct Red 239 (20)	EtOH/H_2O	16.0/8.2	4	35.24	60.67	56.96
MMCS-2	Acid Yellow 76 (20)/Reactive Blue 38 (20)	EtOH/H_2O	16.9/15.3	4	10.86	−24.02	14.92

[a] Weight percent per mass of MMCS.

For instance, NPs based on yellow and blue dyes, red and violet dyes, and red and yellow dyes produce green, vinous, and orange tints, respectively. This expected result can be considered to be the first advantage of NPs based on dye mixtures over NPs based on individual dyes.

NPs based on dye mixtures were also found to possess an additional, unexpected feature: In a number of cases, when the dye mixtures were applied on MMCS, their total content in the NPs significantly exceeded the contents of individually applied dyes under the same NP production conditions (Table 1). This important feature has a potentially valuable practical application: the possibility of achieving strong hues without increasing the NP concentration in the dye composition. This phenomenon is general, being observed for both mixtures of dyes of the same class and for mixtures of various classes, including dyes of the same colour (Table 1). This feature can be considered as a second important advantage of NPs based on dye mixtures.

A more sophisticated analysis of the effect of total dye content in NPs on their ratio in the mixture of Acid Yellow 76 and Reactive Blue 38 dyes (Figure 1) reveals a parabolic shape with downward-directed branches, where the top is usually located in the area of comparable quantities of each dye. Such plots are typical for well-known synergistic effects of the oxidation processes of organic compounds and polymers [22]. It is interesting that the relationships for dye loadings are in close agreement with the relationships for dye concentrations in solutions used for the NP preparation process (1, 0.70, 0.48, and 0.24 versus 1.0, 0.75, 0.50, and 0.25, respectively). Among these effects, a special group is composed of those explained by trivial mechanisms, in accordance with which the synergistic effect is associated with the limited solubility of one or both of the additives [23]. This synergistic effect is due to the interactions of dyes when the adsorption of one dye does not interfere with, but rather promotes the adsorption of another. Formally, a similar effect is observed for polymer light stabilisers when one stabiliser improves the distribution (solubility) of the other in the polymer [23]. In the first case, each of the additives dissolves independently, and their total concentrations increase relative to their individual concentrations. In the second case, one of the additives increases the solubility of another additive, increasing its concentration in the mixture. In the case of dye adsorption, possible synergistic mechanisms may also be trivial. If each of the dyes is mostly adsorbed on centres of different types, the total content will be higher than that in the case of the adsorption of individual substances. Correspondingly, in the case of multilayer adsorption, one of the dyes may form new centres for the adsorption of another dye. Particular ratios of dye concentrations at which the synergistic effect is strongest can only be determined empirically at present, because of the lack of experimental data allowing the determination of the mechanism.

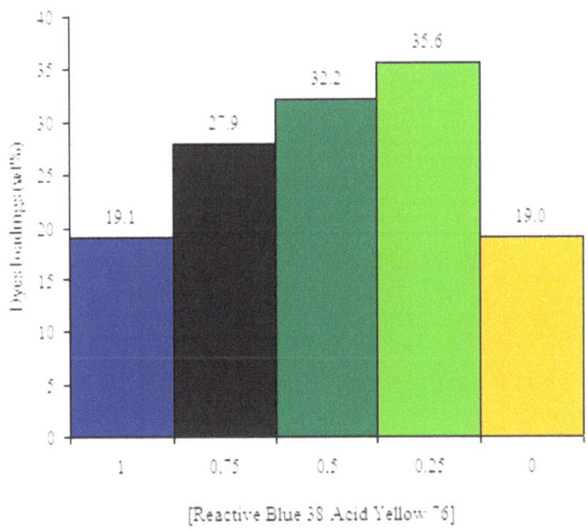

Figure 1. The effect of limited dye loading in MMCS-2-based NPs on the relationship of concentrations of Reactive Blue 38 and Acid Yellow 76 in the ethanol/water solutions used for NP preparation. (Blue is Reactive Blue 38 only, dark green, medium green and light green is Reactive Blue 38/ (Reactive Blue 38 + Acid Yellow 76) relation 0.75, 0.5 or 0.25, and yellow is Acid Yellow 76 only).

The use of NPs based on dye mixtures was also found to produce increasing colourfastness to light (photostability) (Table 2), evaluated in points (i.e., ISO 105-B04: 1997); colour resistance to artificial weathering; xenon arc lamp attenuation test).

In some cases, the effect was rather significant: the colourfastness of the mixed pigment not only reached the colourfastness of the most stable component, but even noticeably exceeded it (Table 2). For such systems, the phenomenon can be considered as truly synergistic one, because the effect observed exceeds not only the sum of the effects of the components with regard to their ratio in the mixture, but even the value that might be provided by the most efficient component. This can be considered as one other advantage of NPs based on dye mixtures compared with NPs based on individual dyes.

Table 2. Colourfastness of PMMA films [a] with nanopigments.

Nanopigment	MMCS	Dye Loading in NP (wt%)	NP Content in PMMA (wt%)	Colourfastness
Direct Violet 9	MMCS-2	19.8	4	4–5
Direct Red 239/Direct Violet 9	MMCS-2	10.5/16.4	4	5–6
Direct Red 239/Direct Violet 9	MMCS-2	16.9/7.8	4	3–4
Direct Red 239	MMCS-2	18	4	2–3
Direct Blue 199	MMCS-1	10.5	4	4
Direct Blue 199/Direct Yellow 1	MMCS-1	13.3	10	7
Direct Blue 199/Direct Yellow 1	MMCS-1	13.4	10	6–7
Direct Yellow 1	MMCS-1	9.8	8	6–7
Basic Yellow 24	MMCS-1	20.4	4	5
Basic Yellow 24/Direct Yellow 1	MMCS-1	20.3/9.8	8	5

[a] The thickness of the PMMA films was ~400 μm.

The mechanism of this synergistic stabilising action is rather complicated. It is necessary to take into account the colourfastness (photostability) of the dyes themselves, estimated based on the relative rate of their consumption, as measured by the decrease in absorption in the visible region. The simplest explanation is based on experimental facts indicating a (frequently significant) reduction in the consumption rate of the less-stable dye, caused by light in the presence of the more stable dye (Figure 2, curves 1 and 3).

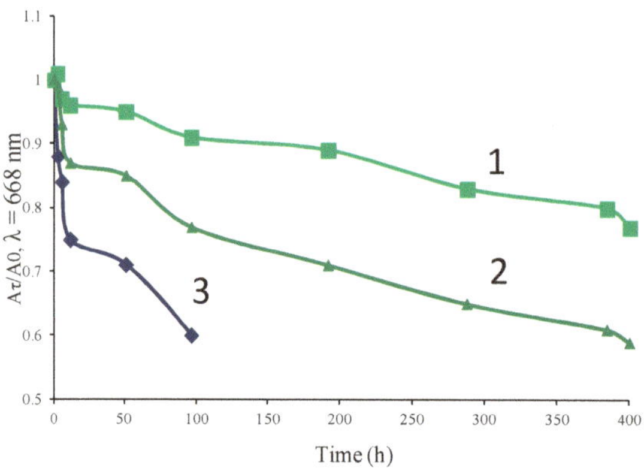

Figure 2. Kinetic curves of the photodegradation of Reactive Blue 38 in PS films coloured by MMCS-2-based NPs with a mixture of Reactive Blue 38 (8.6 wt%) and Acid Yellow 76 (27.0 wt%) (1) and in mixtures of NPs with Reactive Blue 38 (19.1 wt%) and NPs with Acid Yellow 76 (19.0 wt%) only (2). Curve 3 shows the kinetics of Reactive Blue 38 consumption in the film coloured by NPs with Reactive Blue 38 (19.1 wt%) only. The NP contents in the films are 1.5 (1), 0.7 (2), and 0.8 wt% (3). The film thickness is 400 μm. The measured parameter is the optical density at time (h) after photolysis (Aτ) and initial one (A0) at a wavelength of 668 nm, corresponding to the absorption maximum of Reactive Blue 38, as determined experimentally based on the UV–Vis absorption spectrum of PS film containing NPs with Reactive Blue 38 only.

Figure 2 describes the changes in optical density at a wavelength of 668 nm, corresponding to the maximum of the absorption spectrum of Reactive Blue 38. As can be seen in Figure 2, the kinetic curves have a complex shape—a fast initial section is followed by a slower stage. The criterion of stability in this case is a decrease in optical density by 40%, corresponding to a pronounced change in the colour of the composition; therefore, further continuation of irradiation is of no practical importance.

In the example under consideration for NPs based on Acid Yellow 76 and Reactive Blue 38, the screening effect contributes significantly. This screening action is due to the fact that the light-resistant yellow dye strongly absorbs light in the short-wavelength visible region and, thus, reduces the amount of light absorbed by the blue dye in this—the most dangerous area for it. In this case, Acid Yellow 76, which is a more light-stable dye, protects (screens) Reactive Blue 38—the less stable dye—against the most dangerous UV light. The contribution of the screening effect can be assessed by analysing the consumption kinetics of Reactive Blue 38 in the mixture of NPs based on dyes individually applied on MMCS (Figure 2, curve 2). Importantly, in this experiment, for the samples that include a pigment based on a mixture of dyes and a mixture of pigments containing individual dyes, equal optical density in the absorption range of every dye was required. Comparing curves 1–3 in Figure 2, the screening effect contributes significantly to the retardation of the consumption of Reactive Blue 38—the less light-stable dye—in both the initial fast stage and the later slower stages.

It is worth mentioning that the presence of clearly defined bands in the absorption spectra and the absence of noticeable light scattering are important proof of the adsorption of the dyes at the molecular level. It is essential that absorption spectra are also recorded for ionic dyes that are not soluble in organic solvents and cannot be directly embedded in organic polymers and photopolymerising systems.

When NPs with extremely low colourfastness dyes were used (Figure 3), a significant increase in the consumption rate for the more stable dye was observed (Figure 4), leading to low colourfastness. For instance, in the presence of NPs based on a mixture of Disperse Blue 56 and Disperse Yellow 13, the colourfastness of the PMMA films was only 2, whereas that for NPs based on individual Disperse Blue 56 dye was 3. However, even in this specially selected example, where a photochemically active dye predominates over a stable dye, the antagonism for dye mixtures in NPs was not observed. This is because the more stable Disperse Blue 56 dye retards the consumption of the Disperse Yellow 13 dye. Consequently, the colourfastness for NPs based on dye mixtures is much higher than for NPs based on Disperse Yellow 13, the colourfastness of which is only 1.

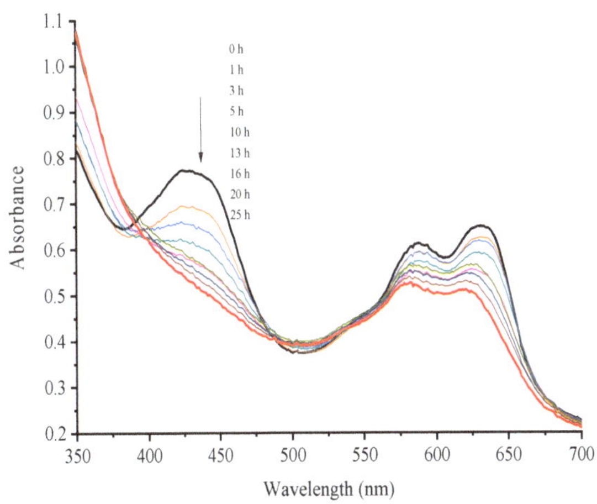

Figure 3. Variation in the near-UV and visible absorption spectra of PMMA films coloured by MMCS-2-based NPs with Disperse Blue 56 and Disperse Yellow 13 with the duration of photolysis, as determined by Suntest. The NP content in the film was 1.04 wt%, and the loadings of Disperse Blue 56 and Disperse Yellow 13 in the mixture were 16.1 wt%. The film thickness was 400 μm. The numbers in the figure next to the arrow indicate the irradiation time, hours.

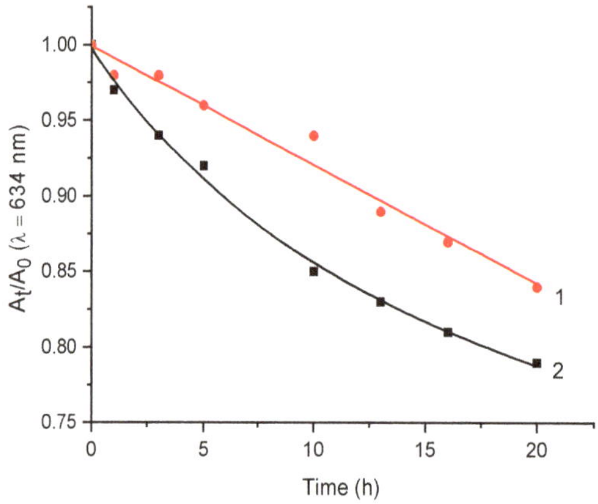

Figure 4. Kinetic curves of the photodegradation of Disperse Blue 56 in PMMA films coloured by MMCS-2-based NPs with Disperse Blue 56 (17.5 wt%) only (1) or a mixture of Disperse Blue 56 and Disperse Yellow 13 (16.1 wt%) during irradiation in Suntest. The NP contents in the film were 2.45 wt% (1) and 3.1 wt% (2), and the film thickness was 300 µm. The measured parameter was the optical density at a wavelength of 634 nm, corresponding to the maximum absorption of Disperse Blue 56, as determined experimentally by the UV–Vis absorption spectrum of a PS film containing NPs with Reactive Blue 56 only.

3.3. UV Inks Based on NPs with Dye Mixtures

NPs based on mixtures, similar to those based on individual dyes [21], may be used as the colouring components of photopolymerisable oligomer/monomer compositions (UV inks). These inks preserve all of the advantages in comparison with UV inks based on common pigments, among which higher colourfastness to rubbing and reduced UV radiation time required for the fast fixing of colouring compositions are of the highest importance (Table 3). In this respect, compared with NPs based on individual dyes, NPs based on dye mixtures used in UV inks naturally preserve all of the advantages associated with the possibility of varying colour and hue over a broad range due to the increasing concentration of the colouring agent, and they also exhibit increased colourfastness.

Table 3. Colourfastness of textile materials coloured by UV inks [a] with nanopigments or pigments to physicochemical impacts.

Nanopigment (wt%)	MMCS	Dye Loading in NP [a] (wt%)	Content of Colouring Component (wt%)	Physicochemical Impacts				Irradiation Time (min)
				Dry Rubbing	Distilled Water	Perspiration	Washing	
Acid Yellow 76/Reactive Blue 38	MMCS-2	16.9/15.3	4	5/4	5/5/5	5/5/5	5/5/5	2
Acid Yellow 76/Reactive Blue 38	MMCS-2	8.5/19.4	4	5/4	5/5/5	5/5/5	5/5/5	2
Direct Yellow 1/Direct Red 239	MMCS-2	16.0/8.2	4	5/3	5/5/5	5/5/5	5/5/5	2
Disperse Blue 56/Disperse Yellow 13	MMCS-2	17.0	7	5/5	5/5/5	5/5/5	5/5/5	2
Direct Blue 199/Direct Yellow 1	MMCS-1	13.3	7	5/5	5/5/5	5/5/5	5/5/5	2
Direct Blue 199/Direct Yellow 1	MMCS-1	13.4	7	5/5	5/5/5	5/5/5	5/5/5	2
Acid Violet 49/Disperse Blue 56	MMCS-2	2.4/18	7	5/4–5	5/5/5	5/5/5	5/5/5	3
Direct Red 239/Direct Violet 9	MMCS-2	10.5/16.4	4	5/4	5/5/5	5/5/5	5/5/4	2
Direct Red 239/Direct Violet 9	MMCS-2	16.9/7.8	4	5/4–5	5/5/5	5/5/5	5/5/4	2
Pigment Irgalite Blue NGA	-	-	2	4/2–3	5/5/5	5/5/5	5/5/5	12

[a] OUMA (67–70 wt%), TMPTA (23 wt%), 2,6-di-tert-butylanthraquinone (3 wt%).

3.4. Nano-Sized Structure of Printing Composition with NPs Based on Dye Mixtures

Similar to the case with NPs based on individual dyes [21], the advantages of UV inks with NPs based on dye mixtures are likely the result of their nano-sized structure. One important source of support for this hypothesis is provided by the AFM data (Figure 5), which indicate the absence of structures higher than 100 nm and wider than 500 nm at the base from the film surface produced by curing UV inks with NPs based on a mixture of Acid Violet 49 and Disperse Blue 56 dyes. As mentioned above [21], inks with NPs are fundamentally different from UV inks based on common pigments. In the latter case, the surface has heterogeneities with typical sizes of 0.8–1.2 μm in height and 1–2 μm in width at the base.

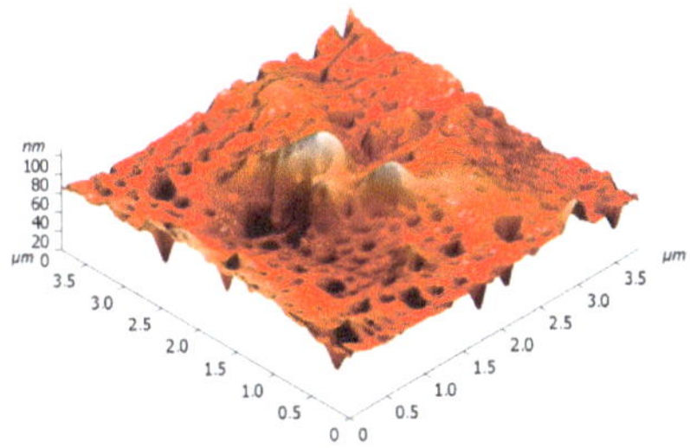

Figure 5. AFM image of a UV-cured film containing OUMA (67 wt%), TMPTA (23 wt%), 2,6-di-tert-butylanthraquinone (3 wt%), and MMCS-2-based NPs with a mixture of Acid Violet 49 and Disperse Blue 56. The NP content is 7 wt%, and the loadings of Acid Violet 49 and Disperse Blue 56 on the MMCS-2 are 2.4 and 18.0 wt%, respectively.

Figure 3 shows that, compared with films coloured by common pigments, films containing NPs based on dye mixtures have an additional feature: they are transparent in the visible range beyond the absorption bands of the dyes. Thus, the NP sizes do not fundamentally exceed the light wavelength (400–700 nm). Therefore, the spectrophotometric analysis corroborates the conclusion about the nano-sized structure of NPs and materials based on these NPs, supporting the AFM results.

A more in-depth study of the films by small-angle X-ray diffraction analysis, however, indicates that some of the NP nanoparticles are present in the form of aggregates (packs), each comprising several tens of MMCS plates with applied dyes (Figure 6). The presence of clearly expressed basal reflexes, which are absent in similar films without NPs and in films with a common pigment, proves the presence of regulation zones provided by the formation of MMCS plate aggregates (30 nm thick) with a characteristic distance between layers (plates) of 3.5 nm. These assessments correlate well with the data obtained for analogous samples containing MMCS in the absence of a dye (i.e., an aggregate thickness of ~35 and an interlayer distance of 3.9 nm). Because the interlayer distance in these aggregates exceeds the MMCS interlayer distance (1.9 nm), the data obtained indicate intercalation of the oligomer/monomer binder to almost the same extent in both NPs and MMCS without the dye. These results are fully consistent with what is known about the structure of polymer mixtures with montmorillonite nanoparticles, in accordance with which both polymers intercalate into the interlayer space when mixed, and nanoparticles exfoliate [23]. The local microsurroundings of the dye molecules in such aggregates (i.e., oligomer/monomer binder and a surfactant fragment) do not significantly differ from their surroundings when they are located in individual MMCS particles. Consequently, the difference in spectral characteristics should not be significant.

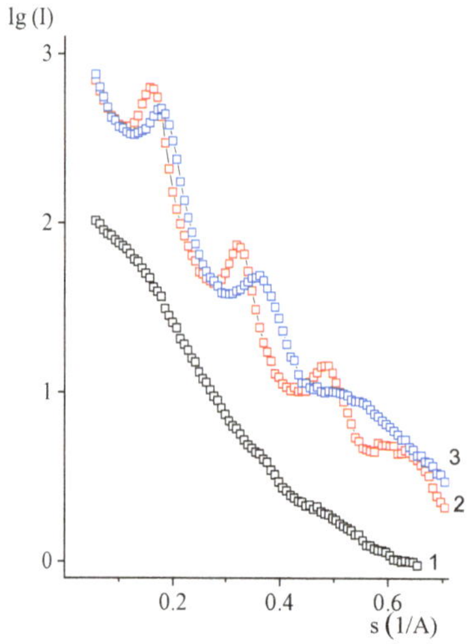

Figure 6. The small-angle X-ray scattering of UV-cured films containing OUMA (67 wt%), TMPTA (23 wt%), and 2,6-di-tert-butylanthraquinone (3 wt%) without NPs (1), with MMCS-2 (7.0 wt%) only (2), and with MMCS-2-based NPs (7.0 wt%) containing a mixture of Reactive Blue 38 (21.7 wt%) and Acid Yellow 76 (20.0 wt%) (3).

4. Conclusions

A new approach to the production of coloured coatings, with nanopigments (NPs) prepared by the dye mixture treatment of montmorillonite modified with cationic surfactants, was developed.

The application of dye mixtures on MMCS allows the preparation of NPs with a wide colour range. Additional advantages of these NPs over NPs based on individual dyes include the possibility of increasing the total concentration of the colouring component in the NPs, as well as increased colourfastness.

The use of NPs based on dye mixtures in the composition of UV inks preserves all of the advantages of NPs relative to common organic pigments, among which a decrease in the radiation time required for the fast fixing of colour composition and an increase in colourfastness to rubbing are the most important.

The nano-sized structure of NPs and compositions based on these NPs was proven through a number of physicochemical methods, including UV–Vis spectrophotometry, X-ray diffraction analysis, and atomic force microscopy.

Author Contributions: Conceptualization, V.B.I. and P.P.L.; methodology, V.B.I.; validation, V.B.I. and P.P.L.; formal analysis, E.V.S.; investigation, E.V.S.; resources, V.B.I.; data curation, E.V.S.; writing—original draft preparation, V.B.I.; writing—review and editing, P.P.L.; visualization, V.B.I. and P.P.L.; supervision, P.P.L. All authors have read and agreed to the published version of the manuscript.

Funding: The study was financed as part of the state assignment of the Semenov Federal Research Center of Chemical Physics, Russian Academy of Science.

Institutional Review Board Statement: Not applicable.

Informed Consent Statement: Not applicable.

Data Availability Statement: Not applicable.

Conflicts of Interest: The authors declare no conflict of interest.

References

1. Lebedevaite, M.; Talacka, V.; Ostrauskaite, J. High biorenewable content acrylate photocurable resins for DLP 3D printing. *J. Appl. Polym. Sci.* **2021**, *138*, 50233. [CrossRef]
2. Mihaylova, E.M. Water-Soluble Holographic Photopolymers for a Sustainable Future—A Review. *Coatings* **2022**, *12*, 1765. [CrossRef]
3. Irfan, M.; Martin, S.; Obeidi, M.A.; Miller, S.; Kuster, F.; Brabazon, D.; Naydenova, I. A Magnetic Nanoparticle-Doped Photopolymer for Holographic Recording. *Polymers* **2022**, *14*, 1858. [CrossRef]
4. Tomašegović, T.; Poljaček, S.M.; Jakovljević, M.S.; Urbas, R. Effect of the Common Solvents on UV-Modified Photopolymer and EPDM Flexographic Printing Plates and Printed Ink Films. *Coatings* **2020**, *10*, 136. [CrossRef]
5. Shaikh, S.F. Nanomaterials Applied in Coatings: Synthesis, Structures, Properties, and Applications. *Coatings* **2022**, *12*, 1773. [CrossRef]
6. Mathiazhang, A.; Joseph, R. Nanotechnology-a new prospective in organic coating—Review. *Int. J. Chem. Eng. Appl.* **2011**, *2*, 225–237.
7. Selvam, S.; Rizwan, B.; Kalarical, J.S.; Sriman, N.S.; Balachandran, U.N. Functional pigments from chromium (III) oxide nanoparticles. *Dyes Pigments* **2012**, *94*, 548–552.
8. Dayakar, T.; Rao, K.V.; Chakra, C.S. Synthesis and characterization of $MgFe_2O_4(0.5)/TiO_2(0.5)$ nano ceramic pigment by mechanochemical synthesis. *Int. J. Nano Sci. Technol.* **2013**, *1*, 01–08.
9. Patel, M.A.; Bhanvase, B.A.; Sonawane, S.H. Production of cerium znc molybdate nano pigment by innovative ultrasound assisted approach. *Ultrason. Sonochemistry* **2013**, *20*, 906–913. [CrossRef] [PubMed]
10. Blosi, M.; Albonetti, S.; Gatti, F.; Baldi, G.; Dondi, M. Au-Ag nanoparticles as for digital decoration. *Dyes Pigments* **2012**, *94*, 355–362. [CrossRef]
11. Cavalcante, P.M.T.; Dondi, M.; Guarini, G.; Raimondo, M.; Baldi, G. Color performance of ceramic nano-pigments. *Dyes Pigments* **2009**, *80*, 226–232. [CrossRef]
12. Mara, M.; Csanády, A.; Tolnaj, G.; Németh, P.; Bertóti, I.; Sajo, I.; Papp, K.; Hargitai, H. Nano-micro pigment composites for high performance paints. *Mater. Sci. Forum* **2010**, *659*, 203–208.
13. Wilson, J.H.; Bradbury, R.; Annable, T.; Yeates, S.G. Inkjet printable aqueous dye-polymer nanoparticles. *Dyes Pigments* **2012**, *95*, 201–205. [CrossRef]
14. Zhao, X.; Meng, Q.; Li, Q. Hydrophobic dye/polymer colorants synthesized by miniemulsion solvent evaporation technique. *Dyes Pigments* **2014**, *100*, 41–49. [CrossRef]
15. Jiang, B.; Tao, P.H.; Huang, Y.D. Study of the adsorption performance and preparation of functional nano-silica pigment particles. *Dyes Pigments* **2014**, *104*, 169–174. [CrossRef]
16. Raha, S.; Ivanov, I.; Quazi, N.H.; Bhattacharya, S.N. Photo-stability of rhodamine-B/montmorillonite nanopigments in polypropylene matrix. *Appl. Clay Sci.* **2009**, *42*, 661–666. [CrossRef]
17. Raha, S.; Quazi, N.; Ivanov, I.; Bhattacharya, S. Dye/clay intercalated nanopigments using commercially available non-ionic dye. *Dyes Pigments* **2012**, *93*, 1512–1518. [CrossRef]
18. Tang, P.; Feng, Y.; Li, D. Facile synthesis of multicolor organic-inorganic hybrid pigments based on layered double hydroxides. *Dyes Pigments* **2014**, *104*, 131–136. [CrossRef]
19. Zavodchikova, A.A.; Safonov, V.V.; Ivanov, V.B. Nanopigments on the basis of dyes mixtures. Izvestiya Vysshikh Uchebnykh Zavedenii. *Seriya Teknol. Tekst. Promyshlennosti* **2011**, *4*, 82–85.
20. Zavodchikova, A.A.; Ivanov, V.B.; Safonov, V.V.; Solina, E.V. Photopolymerizable Composition for Printing on Textile. Materials. Patent 2472197, 13 March 2013.
21. Zavodchikova, A.A.; Safonov, V.V.; Solina, E.V.; Ivanov, V.B. Ultraviolet paints based on nanopigments for printing on textile materials. *Text. Res. J.* **2014**, *84*, 50–60. [CrossRef]
22. Ivanov, V.B.; Shlyapintokh, V.Y. Synergism in photostabilization of polymers. In *Developments in Polymer Stabilisation—8*; Scott, G., Ed.; Elsevier Applied Science Publisher: London, UK, 1987; pp. 20–43.
23. Venu, S.S.; Kochupurayil, A.M.; Swapankumar, G. Rhodamine 6G intercalated montmorillonite nanopigments—Polyethelene composites: Facile synthesis and ultraviolet stability study. *Am. Ceram. Soc.* **2011**, *94*, 1731–1736.

coatings

MDPI

Article

Structural and Magnetic Specificities of Fe-B Thin Films Obtained by Thermionic Vacuum Arc and Magnetron Sputtering

Cornel Staicu [1,2,*], Claudiu Locovei [2,3], Andrei Alexandru Dinu [2,3], Ion Burducea [4], Paul Dincă [1], Bogdan Butoi [1], Oana Gloria Pompilian [1], Corneliu Porosnicu [1], Cristian Petrica Lungu [1] and Victor Kuncser [3,*]

[1] Low Temperature Plasma Physics Department, National Institute for Laser, Plasma and Radiation Physics, 077125 Magurele, Romania
[2] Faculty of Physics, University of Bucharest, 077125 Magurele, Romania
[3] Magnetism and Superconductivity Department, Institute of Materials Physics, 077125 Magurele, Romania
[4] Horia Hulubei National Institute of Physics and Nuclear Engineering, 077125 Magurele, Romania
* Correspondence: cornel.staicu@inflpr.ro (C.S.); kuncser@infim.ro (V.K.)

Abstract: Fe-B based compounds are of high interest due to their special properties and the wide range of involved applications. While B is the element that facilitates the increase in the hardness and the degree of wear resistance, it is also an effective glass former, controlling the formation of a much-desired amorphous structure with specific magnetic properties. Major difficulties related to the proper engineering of Fe-B thin films lay especially in their preparation under well-defined compositions, which in turn, should be accurately determined. The present study closely analyzes the morpho-structural and magnetic properties of thin coatings of Fe-B of approximately 100 nm thickness and with the nominal B content ranging from 5 at. % to 50 at. %. The comparison between films obtained by two preparation methods, namely, the thermionic vacuum arc and the magnetron sputtering is envisaged. Morpho-structural properties were highlighted using X-ray diffraction supplemented with X-ray reflectometry and scanning electron microscopy, whereas the elemental investigations were performed by X-ray dispersive spectroscopy and Rutherford back-scattering spectroscopy. The magnetic properties of the Fe-B layers were carefully investigated by the vectorial magneto-optic Kerr effect and conversion electron Mössbauer spectroscopy. The high capability of Mössbauer Spectroscopy to provide the phase composition and the B content in the formed Fe-B intermetallic films was proven, in correlation to Rutherford back-scattering techniques, and to explain their magnetic properties, including the magnetic texture of interest in many applications, in correlation with longitudinal magneto-optic-Kerr-effect-based techniques.

Keywords: TVA; magnetron Sputtering; Fe-B films; magnetic texture; Mössbauer spectroscopy

1. Introduction

Since the 1970s, when the preparation of amorphous Fe-P-C compounds used an ultrafast solidification technique [1], many other amorphous metals containing iron and appropriate glass formers have been developed. Such alloys present a particular interest, especially in the field of electrical engineering (transformers and sensors based on magneto-impedance effects), due to their special soft magnetic characteristics. In addition to P and C, B is also a good glass forming agent helping in the formation of Fe-B amorphous compounds with relatively low concentrations of B. Except in its combination with Fe, the unique properties of boron and its behavior in relation to other metals have led to the production of a large class of borons, from YB_{66} (monochromator for synchrotron radiation), to (MgB_2) [2], known for its superconducting properties, and ($AlFe_2B_2$) [3], known for its magneto-caloric properties. Last but not least, boron is essential in the class of $Nd_2Fe_{14}B$ hard magnetic materials [4–8]. In literature, the interest in iron boride is already well known. Iron–boron compounds are of considerable interest because boride layers on iron and steel have a high wear resistance [9–13]. Iron borons are also used as deoxidizing agents, designed to increase the hardness of steel and coating alloy components applied to welded electrodes. However, despite the widespread application of these borons in industry, their physical properties have not been sufficiently investigated. Although Fe-B based metastable and amorphous systems can be easily obtained by melt spinning methods as ribbons for a narrow range of B content [14,15], more challenging preparation and investigation issues appear in the case of the counterpart thin films. Only a few studies are reported on the magnetic properties of Fe-B based amorphous intermetallic thin films, e.g., containing also rare earth (RE) elements [16]. In the case of thin films, strong variations of both mechanical and magnetic properties are expected versus the preparation conditions, the involved techniques, as well as the B content, all these issues defining finally whether the films are amorphous or not. The accurate determination of not only the average B concentration in the film but also of its distribution throughout the film are other major challenges, as many of the basic techniques are not able to provide reasonable elemental analysis for the light elements (including B).

In light of the above, the present paper aims to make a comparison between the structural and magnetic properties of thin coatings of Fe-B with various concentrations, as obtained by two deposition methods: (i) thermionic vacuum arc (TVA) and (ii) magnetron sputtering (MS). To determine the structural and compositional properties, the obtained thin films were investigated using X-ray diffraction (XRD), X-ray reflectometry (XRR), scanning electron microscopy (SEM), energy dispersion X-ray spectroscopy (EDX), and Rutherford backscattering (RBS) spectroscopy. At the same time, the magnetic properties of the Fe-B layers were carefully investigated by vectorial magneto-optic Kerr effect (MOKE) and conversion electron Mössbauer (CEM) spectroscopy.

2. Experimental Details

Fe-B thin films of different concentrations were prepared by the two methods mentioned above. The nominal concentrations, presented in Table 1, second column, correspond to the imposed effective thicknesses of Fe and B presented in columns 3 and 4. Accordingly, the specific deposition rates of the two elements were calculated and controlled via quartz microbalance measurements used in the cases of each deposition method (TVA and MS, respectively).

Table 1. Prepared samples, their nominal composition and corresponding effective thickness for each element.

No.	Concentration (at. %)	Fe (nm)	B (nm)
1	Fe + B (5%)	96.9	3.3
2	Fe + B (10%)	97.3	7.0
3	Fe + B (20%)	86.5	14.0
4	Fe + B (30%)	79.3	22.0
5	Fe + B (40%)	70.0	30.0
6	Fe + B (50%)	60.0	40.0

2.1. *Preparation by Thermionic Vacuum Arc versus Magnetron Sputtering*

To obtain thin films with well-defined atomic compositions (from 5 at. % to 50 at.% of B) by simultaneous deposition of Fe and B, two specific anode–cathode systems were used in case of TVA. They consist of a heated tungsten filament and a heat-resistant anode made of either titanium diboride (TiB_2) for boron or of carbon for iron (see Figure 1) [17]. The electrons emitted by the incandescent filament of W are subsequently accelerated to the anode by a Wehnelt cylinder acting as an electrostatic lens by applying a positive voltage [18–20]. In the first step, the material of interest is heated by electron bombardment. The second part consists of melting the material, giving rise to a pressure of metal vapors. Subsequently, due to the inelastic collisions between the thermally emitted filament electrons and the Fe and B atoms, a plasma ignition results in the pure metallic vapors of these materials. By carefully controlling the plasma parameters (filament current, discharge current, respectively, the voltage applied on the anode) a constant evaporation rate is obtained for each material. The main advantage of the TVA method with respect to MS (draws of the two discharge configurations are presented in Figure 1) is the high quality of the deposits due to the high vacuum conditions (lack of an additional discharge gas, e.g., Ar). Films characterized by a high degree of purity are formed due to the absence of gaseous inclusions [21]. Another advantage is the compactness of the coating with higher adhesion to the substrate.

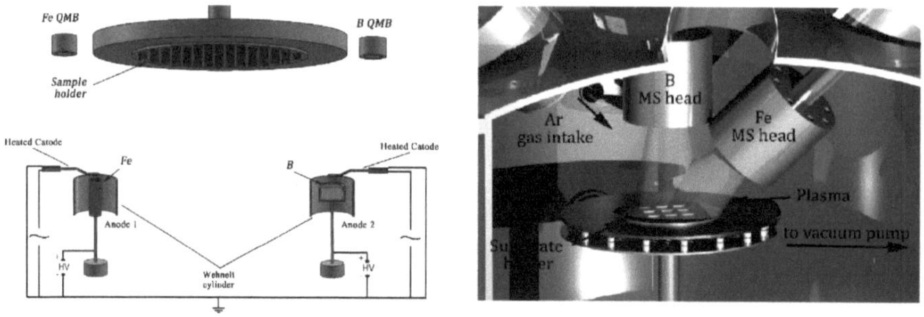

(a) (b)

Figure 1. Experimental set-up for Fe-B deposits by (**a**) TVA method (left hand) and by (**b**) MS (right hand), respectively.

The second method used to obtain Fe-B thin films is MS. For this experiment, Fe-B deposits were made using r.f. magnetron sputtering for B and the dc sputtering mode for Fe. The substrates were mounted in the vacuum chamber on a fixed circular holder placed 10 cm from the sputtering source. The used magnetron system was composed of a single water-cooled cathode, provided with two circular targets of B and Fe (2 mm thick and 50 mm in diameter each). To clean the targets of oxides and other impurities, an argon discharge plasma was initially ignited for approximately 30 min. Throughout the cleaning process, the sample holder was shielded from the plasma with a shutter. The working pressure of Ar was maintained at 1.5×10^{-2} under a continuous Ar flow of 20 sccm. An in situ acceleration voltage of -60 V was also used on the substrates to improve the adhesion and compactness of the deposited layers.

The substrates for the coatings in this study, independent of the deposition method, consisted of silicon wafers with a 12 mm $\times$ 15 mm size, with an optically polished surface and an average roughness of 6 nm. The substrates mounted in the vacuum chamber were ultrasonically cleaned using a solution of acetone as the first cleaning agent and isopropyl alcohol in the second stage and finally dried in atmospheric conditions. Fe-B thin films with a thickness of 100 nm were obtained under real time monitoring conditions via quartz microbalance systems. The base pressure was better than 10^{-5} mbar in each case.

2.2. Characterization

The structural and morphological properties of the obtained samples were analyzed using XRD, XRR and SEM, whereas the compositional measurements were performed by EDX and RBS.

XRD measurements were performed in Bragg–Brentano θ-2θ geometry at room temperature using the Bruker D8 ADVANCE diffractometer (CuKα = 1.54 Å) (Bruker Scientific Instruments, Madison, IA, USA). The data acquired for the determination of the crystalline phases were recorded in the range 2θ = 35–60° at a scanning rate equal to 0.02°/s. Film uniformity and thickness along with Fe-B thin film density were investigated by the same device working under the XRR mode with a scan rate of 0.008°/s. The Z-alignment procedure was initially performed, and the flatness of the sample was calibrated by the maximum of rocking curve.

For the SEM measurements, a scanning electron microscope, FEI Co., Inspect S. model (LabX Media Group, Midland, ON, Canada) was used, with the electron acceleration voltage varying between 0–30 kV to a working distance in the range of 0–30 mm under high vacuum conditions. The parameters used for the measurements were: acceleration voltage set to 25 kV, magnification 10,000X, working distance 13.5 mm, pressure of 10^{-4} Pa. Energy-dispersive X-ray spectroscopy (EDX) (LabX Media Group, Midland, ON, Canada) was carried out for topographic inspection and local chemical analysis. Typically, it is impossible to detect B by EDX. However, O X-ray yields are easily identified. The INCA package software (version 7.3) was used for the EDX elemental analysis. The energy calibration of the EDX facility was carried out with the Kα X-ray emissions of copper (8.05 keV) and of silicon (1.74 keV).

For RBS measurements, the samples were mounted on a goniometric support adjustable on 3 axes with an accuracy of 0.01°. The solid silicon detector with an energy resolution of 18 keV used for the detection of retro-scattered particles of He was positioned at an angle of 165° with respect to the ion beam. A 0.5 mm^2 wide beam of He 2.6 MeV monoenergetic ions was focused on the sample. In addition to RBS, the resonance method, non-Rutherford backscattering spectrometry (NRBS) (High Voltage Engineering Europe B.V., Amsterdamseweg, Netherlands), using 3.042, 3.884 and 4.282 MeV He beams, was also applied for crosschecking the stoichiometry and thickness of the samples.

The magnetic characterizations were performed at room temperature (RT) via a vectorial MOKE device (M&A) (Quantum Design, San Diego, CA, USA) which allows the acquisition of hysteresis loops for various orientations of the applied magnetic field with respect to a fixed direction in the film plane. The in-plane angular distribution of the magnetization easy axis becomes accessible with direct indications on the magnetic texture of the films. Finally, microscopic aspects related to magnetic behavior and local interactions in relation to atomic and electronic configurations around the Fe atoms were analyzed by RT CEM spectroscopy. A ^{57}Co (Rh) radioactive source was installed on a constant acceleration spectrometer. The sample was introduced in a gas flow proportional counter working with a mixture of methane (5%) and He. The γ radiation was incident to the sample plane, and the fitting of the Mössbauer spectra was achieved through the NORMOS software (version 1.0.11) [22]. Isomer shift values were reported relative to metallic α-Fe at RT.

3. Results and Discussion

The results will be comparatively presented on samples obtained using the two deposition methods, in terms of structure and morphology, as well as of the magnetic properties.

3.1. Morphology and Structure of Fe-B Films

The comparative SEM measurements on similar films from the two batches of samples (obtained by TVA and MS, respectively) are presented in Figure 2. Films from the two batches, with 20 at. % and 50 at. % of B, were chosen to be presented due to the clear differences between each other.

Figure 2 shows the SEM images of Fe-B films at a magnification of 10 kX. The morphology is lacking significant roughness on the surface of the films in both deposition batches. However, a clear difference can be seen between the films obtained by TVA and MS. According to Figure 2, the level of mesoscopic uniformity and roughness of the films obtained by TVA is clearly superior to the ones obtained by MS. Lower differences (films of very good uniformity) were observed for other B contents.

XRD patterns on two samples from each batch are presented for comparison in Figure 3.

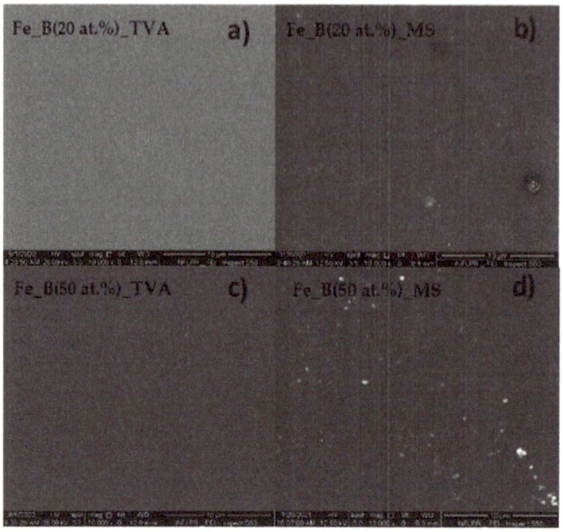

Figure 2. SEM images for samples: (**a**) Fe-B (20 at. % of B) obtained by TVA, (**b**) Fe-B (20 at. % of B) obtained by MS, (**c**) Fe-B (50 at. % of B) obtained by TVA, (**d**) Fe-B (50 at. % of B) obtained by MS.

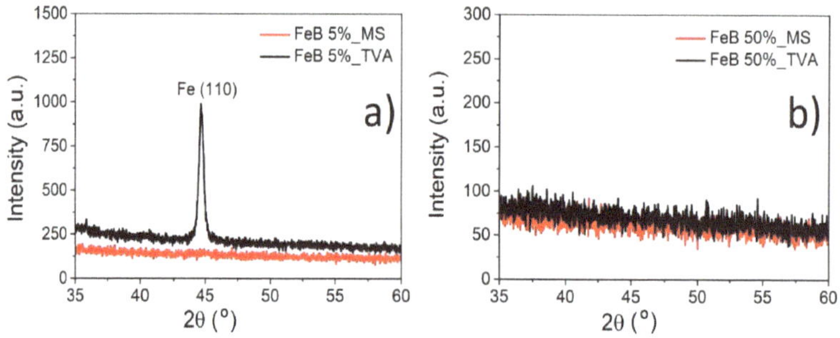

Figure 3. XRD patterns of Fe-B thin films with 5 at. % of B (**a**) and 50 at. % of B (**b**).

First, it is observed that for low concentrations of B (5%), a well crystallized film is obtained by TVA (black line), whereas an amorphous-like or nanocrystalline film (red line) is obtained by the more energetic MS method. For high concentrations of B, the films obtained by both methods are amorphous. However, XRD is not able to determine an accurate estimation of the relative content of crystalline and amorphous phase, as expected at low to intermediate content of B. In this respect, CEM spectroscopy results, as presented subsequently, are more concluding.

The XRR measurements for both films with 20 at. % boron nominal composition obtained by TVA and MS, respectively, are shown in Figure 4. Much better formed oscillations are evidenced in the XRR pattern of the film obtained by TVA, as direct proof of a much better uniformity of the film at nanoscale compared to the case of the similar film obtained by MS. However, from the involved beats, a different thickness is evidenced for the two films: only approximately 71(1) nm in case of the film obtained by TVA and 103(2) nm for the film obtained by MS.

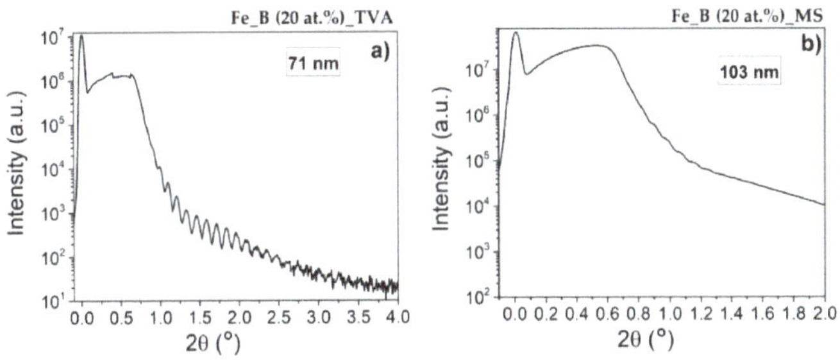

Figure 4. XRR patterns of Fe-B films (20 at. % B) obtained by TVA (**a**) and MS (**b**).

EDX measurements were made on the two films with 50 at. % of B obtained by TVA and MS, respectively. The difficulty of an accurate determination of light elements (e.g., Be, B) by this technique is well known. However, the EDX spectra of the two samples with the highest nominal concentrations of B (50 at. %) prepared by the two methods give evidence for a lower amount of residual oxygen and to a higher boron content in the sample prepared by TVA as compared to the one prepared by MS, and consequently, to a much more proper ratio between Fe and B in the first sample. Nevertheless, this is rather a qualitative result and not an accurate one, especially related to the B amount. A much more proper method for the elemental investigation of such films is RBS. The elemental content was also checked by NRMS, which is even more sensitive to the light elements when compared to RBS, especially when the substrate has a high Z number.

An exemplification of the RBS data for samples of different B content obtained by MS are presented in Figure 5a, whereas NRBS spectra collected on samples obtained by TVA are shown in Figure 5b. The RBS spectra were simulated using SIMNRA software package [23]. Areal density or thin film units (10^{15} atoms/cm^2) are the natural units for ion beam analysis (IBA) since the energy loss is measured in eV/(atoms/cm^2), and one monolayer is of the order of 10^{15} atoms/cm^2 [24,25]. The following tables present the RBS results regarding the stoichiometry of the samples.

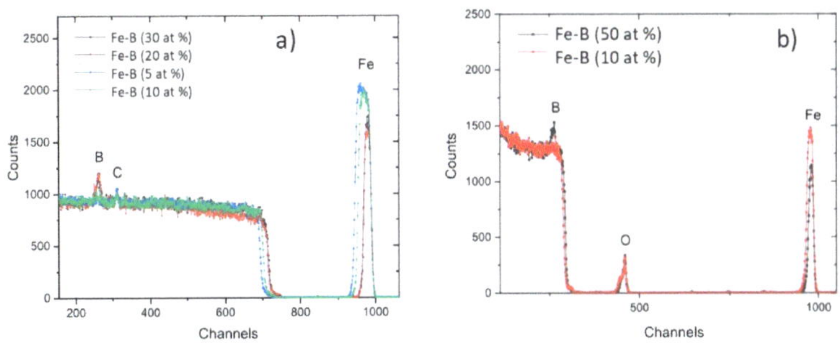

Figure 5. RBS data for Fe-B samples of different B concentrations obtained by MS (**a**) and NRBS spectra for samples obtained by TVA (**b**).

On all the samples, the presence of some impurities that could be due to some surface contamination of the samples is noticeable, e.g., it is well-known that also during ion beam analysis there is a carbon build-up on the sample surface. This is a mixture of C, H and O with thicknesses in the order of 50 TFU (1 TFU = 10^{15} atoms/cm^2). Other contaminants may be present on the sample surfaces (in a very low concentration) due to the handling or from some other reasons. In addition, a small and negligible concentration of W of 0.03% was observed in all TVA samples, probably resulting from the tungsten filament used as a cathode in TVA deposition. However, the most significant observation is related to an order of magnitude higher content of oxygen (tenths of at. %) in samples obtained by MS compared to samples obtained by TVA (order of at. %) showing definitively the much better quality of samples obtained by the second method. Concerning the Fe-B films obtained by MS, the Mössbauer results show the presence of an Fe oxide phase superposed over Fe-B intermetallic phases, and therefore, the average composition of the Fe-B phase cannot be subtracted directly from Table 2 by a simple renormalization between Fe and B. We noted that the ^{57}Fe Mössbauer results presented in the next section definitively prove the lack of Fe oxide phases and the formation of only Fe-B intermetallic phases in the Fe-B samples obtained by TVA. In this case, the few at. % of oxygen in the samples should be seen as inclusions and not making chemical bounds. As a consequence, the oxygen content will influence only the relative content of Fe and B in the Fe-B phases formed in the films, with direct influence on their magnetic properties. The real elemental contents in the Fe-B thin films obtained by TVA (in at. %), to be considered in the following for the interpretation of Mössbauer and magnetic results are presented in Table 3.

Table 2. Atomic composition of samples Fe-B obtained by TVA and MS, respectively.

Sample Name	Samples Obtained by TVA	Samples Obtained by MS
Fe-B (10 at. %)	$Fe_{0.91}B_{0.0797}O_{0.01}W_{0.0003}$	$Fe_{0.51}B_{0.045}C_{0.045}O_{0.37}H_{0.03}$
Fe-B (20 at. %)	$Fe_{0.6997}B_{0.28}O_{0.02}W_{0.0003}$	$Fe_{0.536}B_{0.13}C_{0.05}O_{0.24}H_{0.044}$
Fe-B (30 at. %)	$Fe_{0.6197}B_{0.3}O_{0.07}W_{0.0003}$	$Fe_{0.48}B_{0.2}C_{0.03}O_{0.26}H_{0.03}$
Fe-B (40 at. %)	$Fe_{0.5597}B_{0.38}O_{0.05}W_{0.0003}$	$Fe_{0.43}B_{0.255}C_{0.01}O_{0.28}H_{0.025}$
Fe-B (50 at. %)	$Fe_{0.4497}B_{0.45}O_{0.05}W_{0.0003}$	$Fe_{0.33}B_{0.3}C_{0.02}O_{0.32}H_{0.03}$

Table 3. Atomic composition of samples Fe-B obtained by TVA.

Sample	TVA
Fe-B (10 at. %)	$Fe_{0.91}B_{0.09}$
Fe-B (20 at. %)	$Fe_{0.71}B_{0.29}$
Fe-B (30 at. %)	$Fe_{0.67}B_{0.33}$
Fe-B (40 at. %)	$Fe_{0.60}B_{0.40}$
Fe-B (50 at. %)	$Fe_{0.50}B_{0.50}$

It is worth noticing that the real compositions are almost similar to the nominal ones in cases with low B content (i.e., less than 10 at. %) as well as in cases with higher B content (i.e., higher than 40 at. %), whereas in cases with intermediate B content, the real compositions are higher than the nominal ones.

3.2. Mössbauer Spectroscopy

The Mössbauer spectra of the Fe-B thin films obtained by TVA are presented in Figure 6. The emission spectrum of sample with 5 at. % B nominal composition consists in a sextet with narrow lines and hyperfine parameters close to metallic Fe (body centered cubic structure –bcc). Hence, the specific isomer shift (IS) is 0.01 mm/s and the hyperfine magnetic field (B_{hf}) is 33.1 T (0.00 mm/s and 33.15 T, respectively, for metallic Fe). The corresponding quadrupole splitting (QS) is 0.00 mm/s. Two superposed magnetic components, i.e., an external narrow and an inner broad magnetic sextet, respectively, are evidenced in the spectrum of the Fe-B film with 10 at. % B nominal composition. The first magnetic pattern with IS = 0.01(1) mm/s and B_{hf} = 33.1 T and with a relative spectral contribution of 29% corresponds again to a metallic Fe-like phase. The broad sextet was fitted via a hyperfine magnetic field distribution (presented on the right side of the spectrum), corresponding to a continuous distribution of local atomic configurations around the central Fe, specific to an amorphous Fe-B intermetallic phase.

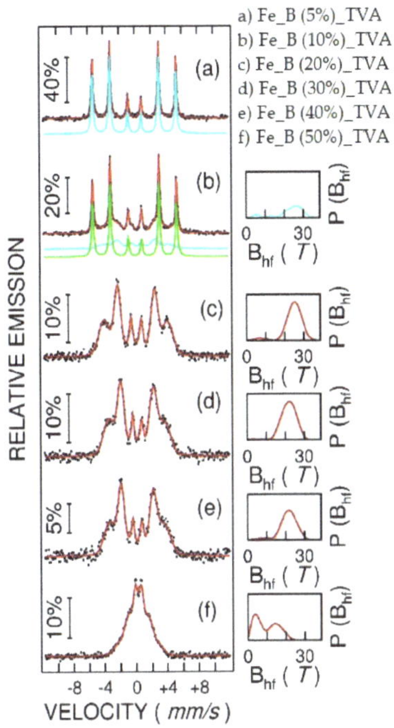

Figure 6. RT CEM spectra of Fe-B thin films obtained by TVA, with the following nominal compositions of B (in at. %): 5% (**a**), 10% (**b**), 20% (**c**), 30% (**d**), 40% (**e**), 50% (**f**). The vertical bars indexes the effect of the spectra (in %). The probability distributions of the hyperfine magnetic field are shown on the right hand of the spectra. The black dots are experimental points, the blue and green continuous lines provides the spectral components discussed in the text whereas the red lines represent the total spectral envelope given by the fit.

The specific average hyperfine parameters of this Fe-B intermetallic phase are: <IS> = 0.14 mm/s, <Bhf > = 24 T and QS almost 0.0 mm/s. An average of the hyperfine parameters over the two main local configurations corresponding to the two spectral components gives: 29.4 T for the average hyperfine magnetic field and 0.04 mm/s for the average Isomer Shift. Further on, at higher B contents, the Mössbauer spectra consist only in broad magnetic sextets, specific to amorphous Fe based phases [26–29], which start to collapse at higher B or RE content due to enhanced magnetic relaxation. We noted the uniform distribution of B_{hf} with nominal B content between 20 and 40 at. % and the tendency of forming two average local configurations for B content of 50 at. % simultaneously with enhanced magnetic relaxation phenomena. On the other hand, the intensity ratio between the second and the third absorption line, R_{23}, in either the elemental sextets under the hyperfine magnetic field distributions at higher B concentrations as well as in the narrow sextet of samples of low B concentrations is 4.0, giving a direct proof for an in-plane orientation of the Fe spins [26] in the Fe-B films obtained by TVA.

The evolution of the average hyperfine magnetic field and average isomer shift versus the real B content in such intermetallic compounds in case of reduced magnetic relaxation (i.e., for B nominal content lower than 40 at. %) is given in Figure 7. An almost linear decrease in the average hyperfine magnetic field and an almost linear increase in the isomer shift is observed. Similar tendencies were also previously reported for Fe-B compounds [12].

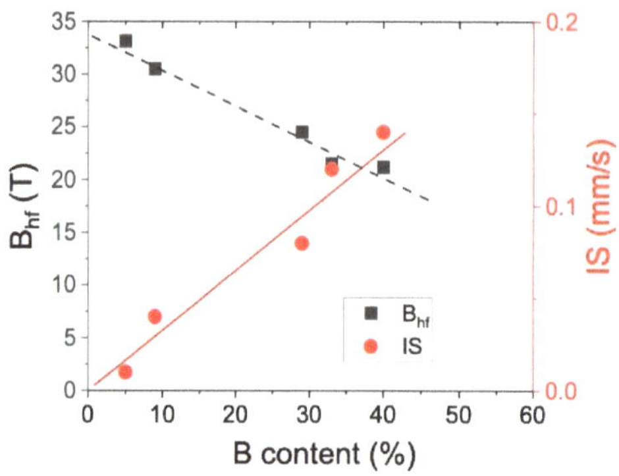

Figure 7. The dependence of average B_{hf} (black points) and IS (red points) versus the real B content in the intermetallic Fe-B thin films obtained by TVA.

The Mössbauer spectra of Fe-B thin films obtained by MS are shown in Figure 8. Two distinct features from the RT CEM spectra of films obtained by TVA are to be mentioned in this case: (i) the presence of a relative intense central component, and (ii) the R_{23} ratio between the second and the third emission line of the external sextet in the spectra of low B content is 0, giving a direct indication for a perpendicular to the film plane orientation for the Fe spins in the intermetallic compounds. In this context, the following fitting procedures were considered for the films obtained by MS. The central components were always fitted by a narrow hyperfine field distribution extending from 0 up to maximum 20 T, whereas depending on the nominal B content, either an external relative broad sextet (at low B concentration) or an external hyperfine field distribution spreading out from 10 to 35 T (at high B content) were considered. The corresponding hyperfine magnetic field distributions are shown on the right side of each spectrum.

We noted that <IS> values belonging to the central component accounted by the distribution at low fields (e.g., with $<B_{hf}>$ from 6 to 11 T, depending on the sample composition) range between 0.4 and 0.5 mm/s, which are specific values for Fe^{3+} ions, whereas <IS> values belonging to the broad external sextet accounted by the distribution at high fields (e.g., with $<B_{hf}>$ from 20 to 29 T, depending on the sample composition) range between 0.1 and 0.2 mm/s, as specific to Fe-B intermetallic phases with a B content ranging from approximately 25 at. % to 45 at. % (see Figure 7). As a consequence, and also based on the RBS results presented in Table 2, the central component was assigned to very disordered Fe_2O_3 clusters under a strong magnetic relaxation regime. The distributed sextet of higher average hyperfine magnetic fields to an amorphous Fe-B phase of higher amount of B and the most external sextet is evidenced mainly for films of low B content, to a disordered nanosized metallic Fe phase with inclusions of B.

In cases with samples with a nominal content of B of 5 at. %, the external sextet is characterized by B_{hf} = 32.3(2) T and IS = 0.01(1) mm/s, suggesting the formation of an intermetallic Fe-B phase with 7(2) at. % of B with the Fe spins oriented perpendicularly to the film plane. Its relative spectral contribution is 50(1) %, meaning that almost half of the Fe atoms enter in this phase. The rest of the Fe atoms/ions give rise to the Fe oxide clusters (50(1) % spectral contribution).

A quite similar spectral decomposition results also for the film with a nominal content of 10 at. % of B. A higher IS value for the intermetallic component (IS = 0.04(1) mm/s) suggests a higher B content in the Fe-B phase (approximately 10(2) at. % of B) of this sample. The relative spectral area of the Fe oxide phase is approximately 53(1) %, meaning that approximately 27 Fe^{3+} ions and 40 O^{-2} ions (from 100 atoms) participate in the Fe oxide phase and approximately 24 Fe atoms and 3 B atoms participate in the Fe-B phase, in reasonable agreement with RBS data in Table 2. The same perpendicular to plane orientation of the Fe spins in the Fe-B phase is evidenced by Mössbauer spectroscopy (R_{23} = 0) for this sample as well.

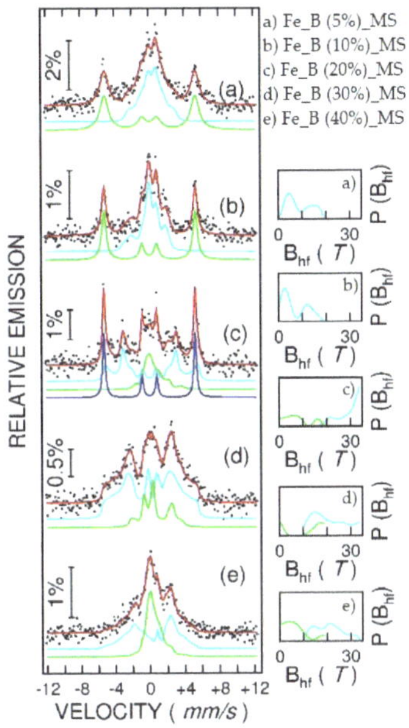

Figure 8. RT CEM spectra of Fe-B thin films obtained by MS, with the following nominal compositions of B (in at. %): 5% (**a**), 10% (**b**), 20% (**c**), 30% (**d**), 40% (**e**). The vertical bars indexes the effect of the spectra (in %). The probability distributions of the hyperfine magnetic field are shown on the right hand of the spectra The black dots are experimental points, the blue and green continuous lines provide the spectral components discussed in the text, whereas the red lines represent the total spectral envelope given by the fit.

In cases with samples with nominal content of B of 20 at. %, except the external sextet characterized bt B_{hf} = 32.9(1) T and IS = 0.01(1) mm/s and associated to a local configuration close to nanosized metallic Fe (i.e., of very low B content), there is a second component fitted by a distribution of hyperfine magnetic fields with $<B_{hf}>$ =29.2(2) T and $<IS>$ =0.12(2) mm/s, easily associated via Figure 7 to a Fe-B phase with about 30 at. % of B (i.e., $Fe_{70}B_{30}$). The relative spectral area of the last component is 44(1) % meaning, according to Table 2, 23 atoms of Fe and 11 atoms of B (from 100 atoms) in this Fe-B phase. The most inner spectral component, fitted this time via a distribution of hyperfine magnetic fields with $<B_{hf}>$ =7.2(2) T and $<IS>$ =0.45(2) mm/s, is associated with the Fe_2O_3 phase. By the relative spectral area of this component of 26(1) %, a number of 14 ions of Fe^{3+} and 21 atoms of O^{2-} from 100 atoms contribute to the Fe oxide phase, again in reasonable agreement with the RBS data in Table 2. We mention the perpendicular to the film plane orientation of the Fe spins in the metallic phase poor in B (similar to the one of spectra corresponding to films with less than 10 at. % of B) and the in plane orientation of the Fe spins in the amorphous intermetallic phase of higher B content ($Fe_{70}B_{30}$).

Only two spectral components were considered for samples of higher B content. In cases with samples with nominal content of B of 30 at. %, the outer magnetic component, fitted via a distribution of hyperfine magnetic fields with $<B_{hf}>$ = 20.5(2) T and $<IS>$ = 0.15(2) mm/s, was assigned to an amorphous intermetallic Fe-B phase of type $Fe_{60}B_{40}$ (see Figure 7), whereas the inner magnetic component was fitted by a distribution of hyperfine magnetic fields characterized by $<B_{hf}>$ = 10.5(2) T and $<IS>$ = 0.53(2) mm/s to the Fe_2O_3 phase. The relative spectral areas of 70(2) % and 30(2) %, respectively, infer 33 atoms of Fe and 22 atoms of B in the Fe-B phase and 15 ions of Fe^{3+} and 22 ions of O^{2-} in the Fe oxide phase from 100 atoms, again in reasonable agreement with Table 2. A similar reasoning is applied for the film with nominal composition of 40 at. % of B. Accordingly, the Mössbauer results support the formation of an amorphous intermetallic phase of type $Fe_{52}B_{48}$ ($< B_{hf} >$ = 19.5(2) T and $< IS >$ =0.19(2) mm/s) with relative spectral area of 54(2) % and an Fe oxide phase ($<B_{hf}>$ = 6.5(2) T and $<IS>$ = 0.51(2) mm/s) with a relative spectral area of 46(2) %. The corresponding atomic contribution at 100 atoms in the film is 23 Fe and 21 B in the Fe-B phase and 18 Fe and 28 O in the Fe oxide phase, in reasonable agreement with RBS results. The Fe spins of the intermetallic phases reached in B are in the film plane (R_{23} = 4).

3.3. Magnetic Characterization

The large capabilities of the MOKE method to investigate the magnetic behavior in such thin films, including in plane anisotropies, are illustrated by the hysteresis loops presented in Figures 9 and 10. The loops, consisting of the dependence of the Kerr angle on the applied magnetic field, were collected on the Fe-B thin films with thicknesses ranging from 70 to 100 nm, however, much thicker than the penetration depth of laser radiation (wavelength of 632 nm). Specific loops collected on films prepared by TVA, with different concentrations of B are shown in Figure 9. Three orientations of the magnetic field relative to the (110) direction of the Si substrate, 0°, 45° and 90°, were considered for each sample in order to investigate the in-plane anisotropy, as previously reported in [30,31]. Briefly, the idea is that in the case of uniaxial anisotropy, the hysteresis loops evolve from a rectangular shape, specific to a field applied along the easy axis of magnetization to a linear shape, specific to a field applied along the hard axis of magnetization (which is perpendicular to the easy axis in this case). If the saturation field coincides with the switching field the magnetization reversal is achieved by a coherent rotation of the spins (e.g., for a magnetic monodomain), whereas in the case of a saturation field higher than the switching field, the magnetization reversal is via domain wall displacements. If the evolution of the loops is not between these two limits, then we deal with an angular distribution of the easy axis (the broader the distribution, the lower the variation of the loop shape, when the film is rotated, if the loops remain constant at different angular rotations of the samples (from 0 to 90° in the present experiments) a lack of in-plane anisotropy is demonstrated.

In this context, the following observation can be drawn from Figure 9 for the films obtained by TVA: (i) the films with a B content lower or equal to 40 at. % are all ferromagnetic, (ii) the magnetic anisotropy is of uniaxial type and increases progressively with the B content starting from the case of isotropic behavior in those samples with very low B content, where mainly crystalline intermetallic phases are evidenced by Mössbauer spectroscopy (iii) magnetic domains are formed in the films and the magnetization reversal is through displacement of magnetic walls. (iv) The films are soft magnetic, with coercive fields ranging between 20 and 30 Oe, depending on the B content and reaching a maximum for the sample with 30 at. % of B which is completely amorphous and presents negligible magnetic relaxation.

The specific loops collected on films with equivalent nominal compositions of B, but prepared by MS, are shown in Figure 10. The MOKE signal of the samples with the lowest B content (i.e., with nominal compositions of B lower than 10 at. %) is more than one order of magnitude lower than in the case of the films obtained by TVA. This observation, correlated also with the linear shape of negligible coercivity of the loop, gives support for the assignment of the MOKE signal to the Fe oxide phase rather than to the Fe-B intermetallic one. Nevertheless, this assignment is also supported by the Mössbauer results, proving the perpendicular to the film plane alignment of the Fe spins in the Fe-B phase of low B content with no influence on the in-plane magnetization reversal observed by the present longitudinal MOKE geometry. We noted that this intermetallic phase with a relative amount of 50% pointed by Mössbauer spectroscopy would be the only one susceptible to contribute to the magnetization reversal if the Fe spins had in-plane components. The situation is drastically changed in cases of the films with higher nominal B content (i.e., 30 or 40 at. %) where the MOKE signal can be assigned to the Fe-B amorphous intermetallic phase of the higher B content with the Fe spins in the sample plane. This time, the MOKE signal increased by more than one order of magnitude, however, remaining still lower than in the case of films obtained by TVA of equivalent nominal B content. This is due to a relatively lower content of the intermetallic phase in the film (more than 40% of Fe oxide are formed).

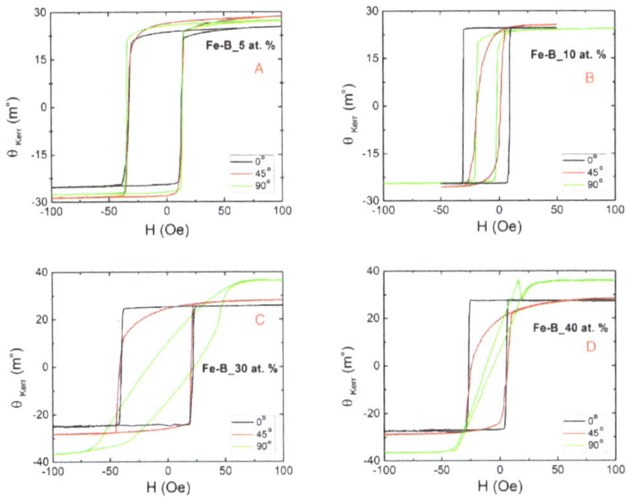

Figure 9. MOKE hysteresis loops for Fe-B films obtained by TVA. Different orientations of the magnetic field with respect to a reference in plane direction are considered. (**A**) is Fe-B 5 at. %, (**B**) is Fe-B 10 at. %, (**C**) is Fe-B 30 at. %, and (**D**) is Fe-B 40at. %.

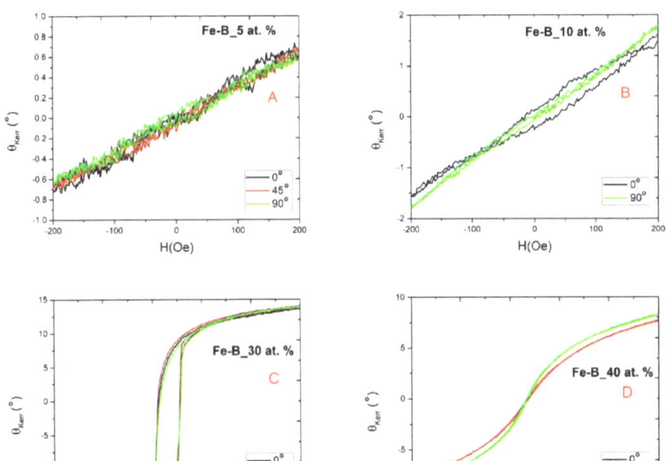

Figure 10. MOKE hysteresis loops for Fe-B films obtained by MS. Different orientations of the magnetic field with respect to a reference in plane direction are considered. (**A**) is Fe-B 5 at. %, (**B**) is Fe-B 10 at. %, (**C**) is Fe-B 30 at. %, and (**D**) is Fe-B 40at. %.

As a last observation, the loop of the sample with nominal content of B of 40 at. % is almost closed, with almost negligible coercivity, resembling rather a loop specific to a paramagnetic material, most probably due to the high content of B in the real $Fe_{52}B_{48}$ phase, where magnetic relaxation phenomena are substantially enhanced.

4. Conclusions

The paper aimed to analyze Fe-B films with various concentrations of B (5 at. % −50 at. %) in terms of morpho-structural and magnetic properties. Thin films (70–100 nm) were successfully obtained by two deposition methods, TVA and MS. The morpho-structural investigations performed by SEM, XRD and XRR have shown a roughness-free morphology of all samples, but with a much better uniformity of samples obtained by TVA. The elemental analysis was surveyed by EDX for samples of high B content and performed in detail by RBS based techniques. The formation of relative pure Fe-B intermetallic films by TVA and a high content of oxygen in films obtained by MS were proven. The real phase composition of the films was obtained by Mössbauer Spectroscopy. The obtained results on the elemental composition are in reasonable agreement with the RBS data. Mössbauer results give evidence for no Fe oxide phase in the films obtained by TVA but the formation of disordered clusters of Fe_2O_3 in the films obtained by MS, in a relative amount between 20 to 50 %, depending on the B content. It was proven that in the films of low B content (less than 10 at. % of B nominal composition), mainly a defect crystalline phase approaching the body centered cubic structure of metallic Fe with B inclusions is formed whereas at increased B concentrations, an amorphous Fe-B phase of high B content is formed. The Fe spins are always oriented in the sample plane for the films obtained by TVA whereas in the case of the films obtained by MS, the Fe spins are perpendicular to the sample plane for the crystalline-like phase and in the film plane in the case of the amorphous-like intermetallic compounds. These results give direct explanation for the different magnetic behavior of the films, as evidenced by the longitudinal MOKE investigations. The highest magnetic anisotropy and saturation magnetization, as well as the lowest coercivity, is obtained for the $Fe_{60}B_{40}$ amorphous films obtained by TVA. The formation of nano-crystalline Fe-B films with perpendicular to plane magnetic anisotropy by MS is another direction of interest for magneto-functional applications.

Author Contributions: Conceptualization, C.S. and V.K.; methodology, C.S. and V.K.; validation V.K. and C.P.; investigation, C.S., P.D., B.B., I.B., C.L., A.A.D. and O.G.P.; data curation, V.K., C.S., I.B., C.L., A.A.D. and O.G.P.; writing—original draft preparation, C.S.; writing—review and editing, V.K. and C.P.; Visualization, C.S.; Supervision, C.P., V.K. and C.P.L.; Project administration; C.P.L., V.K. and C.P.; Funding acquisition, C.P.L., V.K. and C.P. All authors have read and agreed to the published version of the manuscript.

Funding: This work was supported by grants of the Romanian Ministry of Education and research, Core Program PN-19 (Contract No. 21N/2019) and project Supported Institutional Excellence Contract 35 PFE/2022 at NIMP and the Core Program LAPLAS IV—contract 16N/2019 at NILPRP. The research leading to these results has also received funding from the NO-RO grants 2014–2021, under Project contract No 39/2021 and the POC Project REBMAT, ID P_37_697. Ion beam experiments were performed at 3 MV Tandetron accelerator from NIPNE-HH and were supported by the Romanian National Program "Instalații și Obiective de Interes Național" (Grant No. 786/2014).

Institutional Review Board Statement: Not applicable.

Informed Consent Statement: Not applicable.

Data Availability Statement: Not applicable.

Conflicts of Interest: The authors declare no conflict of interest.

References

1. Lin, C.-Y.; Tien, H.-Y.; Chin, T.-S. Soft magnetic ternary iron-boron-based bulk metallic glasses. *Appl. Phys. Lett.* **2005**, *86*, 162501. [CrossRef]
2. Nagamatsu, J.; Nakagawa, N.; Muranaka, T.; Zenitani, Y.; Akimitsu, J. Superconductivity at 39 K in magnesium diboride. *Nature* **2001**, *410*, 63–64. [CrossRef]
3. Tan, X.; Chai, P.; Thompson, C.M.; Shatruk, M. Magnetocaloric Effect in $AlFe_2B_2$: Toward Magnetic Refrigerants from Earth-Abundant Elements. *J. Am. Chem. Soc.* **2013**, *135*, 9553–9557. [CrossRef]
4. Fokwa, B.P.T. Borides: Solid-state chemistry. In *Encyclopedia of Inorganic and Bioinorganic Chemistry*; Scott, R.A., Ed.; John Wiley & Sons, Ltd.: Hoboken, NJ, USA, 16 June 2014; Available online: https://onlinelibrary.wiley.com/doi/10.1002/9781119951438.eibc0022.pub2 (accessed on 25 May 2022).
5. Fokwa, B.P.T. Transition-metal-rich borides—fascinating crystal structures and magnetic properties. *Eur. J. Inorg. Chem.* **2010**, *2010*, 3075–3092. [CrossRef]
6. Albert, B.; Hillebrecht, H. Boron: Elementary Challenge for Experimenters and Theoreticians. *Angew. Chem. Int. Ed.* **2009**, *48*, 8640–8668. [CrossRef]
7. Mori, T. *Boron-Based Materials in Materials Aspect of Thermoelectricity*; Uher, C., Akopov, G., Yeung, M.T., Kaner, R.B., Eds.; CRC Press: Boca Raton, FL, USA, 2016; Available online: https://www.routledge.com/Materials-Aspect-of-Thermoelectricity/Uher/p/book/9781498754903 (accessed on 25 May 2022).
8. Herbst, J.F.; Croat, J.J.; Pinkerton, F.E.; Yelon, W.B. Relationships between crystal structure and magnetic properties in $Nd_2Fe_{14}B$. *Phys. Rev. B* **1984**, *29*, 4176–4178. [CrossRef]
9. Chung, H.Y.; Weinberger, M.B.; Levine, J.B.; Cumberland, R.W.; Kavner, A.; Yang, J.M.; Tolbert, S.H.; Kaner, R.B. Synthesis of ultra-incompressible superhard rhenium diboride at ambient pressure. *Science* **2007**, *316*, 436–439. [CrossRef]
10. Chung, H.-Y.; Weinberger, M.B.; Yang, J.-M.; Tolbert, S.H.; Kaner, R.B. Correlation between hardness and elastic moduli of the ultraincompressible transition metal diborides RuB_2, OsB_2, and ReB_2. *Appl. Phys. Lett.* **2008**, *92*, 261904. [CrossRef]
11. Gu, Q.; Krauss, G.; Steurer, W. Transition Metal Borides: Superhard versus Ultra-incompressible. *Adv. Mater.* **2008**, *20*, 3620–3626. [CrossRef]
12. Weinberger, M.B.; Levine, J.B.; Chung, H.Y.; Cumberland, R.W.; Rasool, H.I.; Yang, J.M.; Kaner, R.B.; Tolbert, S.H. Incompressibility and Hardness of Solid Solution Transition Metal Diborides: $Os_{1-x}Ru_xB_2$. *Chem. Mater.* **2009**, *21*, 1915–1921. [CrossRef]
13. Levine, J.B.; Tolbert, S.H.; Kaner, R.B. Advancements in the Search for Superhard Ultra-Incompressible Metal Borides. *Adv. Funct. Mater.* **2009**, *19*, 3519–3533. [CrossRef]
14. Sanchez, F.H.; Budnick, J.I.; Zhang, Y.D.; Hines, W.A.; Choi, M.; Hasegawa, R. Mössbauer study of the local atomic environments in metastable crystalline Fe-B alloys. *Phys. Rev. B* **1986**, *34*, 4738–4743. [CrossRef]
15. Hesse, J.; Rubartsch, A.J. Model independent evaluation of overlapped Mossbauer spectra. *J. Phys. E Sci. Instrum.* **1974**, *7*, 526. [CrossRef]
16. Kuncser, V.; Filoti, G.; Serbanescu, M.D.; Jianu, A.; Cristea, L.; Dormann, J.L.; Noguès, M.; Maknani, J. Hyperfine fields and magnetic anisotropy in $Fe_{1000-x}Gd_x$ thin films and $Fe_{80-x}Gd_xB_{20}$ ($0 \leq x \leq 24$) thin films and ribbons. *Hyperfine Interact.* **1994**, *94*, 2323–2326. [CrossRef]
17. Marcu, A.; Ticoş, C.M.; Grigoriu, C.; Jepu, I.; Porosnicu, C.; Lungu, A.M.; Lungu, C.P. Simultaneous carbon and tungsten thin film deposition using two thermionic vacuum arcs. *Thin Solid Film.* **2011**, *519*, 4074–4077. [CrossRef]
18. Kuncser, V.; Schinteie, G.; Palade, P.; Jepu, I.; Mustata, I.; Lungu, C.P.; Miculescu, F.; Filoti, G. Magnetic properties of Fe-Co ferromagnetic layers and Fe-Mn/Fe-Co bilayers obtained by thermionic vacuum arc. *J. Alloy. Compd.* **2010**, *499*, 23–29. [CrossRef]
19. Jepu, I.; Porosnicu, C.; Lungu, C.P.; Mustata, I.; Luculescu, C.; Kuncser, V.; Iacobescu, G.; Manin, A.; Ciupina, V. Combinatorial Fe-Co thin film magnetic structures obtained by thermionic vacuum arc method. *Surf. Coat. Technol.* **2014**, *240*, 344–352. [CrossRef]
20. Miculescu, F.; Jepu, I.; Stan, G.E.; Miculescu, M.; Voicu, S.I.; Cotrut, C.; Pisu, T.M.; Ciuca, S. Tailoring the electric and magnetic properties of submicron-sized metallic multilayered systems by TVA atomic inter-diffusion engineered processes. *Appl. Surf. Sci.* **2015**, *358*, 619–626. [CrossRef]
21. Jepu, I.; Porosnicu, C.; Mustata, I.; Lungu, C.P.; Kunkser, V.; Osiac, M.; Iacobescu, G.; Ionescu, V.; Tudor, T. Simultaneously Thermionic Vacuum Arc Discharges In Obtaining Ferromagnetic Thin Films, Romanian Reports. *Physics* **2011**, *63*, 804–816.
22. Brand, R.A. Improving the validity of hyperfine field distributions from magnetic alloys: Part I: Unpolarized source. *Nucl. Instr. Meth. Phys. Res. B* **1987**, *28*, 398–416. [CrossRef]
23. Mayer, M. *SIMNRA User's Guide Technical Report IPP 9/133*; MPI für Plasmaphysik: Garching, Germany, 1997.
24. Nastasi, M.; Mayer, J.W.; Wang, Y. *Ion Beam Analysis: Fundamentals and Applications*; Taylor & Francis Group, LLC.: Boca Raton, FL, USA, 2015.
25. IBANDL. Available online: https://www-nds.iaea.org/exfor/ibandl.htm (accessed on 25 May 2022).
26. Seqqat, M.; Nogués, M.; Crisan, O.; Kuncser, V.; Cristea, L.; Jianu, A.D.; Filoti, G.; Dormann, J.L.; Sayah, D.; Godinho, M. Magnetic properties of $Fe_{100-x}Sm_x$ thin films and $Fe_{80-x}Sm_xB_{20}$ thin films and ribbons. *J. Magn. Magn. Mater.* **1996**, *157*, 225–226. [CrossRef]
27. Stanciu, A.E.; Schinteie, G.; Kuncser, A.; Iacob, N.; Trupina, L.; Ionita, I.; Crisan, O.; Kuncser, V. Unexpected magneto-functionalities of amorphous Fe-Gd thin films crossing the magnetization compensation point. *J. Magn. Magn. Mater.* **2020**, *498*, 166173. [CrossRef]

28. Miglierini, M.B. Radiation Effects in Amorphous Metallic Alloys as Revealed by Mössbauer Spectrometry: Part I. Neutron Irradiation. *Metals* **2021**, *11*, 845. [CrossRef]
29. Kuncser, V.; Palade, P.; Kuncser, A.; Greculeasa, S.; Schinteie, G. *Size Effects in Nanostructures, Basics and Applications*; Kuncser, V., Miu, L., Eds.; Springer: Berlin/Heidelberg, Germany, 2014; pp. 169–237.
30. Locovei, C.; Radu, C.; Kuncser, A.; Iacob, N.; Schinteie, G.; Stanciu, A.; Iftimie, S.; Kuncser, V. Relationship between the Formation of Magnetic Clusters and Hexagonal Phase of Gold Matrix in Au_xFe_{1-x} Nanophase Thin Films. *Nanomaterials* **2022**, *12*, 1176. [CrossRef] [PubMed]
31. Stanciu, A.E.; Schinteie, G.; Kuncser, A.C.; Locovei, C.; Trupina, L.; Iacob, N.; Leca, A.; Borca, B.; Kuncser, V. Magnetic Properties of Nanosized Fe and FeCo Systems on Trenched Mo Templates. *Coatings* **2022**, *12*, 1366. [CrossRef]

coatings

MDPI

Article

Memory Properties of Zr-Doped ZrO_2 MOS-like Capacitor

Catalin Palade *, Adrian Slav, Ionel Stavarache, Valentin Adrian Maraloiu, Catalin Negrila and Magdalena Lidia Ciurea

National Institute of Materials Physics, 405A Atomistilor Street, 077125 Magurele, Romania
* Correspondence: catalin.palade@infim.ro

Abstract: The high-k-based MOS-like capacitors are a promising approach for the domain of non-volatile memory devices, which currently is limited by SiO_2 technology and cannot face the rapid downsizing of the electronic device trend. In this paper, we prepare MOS-like trilayer memory structures based on high-k ZrO_2 by magnetron sputtering, with a 5% and a 10% concentrations of Zr in the Zr–ZrO_2 floating gate layer. For crystallization of the memory structure, rapid thermal annealing at different temperatures between 500 °C and 700 °C was performed. Additionally, Al electrodes were deposited in a top-down configuration. High-resolution transmission electron microscopy reveals that ZrO_2 has a polycrystalline–columnar crystallization and a tetragonal crystalline structure, which was confirmed by X-ray diffraction measurements. It is shown that the tetragonal phase is stabilized during the crystallization by the fast diffusion of oxygen atoms. The capacitance–voltage characteristics show that the widest memory window (ΔV = 2.23 V) was obtained for samples with 10% Zr annealed at 700 °C for 4 min. The charge retention characteristics show a capacitance decrease of 36% after 10 years.

Keywords: non-volatile memory; high-k ZrO_2; MOS-like capacitor; charge retention

1. Introduction

In recent decades, the miniaturization trend of electronic devices has been the mainstream approach for increasing computational power and data storage capabilities. The usage of SiO_2 as a dielectric oxide for classical metal–oxide–semiconductor (MOS) devices [1,2] was made possible due to SiO_2's very good electrical stability and native interface with Si. The scaling down of the SiO_2 tunneling oxide was the preferred method used for increasing the density of the memory structures on the surface unit. However, this approach has some limitations as the SiO_2 tunneling oxide has to be thinned down to 2 nm [3]. If the value is below this, the process of direct tunneling of the charge carriers between the Si substrate and the floating gate layer through the SiO_2 tunneling layer leads to a high-leakage current and a low-charge retention capability of the MOS-like capacitor [4]. A solution to this problem is to replace the SiO_2 with a high-k oxide, such as ZrO_2, HfO_2 or Al_2O_3 [5–11], that has a dielectric constant of $k > 20$, which is sufficiently high to have a low equivalent oxide thickness (EOT) [12] and low enough to avoid the lateral fringing-field effect [13,14] to be used in MOS-like capacitors. ZrO_2 is characterized by a good interface with Si [15] and being thermally stable [16], and the temperature of the phase transformation from monoclinic to tetragonal is almost 500 °C, which is lower than that for HfO_2 [17,18]. Additionally, the dielectric constant, k, of the tetragonal phase is higher than 40 [19]. It is important to have a lower phase transition temperature, as using lower temperature processing together with the magnetron sputtering (MS) deposition method (which is a cheap and reproducible method) means cost-effective production and equipment investments.

In the literature, there are very few articles related to the memory properties of the ZrO_2-based trilayer MOS-like capacitors. Some of these articles describe the charge storage properties of Ge or Zr nanocrystals embedded in the ZrO_2 matrix as a floating gate layer [20–22] located at a tunnelable distance from the Si substrate. Other articles describe the ferroelectric properties of ZrO_2 in association with HfO_2 [23–25].

In this paper, we use Zr-doped ZrO_2 for the formation of charge storage centers in a ZrO_2-based MOS-like trilayer structure. The main advantage of using an oxygen-deficient ZrO_2 layer as a floating gate layer compared with Zr nanocrystals embedded in ZrO_2 is related to the high density of the storage centers. This is translated into a very stable memory structure with a long charge retention time. For this, we prepared a MOS-like memory structure of *ZrO_2 [gate oxide]/Zr–ZrO_2 [floating gate]/ZrO_2 [tunneling oxide]/p-Si* by MS, followed by rapid thermal annealing (RTA) between 500 °C and 700 °C and a metallization process. In the Zr–ZrO_2 floating gate layer, we used 5% and 10% Zr. The broadest memory window (ΔV = 2.23 V) was obtained from the samples with 10% Zr annealed at 700 °C.

2. Materials and Methods

2.1. Sample Preparation

The memory structures *ZrO_2 [gate oxide]/Zr–ZrO_2 [floating gate]/ZrO_2 [tunneling oxide]/p-Si* were prepared by using an MS deposition technique (Surrey Nanosystems Gamma 1000 equipment, East Sussex, UK) on a *p*-type Si substrate (1–10 Ω·cm resistivity). Before the deposition, the Si wafers were chemically cleaned by using a two-step procedure. In the first step, the organic contaminants were removed in a piranha solution (H_2O_2:H_2SO_4 = 1:3), followed by the second step of native SiO_2 removal in an aqueous solution of HF (2%). The preparation of the trilayer MOS-like capacitor starts with the deposition of a 13-nm tunneling ZrO_2 layer from a ZrO_2 target on which the power of P = 60 W RF was applied. Then, a 13-nm Zr-doped ZrO_2 floating gate layer was deposited by co-sputtering the Zr and the ZrO_2 (P = 60 W RF) from separate targets. For the Zr, we used two concentrations of 5% and 10% (P = 10–30 W DC) for obtaining 2 types of samples. Finally, the gate oxide layer of 30 nm was also deposited at a P = 60 W RF. The memory structure deposition was performed in an Ar atmosphere at a 25-sccm (standard cubic centimeters per minute) flow rate and a working pressure of p = 3 mTorr. To ensure the uniformity of the layer deposition, the Si substrate was rotated at 15 rpm. Additionally, a 56-nm ZrO_2 control structure was deposited under similar conditions as the trilayer one. After the deposition, the trilayer memory structure was nanostructured by RTA at different temperatures between 500 °C and 700 °C for 4 min in an N_2 atmosphere.

Table 1 presents the structure sketch and the preparation details of the trilayer MOS-like capacitor and the control structure.

Table 1. Structure configuration and preparation details of the trilayer *ZrO_2 [gate oxide]/Zr–ZrO_2 [floating gate]/ZrO_2 [tunneling oxide]/p-Si* and the control structure.

	Name	Floating Gate Layer Composition	Structure	Layers Thickness [nm]	Thermal Annealing
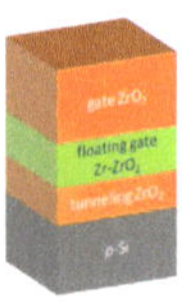	Zr5-500 Zr5-600 Zr5-700 Zr10-500 Zr10-600 Zr10-700	5% Zr + 95% ZrO_2 10% Zr + 90% ZrO_2	ZrO_2/Zr-ZrO_2/ZrO_2/Si	30/13/13	T = 500 °C, t = 4 min T = 600 °C, t = 4 min T = 700 °C, t = 4 min
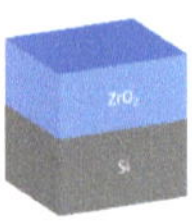	Control	-	ZrO_2/Si	56	T = 700 °C, t = 4 min

For the electrical characterization, Al electrodes with a 1-mm^2 area were deposited by thermal evaporation in a top-down configuration, forming a MOS-like capacitor.

2.2. Characterization Methods

The trilayer MOS-like capacitor structure and morphology were characterized by high-resolution transmission electron microscopy (HRTEM) in the Jeol JEM-ARM 200F electron microscope (JEOL, Tokyo, Japan). The crystalline structure was determined by X-ray diffraction measurements using the Rigaku SmartLab diffractometer (Rigaku, Tokyo, Japan), and the oxidation states of Zr were determined by X-ray photoelectron spectroscopy (XPS). A SPECS XPS spectrometer (SPECS Surface Nano Analysis GmbH, Berlin, Germany) with a PHOIBOS 150 analyzer was used for this. This uses a 300-W monochromatic RX source (Al Kα—1486.61 eV) and a SPECS FG15/40 (SPECS Surface Nano Analysis GmbH, Berlin, Germany) flood gun for charge neutralization.

The memory characteristics of each trilayer structure were determined by capacitance–voltage (*C–V*) measurements using an Agilent 4980A LCR meter (Agilent Technologies, Santa Clara, CA, USA). The frequency dependence of the capacitance (*C-f*) was also measured between 10 kHz and 1 MHz with different voltages applied to the gate electrode. The capacity of the MOS-like trilayer capacitor to retain electrical charges was determined by the time dependence of the capacitance (*C-t*), measured at a frequency of 1 MHz.

3. Results and Discussion

The structure and morphology of the sample with the 10% Zr content in the floating gate layer annealed at 700 °C for 4 min (Zr10-700) were determined by cross-section HRTEM analyses (Figure 1a), where the columnar crystallization of ZrO_2 can be observed.

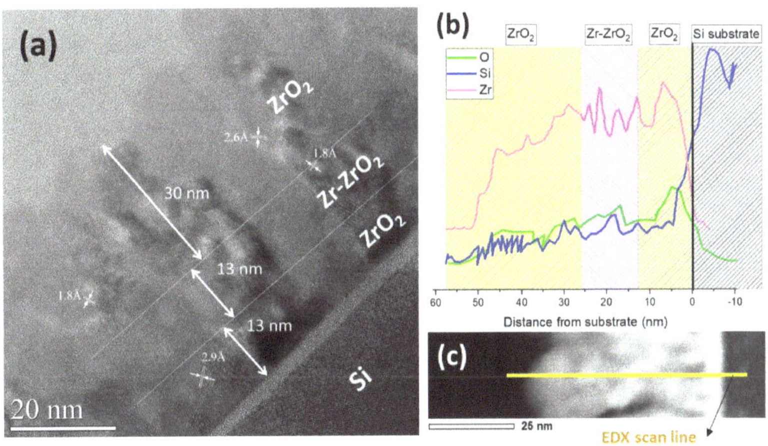

Figure 1. (**a**) HRTEM image obtained of the Zr10-700 structure; (**b**) EDX analysis with relative concentrations of Zr (magenta), Si (blue), and O (green), measured on the cross-section of the sample; (**c**) High-angle annular dark-field (HAADF) cross-section image.

In the HRTEM image, it is possible to measure some lattice fringe distances of 0.29 nm, 0.18 nm, and 0.16 nm that correspond to the planes (111), (002), and (103) of the ZrO_2 tetragonal phase, respectively.

The contrast between the deposited layers is low, and the trilayer configuration of the memory capacitor cannot be observed as distinct individual layers. The reason for this is the presence of the columnar crystallization process. The fast diffusion of oxygen atoms leads to uniform nonstoichiometric suboxides in the whole trilayer structure, not only in the middle layer. This will stabilize the tetragonal phase during the crystallization. In our opinion, the crystallization process begins with the nucleation on the tunneling layer, and the crystal growth continues normally to the free surface in a columnar morphology with a growth rate that depends on the nuclei orientations.

Some information about the structural configuration of the trilayer structure is obtained by analyzing the energy-dispersive X-ray spectroscopy (EDX) data. Figure 1b presents the relative concentration of Zr (magenta), Si (blue), and O (green) measured on the cross-section of the Zr10-700 sample (Figure 1c).

The maximum concentration of Zr is located near the floating gate layer, as expected from the preparation conditions. These results confirm the trilayer structure configuration of the MOS-like memory capacitor.

The polycrystallinity of the trilayer memory structure observed in the HRTEM analysis (Figure 1a) was also confirmed by the two-dimensional fast Fourier transform (FFT) (Figure 2b) applied to the HRTEM image (Figure 2a), followed by an inverse FFT (IFFT) (Figure 2c). The colored regions are obtained by filtering the spatial frequencies and orientations in the FFT image, which are labeled with numbers and colors in Figure 2b. Each colored region in Figure 2c represents a ZrO_2 nanocrystallite or a group of them having the same lattice fringe spacing and direction, giving the same spot in the FFT image. For example, the green area shows a group of four nanocrystallites that have the same crystalline orientation and similar lattice fringes in the HRTEM image. Each nanocrystallite resembles a coherent domain of ZrO_2.

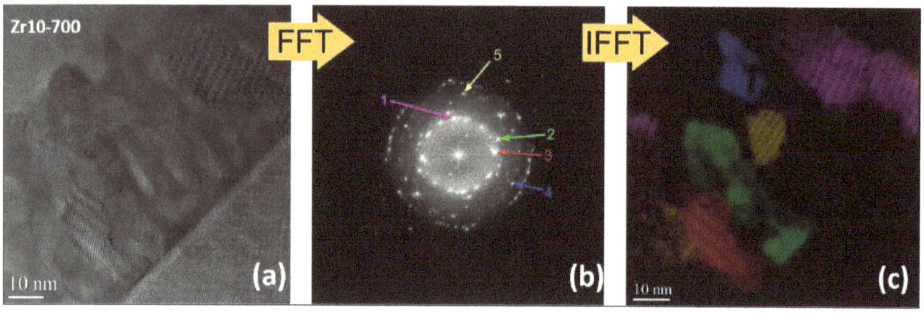

Figure 2. (**a**) HRTEM image obtained on the Zr10-700; (**b**) FFT algorithm applied on the HRTEM image. Each number correspond to a different lattice fringe spacing and direction; (**c**) IFFT image with colored regions that represents ZrO_2 nanocrystallites with different crystalline lattice fringe spacings and orientation.

The crystalline structure of the MOS-like capacitor was also evidenced by XRD measurements (Figure 3).

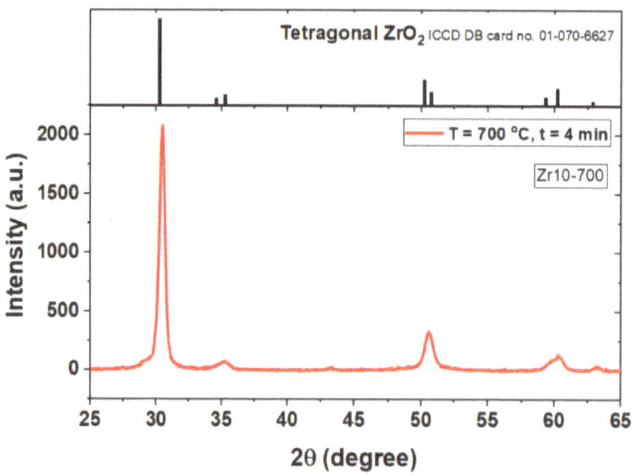

Figure 3. XRD spectra of the sample obtained after 4 minutes at 700 °C RTA (Zr10-700).

The XRD diffractogram presents high intensity peaks positioned at 2θ = 30.48 deg, 2θ = 35.2 deg, 2θ = 50.6 deg, and 2θ = 60.32 deg. These peaks correspond to a high-pressure tetragonal phase of the ZrO_2, indexed in the ICCD DB card no. 01-070-6627, for the crystallographic planes of (111), (002), (200), and (211), respectively. The 30.48 deg peak corresponds to the (111) plane with a lattice spacing of 0.293 nm, which is smaller than the 0.295 nm (for 30.28 deg) of the tetragonal lattice phase. However, the experimental (111) peak has a very large base (from 28.4 to 31.5 deg) covering the positions of the (-111) and (111) peaks of the monoclinic phase of the ZrO_2. This monoclinic phase is present in several percentages in the structure and explains the small coherence size of the ZrO_2 crystallites.

At room temperature and atmospheric pressure, the most stable phase of ZrO_2 is the monoclinic phase, which is characterized by a lower value of the dielectric constant compared with the tetragonal phase. In our case, by co-depositing between 10% Zr–90% ZrO_2 for the floating gate layer, some amorphous zones of the ZrO_2 at the nanometric scale will probably appear.

A possible explanation for the tetragonal phase being the majority phase in the trilayer structure is that during the RTA at 700 °C, the crystallization starts in both the tunneling ZrO_2 and the gate ZrO_2 and continues in the floating gate layer, similarly to the case of HfO_2-based trilayers. [26]. In the present case, the internal strain appears by oxygen vacancy formation and is also due to the 10% Zr excess in the floating gate.

We also performed X-ray photoelectron spectroscopy (XPS) measurements at the surface of the trilayer structure (Figure 4).

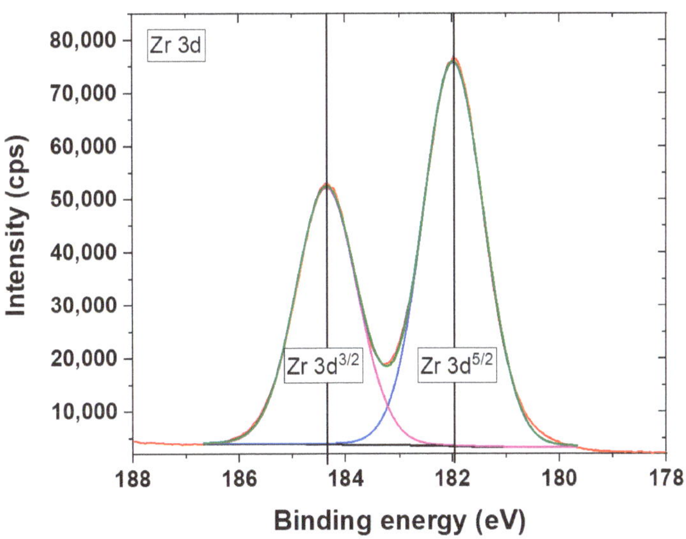

Figure 4. XPS spectra of the Zr 3d obtained on the Zr10-700. The colored lines represent components of spectral deconvolution (Voigt profiles) were each maximum corresponds to Zr 3d doublets.

As it can be seen, the Zr 3d doublets Zr $3d^{5/2}$ and Zr $3d^{3/2}$ have a splitting energy of 2.4 eV and an area ratio of 3/2. The binding energy clearly indicates that Zr is in an oxidation state; therefore, there is no metallic Zr in the ZrO_2 gate oxide layer contributing to the formation of the tetragonal phase of ZrO_2.

The memory properties of the MOS-like capacitor with Zr concentrations of 5 and 10% in the floating gate layer and annealing temperatures between 500 °C and 700 °C (for 4 min) were investigated by *C–V* measurements. The *C–V* characteristics are presented in Figure 5 together with the *C–V* curve taken on the control structure, for comparison.

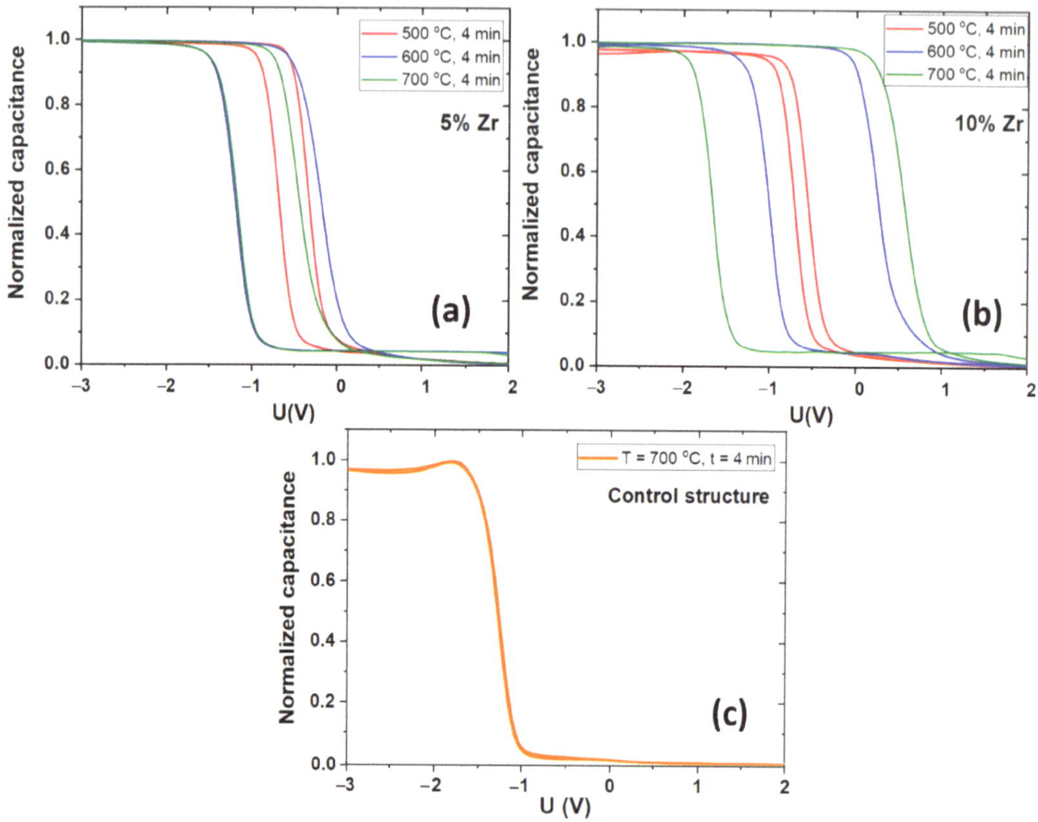

Figure 5. *C*–*V* characteristics obtained on the MOS-like capacitor with (**a**) 5% and (**b**) 10% concentration in the floating gate layer, annealed at 500°C, 600 °C, and 700 °C for 4 min; (**c**) *C*–*V* curves obtained on the control structure of 56 nm of ZrO_2 annealed at 700 °C for 4 min.

The *C-V* measurements were performed by swiping the voltage between +2 V (inversion regime) and −3 V (accumulation regime) with a swiping speed of 1 V/s at 1 MHz frequency [27]. The hysteresis loops in Figures 5a and 5b show that the best memory window of ΔV = 2.23 V is obtained on the capacitors with the 10% Zr content in the floating gate and an RTA at 700 °C (for 4 min in an N_2 atmosphere), in contrast with the *C*–*V* characteristic of the control structure (having no hysteresis).

Therefore, we can conclude that the memory properties of the trilayer MOS-like capacitor are due to both oxygen vacancies present in the trilayer structure and to Zr excess in the Zr–ZrO_2 floating gate layer acting as charge storage nodes.

The sample with the best memory characteristics (Zr10-700) was chosen for further investigations of the frequency dependence of the capacitance (Figure 6a) and the resistance (Figure 6b). The *C-f* and *R-f* characteristics were measured at different voltage values (−3 V–+2 V) applied to the ZrO_2 gate oxide electrodes. The capacitance and resistance values were recorded during the frequency swiping between 10 kHz and 1 MHz (Figure 6).

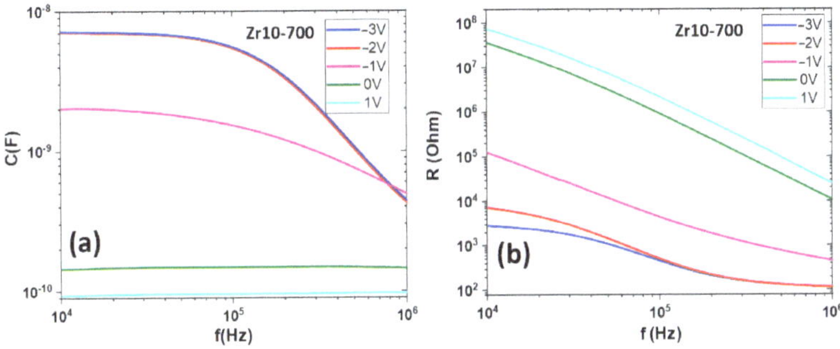

Figure 6. (**a**) *C-f* and (**b**) *R-f* measurements made at different voltages applied to the gate oxide electrode.

The voltage dependence of the *C-f* and *R-f* characteristics can be explained by the voltage dependence of the depletion region at the interface between the Si substrate and the ZrO_2 tunneling layer. By applying a negative voltage on the ZrO_2 gate oxide, the majority of charge carriers (holes) accumulate at the Si–ZrO_2 interface, meaning that there is no depletion region at the Si surface [27]. In the present case, the total capacitance is at its maximum value and is given only by the capacitance of the trilayer structure. By swiping the applied voltage from negative to positive, the Si–ZrO_2 interface is depleted of charge carriers. In this case, the total capacitance is the series capacitance of the trilayer structure and the capacitance of the depleted region. Since the depleted region capacitance depends on the dielectric constant of the Si and the width of the depleted region, the total capacitance decreases as the voltage approaches the inversion regime, where the minority of charge carriers arrive at the Si–ZrO_2 interface due to the high electric field.

In our previous work [28], we studied the frequency dependence of the capacitance and the resistance of a trilayer structure and developed a model that allows us to simultaneously fit the *C-f* and *R-f* characteristics. From the fit, we extracted the material parameters, such as the dielectric constant, of each layer in the MOS-like capacitor. We applied the same model to this structure (*ZrO_2 [gate oxide]/Zr–ZrO_2 [floating gate]/ZrO_2 [tunneling oxide]/p-Si*) and fitted the frequency dependence of the capacitance and the resistance in the accumulation regime (−3 V) to determine the dielectric constant of each layer. The results of the fit are presented in Figure 7.

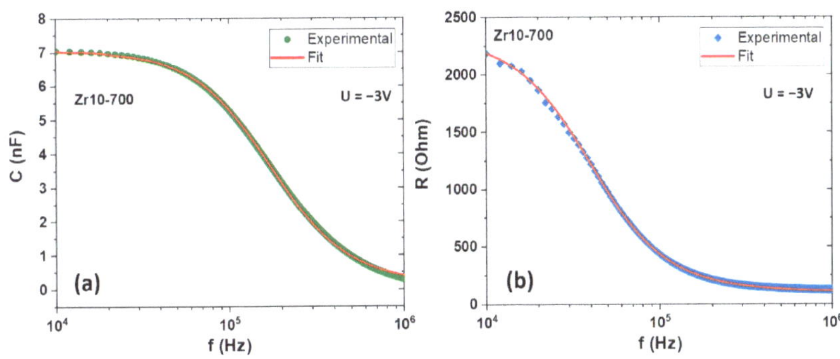

Figure 7. The frequency dependence of the (**a**) capacitance and (**b**) resistance measured on Zr10-700 in the accumulation regime (−3 V).

The values of the dielectric constants for each layer obtained from the *C-f* and *R-f* fits are presented in Figure 8.

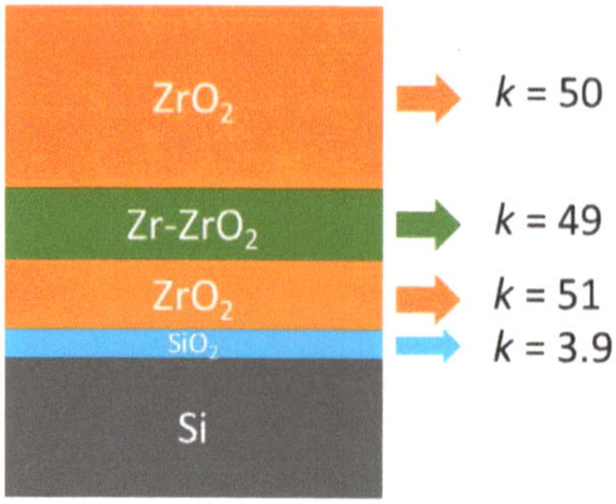

Figure 8. The schematic representation of the as-deposited trilayer structure and the dielectric constants obtained from the *C*-*f* and *R*-*f* fit curves from Figure 7 (the experimental *C*-*f* and *R*-*f* curves were measured on the Zr-10-700 structure).

Figure 8 shows the values of the dielectric constants obtained from the fit curves from Figure 7. These values of the dielectric constant correspond to the layers in the 700 °C RTA sample. One can observe that they are very close to each other and that they are typical for the tetragonal phase range [19]. These results are in good agreement with the XRD and HRTEM results. Additionally, the small variation in the dielectric constant values from one layer to another one shows that there is an internal strain in the trilayer structure due to oxygen vacancy formation and Zr excess.

The charge retention characteristics were also measured on the Zr10-700 sample at a frequency of 500 kHz and at a temperature of 23 °C. In Figure 9, the *C*-*t* characteristics are presented for both the programmed and the erased processes. The programming process was performed by charging the trilayer structure at −3 V for 1 s applied on the gate oxide (accumulation regime) and then the capacitance value was recorded in time at the applied voltage of −1.6 V (flatband voltage). The erasing process was performed by applying +2 V to the gate oxide for 1 s (inversion regime) and then the capacitance value was recorded in time at the applied voltage of +0.6 V (flatband voltage).

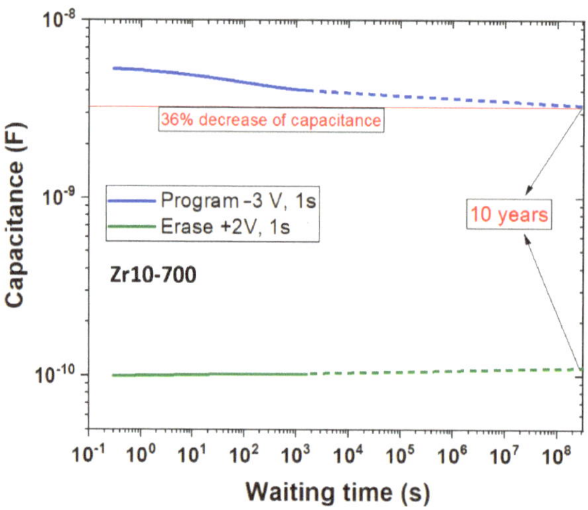

Figure 9. Charge retention characteristics measured at 500 kHz on the Zr10-700 sample. There is a 36% capacitance decrease after 10 years.

The capacitance-decreasing trend was extrapolated for 10 years, and we observed a 36% capacitance decrease, which is in the accepted range of a standard 50% capacitance decrease after 10 years for FLASH memory devices [29].

4. Conclusions

We prepared a MOS-like memory structure of ZrO_2 *[gate oxide]/Zr–*ZrO_2 *[floating gate]/*ZrO_2 *[tunneling oxide]/p-Si* using MS with a 5% and a 10% Zr content in the Zr–ZrO_2 floating gate layer, followed by an RTA process at 500 °C, 600 °C, and 700 °C. The trilayer structure Zr10-700 with the 10% Zr content in the floating gate layer is formed of polycrystalline ZrO_2 that is columnar crystallized in a tetragonal phase (HRTEM, XRD). The polycrystalline character of the trilayer memory structure is confirmed by FFT and IFFT analyses. We consider that the tetragonal phase is stabilized during the crystallization by the fast diffusion of oxygen atoms, leading to uniform nonstoichiometric suboxides in the whole trilayer. Additionally, XRD measurements show that besides the majority of the ZrO_2 tetragonal phase, the MOS-like capacitor structure contains several percentages of monoclinic phase that explains the small coherence size of the ZrO_2 crystallites. The XPS results show that Zr is in an oxidation state; therefore, there is no metallic Zr in the gate ZrO_2 layer to contribute to the ZrO_2 tetragonal phase.

The best memory properties were obtained on trilayer capacitors with 10% Zr annealed at 700 °C. The hysteresis loops taken on them have the broadest memory window of $\Delta V = 2.23$ V, and the charge retention characteristics show a 36% capacitance decrease after 10 years. Additionally, the dielectric constants of each layer in the trilayer, obtained from the fit of *C-f* and *R-f* experimental curves, have values close to each other and they are typical for the tetragonal phase range. We consider that the most important contribution to the memory properties is given by oxygen vacancies with a high enough density, acting as charge storage centers.

Author Contributions: Conceptualization, methodology, software, C.P.; investigation, C.P., A.S., I.S., V.A.M., and C.N.; writing—original draft preparation, C.P.; writing—review and editing, C.P. and M.L.C.; supervision, C.P. All authors have read and agreed to the published version of the manuscript.

Funding: This research was funded by CNCS—UEFISCDI, projects no. PN-III-P1-1.1-PD-2019-1038, PN-III-P2-2.1-PED-2019-0205, PN-III-P2-2.1-PED-2019-4468, PN-III-P1-1.1-TE-2021-1537 within PNCDI III and by the Romanian Ministry of Research, Innovation, and Digitalization.

Institutional Review Board Statement: Not applicable.

Informed Consent Statement: Not applicable.

Data Availability Statement: Not applicable.

Conflicts of Interest: The authors declare no conflict of interest.

References

1. Markov, S.; Sushko, P.V.; Roy, S.; Fiegna, C.; Sangiorgi, E.; Shluger, A.L.; Asenov, A. Si-SiO_2 Interface Band-Gap Transition—Effects on MOS Inversion Layer. *Phys. Status Solidi (a)* **2008**, *205*, 1290–1295. [CrossRef]
2. Kar, S.; Narasimhan, R.L. Characteristics of the Si-SiO_2 Interface States in Thin (70–230 Å) Oxide Structures. *J. Appl. Phys.* **1987**, *61*, 5353–5359. [CrossRef]
3. Majkusiak, B.; Walczak, J. Theoretical Limit for the SiO_2 Thickness in Silicon MOS Devices. *Sci. Technol. Semicond. Insul. Struct. Devices Oper. Harsh Environ.* **2005**, *185*, 309–320. [CrossRef]
4. Timp, G.; Bude, J.; Baumann, F.; Bourdelle, K.K.; Boone, T.; Garno, J.; Ghetti, A.; Green, M.; Gossmann, H.; Kim, Y.; et al. The Relentless March of the MOSFET Gate Oxide Thickness to Zero. *Microelectron. Reliab.* **2000**, *40*, 557–562. [CrossRef]
5. Liu, J.; Li, J.; Wu, J.; Sun, J. Structure and Dielectric Property of High-k ZrO_2 Films Grown by Atomic Layer Deposition Using Tetrakis (Dimethylamido) Zirconium and Ozone. *Nanoscale Res. Lett.* **2019**, *14*, 154. [CrossRef]
6. Huang, K.-W.; Chang, T.-J.; Wang, C.-Y.; Yi, S.-H.; Wang, C.I.; Jiang, Y.-S.; Yin, Y.-T.; Lin, H.-C.; Chen, M.-J. Leakage Current Lowering and Film Densification of ZrO_2 High-k Gate Dielectrics by Layer-By-Layer, In-Situ Atomic Layer Hydrogen Bombardment. *Mater. Sci. Semicond. Processing* **2020**, *109*, 104933. [CrossRef]

7. Kashir, A.; Ghiasabadi Farahani, M.; Hwang, H. Towards an Ideal High-κ HfO_2–ZrO_2-Based Dielectric. *Nanoscale* **2021**, *13*, 13631–13640. [CrossRef]
8. Palade, C.; Slav, A.; Cojocaru, O.; Teodorescu, V.S.; Stoica, T.; Ciurea, M.L.; Lepadatu, A.-M. SiGeSn Quantum Dots in HfO_2 for Floating Gate Memory Capacitors. *Coatings* **2022**, *12*, 348. [CrossRef]
9. Dragoman, M.; Dinescu, A.; Dragoman, D.; Palade, C.; Teodorescu, V.S.; Ciurea, M.L. Graphene/Ferroelectric (Ge-Doped HfO_2) Adaptable Transistors Acting as Reconfigurable Logic Gates. *Nanomaterials* **2022**, *12*, 279. [CrossRef]
10. Dragoman, M.; Dinescu, A.; Dragoman, D.; Palade, C.; Moldovan, A.; Dinescu, M.; Teodorescu, V.S.; Ciurea, M.L. Wafer-Scale Graphene-Ferroelectric HfO_2/Ge–HfO_2/HfO_2 Transistors Acting as Three-Terminal Memristors. *Nanotechnology* **2020**, *31*, 495207. [CrossRef]
11. Stavarache, I.; Cojocaru, O.; Maraloiu, V.A.; Teodorescu, V.S.; Stoica, T.; Ciurea, M.L. Effects of Ge-Related Storage Centers Formation in Al_2O_3 Enhancing the Performance of Floating Gate Memories. *Appl. Surf. Sci.* **2021**, *542*, 148702. [CrossRef]
12. Ni, K.; Saha, A.; Chakraborty, W.; Ye, H.; Grisafe, B.; Smith, J.; Rayner, G.B.; Gupta, S.; Datta, S. Equivalent Oxide Thickness (EOT) Scaling with Hafnium Zirconium Oxide High-κ Dielectric near Morphotropic Phase Boundary. In Proceedings of the 2019 IEEE International Electron Devices Meeting (IEDM), San Francisco, CA, USA, 7–11 December 2019. [CrossRef]
13. Zhao, H.; Yeo, Y.-C.; Rustagi, S.C.; Samudra, G.S. Analysis of the Effects of Fringing Electric Field on FinFET Device Performance and Structural Optimization Using 3-D Simulation. *IEEE Trans. Electron. Devices* **2008**, *55*, 1177–1184. [CrossRef]
14. Yang, C.-F.; Hwu, J.-G. Role of Fringing Field on the Electrical Characteristics of Metal-Oxide-Semiconductor Capacitors with Co-Planar and Edge-Removed Oxides. *AIP Adv.* **2016**, *6*, 125017. [CrossRef]
15. Peacock, P.W.; Xiong, K.; Tse, K.; Robertson, J. Bonding and Interface States of Si:HfO_2 and Si:ZrO_2 Interfaces. *Phys. Rev. B* **2006**, *73*, 075328. [CrossRef]
16. Namavar, F.; Wang, G.; Cheung, C.L.; Sabirianov, R.F.; Zeng, X.C.; Mei, W.N.; Bai, J.; Brewer, J.R.; Haider, H.; Garvin, K.L. Thermal Stability of Nanostructurally Stabilized Zirconium Oxide. *Nanotechnology* **2007**, *18*, 415702. [CrossRef]
17. Zhang, Y.; Zhang, J. First Principles Study of Structural and Thermodynamic Properties of Zirconia. *Mater. Today Proc.* **2014**, *1*, 44–54. [CrossRef]
18. Haggerty, R.P.; Sarin, P.; Apostolov, Z.D.; Driemeyer, P.E.; Kriven, W.M. Thermal Expansion of HfO_2 and ZrO_2. *J. Am. Ceram. Soc.* **2014**, *97*, 2213–2222. [CrossRef]
19. Panda, D.; Tseng, T.-Y. Growth, Dielectric Properties, and Memory Device Applications of ZrO_2 Thin Films. *Thin Solid Film.* **2013**, *531*, 1–20. [CrossRef]
20. Lehninger, D.; Khomenkova, L.; Röder, C.; Gärtner, G.; Abendroth, B.; Beyer, J.; Schneider, F.; Meyer, D.C.; Heitmann, J. Ge Nanostructures Embedded in ZrO_2 Dielectric Films for Nonvolatile Memory Applications. *ECS Trans.* **2015**, *66*, 203–212. [CrossRef]
21. Wu, Y.-H.; Wu, J.-R.; Wu, M.-L.; Chen, L.-L.; Lin, C.-C. Ge-Based Nonvolatile Memory Formed on Si Substrate with Ge-Stabilized Tetragonal ZrO_2 as Charge Trapping Layer. *J. Electrochem. Soc.* **2011**, *158*, H410. [CrossRef]
22. Palade, C.; Slav, A.; Ciurea, M.L. Memory Properties of $GeZrO_2$ Based Trilayer Structure. In Proceedings of the 2021 International Semiconductor Conference (CAS), online, 6–8 October 2021. [CrossRef]
23. Müller, J.; Böscke, T.S.; Schröder, U.; Mueller, S.; Bräuhaus, D.; Böttger, U.; Frey, L.; Mikolajick, T. Ferroelectricity in Simple Binary ZrO_2 and HfO_2. *Nano Lett.* **2012**, *12*, 4318–4323. [CrossRef] [PubMed]
24. Li, Y.; Li, J.; Liang, R.; Zhao, R.; Xiong, B.; Liu, H.; Tian, H.; Yang, Y.; Ren, T.-L. Switching Dynamics of Ferroelectric HfO_2-ZrO_2 with Various ZrO_2 Contents. *Appl. Phys. Lett.* **2019**, *114*, 142902. [CrossRef]
25. Chen, Y.; Wang, L.; Liu, L.; Tang, L.; Yuan, X.; Chen, H.; Zhou, K.; Zhang, D. Thickness-Dependent Ferroelectric Properties of HfO_2/ZrO_2 Nanolaminates Using Atomic Layer Deposition. *J. Mater. Sci.* **2021**, *56*, 6064–6072. [CrossRef]
26. Palade, C.; Lepadatu, A.-M.; Slav, A.; Cojocaru, O.; Iuga, A.; Maraloiu, V.A.; Moldovan, A.; Dinescu, M.; Teodorescu, V.S.; Stoica, T.; et al. A Nanoscale Continuous Transition from the Monoclinic to Ferroelectric Orthorhombic Phase inside HfO_2 Nanocrystals Stabilized by HfO_2 Capping and Self-Controlled Ge Doping. *J. Mater. Chem. C* **2021**, *9*, 12353–12366. [CrossRef]
27. Lepadatu, A.M.; Palade, C.; Slav, A.; Maraloiu, A.V.; Lazanu, S.; Stoica, T.; Logofatu, C.; Teodorescu, V.S.; Ciurea, M.L. Single Layer of Ge Quantum Dots in HfO_2 for Floating Gate Memory Capacitors. *Nanotechnology* **2017**, *28*, 175707. [CrossRef]
28. Palade, C.; Lepadatu, A.M.; Slav, A.; Lazanu, S.; Teodorescu, V.S.; Stoica, T.; Ciurea, M.L. Material Parameters from Frequency Dispersion Simulation of Floating Gate Memory with Ge Nanocrystals in HfO_2. *Appl. Surf. Sci.* **2018**, *428*, 698–702. [CrossRef]
29. Hong, A.J.; Kim, J.; Kim, K.; Wang, Y.; Xiu, F.; Jeon, J.; Park, J.; Rauda, I.; Chen, L.-M.; Yang, Y.; et al. Cr Metal Thin Film Memory. *J. Appl. Phys.* **2011**, *110*, 054504. [CrossRef]

coatings

MDPI

Article

Magnetic Properties of Nanosized Fe and FeCo Systems on Trenched Mo Templates

Anda Elena Stanciu [1,*], Gabriel Schinteie [1], Andrei Cristian Kuncser [1], Claudiu Locovei [1,2], Lucian Trupina [1], Nicusor Iacob [1], Aurel Leca [1], Bogdana Borca [1] and Victor Kuncser [1,*]

1 National Institute of Materials Physics, Atomistilor 405A, 077125 Magurele, Romania
2 Faculty of Physics, University of Bucharest, Atomistilor 405, 077125 Magurele, Romania
* Correspondence: anda.stanciu@infim.ro (A.E.S.); kuncser@infim.ro (V.K.)

Abstract: The manipulation of magnetic anisotropy represents the fundamental prerequisite for the application of magnetic materials. Here we present the vectorial magnetic properties of nanostructured systems and thin films of Fe and FeCo prepared on linearly trenched Mo templates with thermally controlled periodicity. The magnetic properties of the nanosystems are engineered by tuning the shape, size, thickness, and composition parameters of the thin films. Thus, we control coercivity, magnetization, orientation of the easy axis of magnetization, and the long-range magnetic order of the system in the function of the temperature. We distinguish magnetic components that emerge from the complex morpho-structural features of the undulating Fe or FeCo nanostructured films on trenched Mo templates: (i) assembly of magnetic nanowires and (ii) assembly of magnetic islands/clusters. Uniaxial anisotropy at room temperature was proven, characterized, and explained in the case of all systems. Our work contributes to the understanding of magnetic properties necessary for possible further applications of linear systems and undulated thin films.

Keywords: magnetic anisotropy; linear nanostructures; Fe; FeCo; Mo; structured template

1. Introduction

Modern spintronic applications demand nanosized systems (thin films, nanowires, and nanoparticles) that should satisfy several conditions regarding the magnetic properties and magnetic anisotropy. They should present controllable magnetic properties, such as the value of the magnetic anisotropy energy, saturation magnetization, the relative orientation of the easy axis of magnetization of nano-objects, coercivity, and squareness factor. Also, the possibility to switch the magnetization with low-cost external stimuli and to stabilize the magnetic states upon exposure to thermal fluctuations or external fields is required [1–4].

Depending on the application type, the magnetic anisotropy should be either in plane or out of plane. For example, in the case of high-density magnetic recording and high-frequency magnetic devices perpendicular magnetic anisotropy is needed. The in-plane magnetic anisotropy with uniaxial anisotropy is used in random access memory devices and in the hard disk drive head component [5]. Otherwise, an in-plane isotropically coercive free layer allowing stabilizing the magnetization in any direction between the parallel and the antiparallel magnetic configurations, is proposed for efficient spintronic memristive applications [6–9]. In these cases, the intermediate resistance states in between the maximum and the minimum values corresponding to the parallel and antiparallel alignment of the magnetization of top and bottom layers sandwiching a conductive/isolating thin spacers are obtained via the dependence of the giant/tunneling magnetoresistance on the angle between the magnetization directions of the two magnetic layers of different magnetic anisotropies.

Tuning the magnetic anisotropy can be achieved via a series of different preparation methods and conditions. Morpho-structural aspects and structured topography obtained for example by ion beam etching and wet chemical etching [6], oblique deposition [8–10] or applying a magnetic field during deposition [11], can be exploited with the purpose to pattern the magnetic material. External control of magnetic anisotropy, coercivity, or magnetic relaxation/superparamagnetism can be achieved via the application of an electric field [12,13]. A different choice is to achieve in-plane uniaxial magnetic anisotropy that may be obtained in a common way by organizing matter into linear objects with a small dimensionality for example nanowires, stripes, and nanotubes with regular shape and spacing [2,3] or by depositing magnetic systems on valley's deeps of undulated surfaces [14]. Moreover, nanowires and linear systems can be implemented in a wide range of applications in domains such as electronics [15], photonics [16,17], sensoristics [18], or thermoelectrics [19].

A way of preparing nanowires, stripes, and linear structures is based on using pre-engineered substrates to guide the growth. A major advantage of this method is the possibility of producing ordered arrays with a narrow size distribution of the component entities [20]. Thus, self-assembled magnetic monoatomic wires could be obtained on a vicinal surface [21] or on facetted surfaces with controllable periodicity [22,23]. Besides, the angular distribution of the easy axis of magnetization corresponding to the magnetic system and its overall anisotropy can be tuned from an almost uniaxial (Dirac-like angular distribution) to an almost isotropic one (random-like angular distribution) [24].

In the present work, we are following a method that supposes the use of a nano-structured surface with linearly faceted trenches as a template [22,23]. The trenches can be obtained by homoepitaxy of a refractory metal (W or Mo) with body-centered cubic crystalline structure, deposited on a nominal planar surface.

The lateral periodicity of the trenches, which are mainly oriented in the [001] crystallographic direction, can be controlled kinetically by tuning the deposition temperature [25]. We were using Mo for the formation of trenches because it favors the in-plane anisotropy with an easy axis along the trenches ([001] direction) in the case of Fe films deposited on Mo [25–27]. Magnetic nanostructures were subsequently engineered as undulated thin films of Fe and FeCo by deposition on the faceted trenches of the substrate, with the aim to control the magnetic properties and especially the magnetic anisotropy. Here we compare the magnetic properties of an undulated Fe film with those of a FeCo film with similar effective thicknesses given the possibility to tune the coercive field and the saturation magnetization of Fe intermetallic compounds by Co addition [28]. The influence of composition, effective thickness, and shape effects were considered. The wetting particularities of the deposited materials were proven to have also an important role. Accordingly, different types of nanostructures of Fe and FeCo with a distribution of the easy axis of magnetization from in-plane isotropic to uniaxial anisotropic, without or with out-of-plane components, were obtained. These systems have promising applications for memristive devices [5]. On the other hand, studying the magnetic properties of an undulated FeCo film from a fundamental point of view is motivated by the rich specificities the magnetic FeCo or Fe/Co systems exhibit such as magnetostriction [29], magnetoresistance [30], or hosting topological textures known as skyrmions [31]. We revised the magnetic properties of Fe and FeCo undulated thin films with limiting cases (defined by geometrical structure aspects) such as: (i) assembly of magnetic nanowires and (ii) assembly of magnetic islands/clusters.

2. Materials and Methods

The samples were prepared in an ultra-high vacuum (the base pressure was 1×10^{-9} Torr) by Magnetron Sputtering deposition on mono-crystalline Al_2O_3 sapphire substrates with $(11\bar{2}0)$ crystal orientation, acquired from CrysTec GmbH. The thickness of the deposited materials was estimated in-situ with a quartz crystal microbalance and the results were corroborated with post-deposition interferometry and profilometry measurements realized ex-situ. In order to obtain optimal deposition conditions for the realization of the desired structures the following parameters have been adjusted: the pressure of the working gas (Ar), the value of the power used, applying RF or DC voltage, the substrate temperature during deposition or the thermal treatment temperature after deposition. The Mo trenches were obtained by depositing in the first step a Mo layer under a power of 7–9 W in DC, for 20–25 min at RT. In order to obtain sequences of trenches with different periodicity [22], two sets of temperature intervals were used during the homoepitaxial growth of $Mo/Mo/Al_2O_3$: 500–280 °C and 300–180 °C, respectively. In a second step, a Mo layer was deposited with 9 W in DC for 60 min. Magnetic layers of Fe and, respectively, FeCo were realized by deposition at 175 °C from corresponding unique targets, with a DC power of 20 W for 6 min and 15 W for 6.5 min, respectively. On trenches of higher lateral periodicity, we deposited a FeCo film in the same conditions as the previously described FeCo sample. A thicker sample of FeCo depositing under a DC power of 7 W for 60 min was grown also on Mo trenches of higher lateral periodicity. The low DC power used for Mo, Fe, and FeCo (e.g., 1 Å/min for Mo and 2 Å/min for Fe) sputtering allowed for achieving very low deposition rates that are required to obtain small crystallites [28]. The obtained magnetic nanostructures were covered with a 2–5 nm thick Mo layer to prevent oxidation processes.

Morpho-structural and compositional characterization were mainly realized via microscopic measurements using Atomic Force Microscopy (AFM) and Transmission Electron Microscopy (TEM). Atomic force microscopy (AFM) was performed using an MFP 3D SA (Asylum Research, Oxford Instruments, Santa Barbara, United States) microscope. Surface topography images for 1 mm square scan areas were obtained by tapping (or AC or non-contact) mode using commercial Si cantilevers (AC160TS, spring constant k = 26 N/m and resonance frequency f = 275 kHz, Olympus Co. Ltd., Shinjuku, Tokyo, Japan). Transmission Electron Microscopy (TEM) measurements have been performed using a Cs-corrected JEOL ARM 200F instrument (JEOL Ltd., Akishima, Tokyo, Japan) with Electron Energy Loss Spectroscopy (EELS) capabilities. TEM specimens have been prepared using the standard cross-sectional method, involving mechanical polishing followed by ion polishing using the PIPS (Precision Ion Polishing System) system (GATAN Inc., Pleasanton, United States). TEM was also used in the mass thickness contrast mode. Vectorial magnetic measurements were realized using room temperature longitudinal Magneto-Optical Kerr Effect (MOKE) investigating the magnetic properties at different orientations of the applied field in the sample plane. Temperature-dependent magnetic properties were analyzed by Superconducting Quantum Interference Device (SQUID) magnetometry between 10 and 300 K, applying a magnetic field in two directions in the sample plane (longitudinal and transversal geometry) and perpendicular to the sample plane (perpendicular geometry).

3. Results

3.1. Structural and Morphological Investigations of Fe Linear Structures

As described in Refs. [20,21,23], Mo faceted linear structures with [001] crystallographic orientation can be prepared, this direction being favored at Mo/Fe interfaces as the easy axis of magnetization of Fe. Following similar deposition conditions, even some differences may occur due to the use here of a different deposition method [32–35], firstly a 5 nm thick Mo film was deposited on an Al_2O_3 and it was thermally treated after deposition, in order to obtain a planar flat surface. Then, a Mo layer having a thickness of 15 nm was re-deposited on the planar Mo surface while varying the deposition temperature, in order to obtain the arrays of linearly trenched structures [22]. For all samples, we used these structures as a template for the growth of Fe and FeCo thin films. AFM, TEM images, and sketches of representative samples are presented in Figure 1. The AFM images correspond to planar atomic surfaces of the Al_2O_3 substrate (Figure 1a) and to the first wetting Mo layer (Figure 1b). Linear patterns observed in the approximately diagonal direction of the image represent atomic steps corresponding to the Al_2O_3 substrate with an average periodicity of 35 nm, followed by the first thermal treated Mo layer. Figure 1c shows the profiles of the structures over the lines presented in the two images. Beyond the AFM resolution, trenches were evidenced by TEM. In Figure 1d the TEM images (inset: chemical contrast) in the profile plane of a Fe sample with a nominal thickness of approximately

3 nm and lateral periodicity of trenches of approximately 15 nm are shown. Fe molds on the structured Mo surface and accumulates deep inside the trenches.

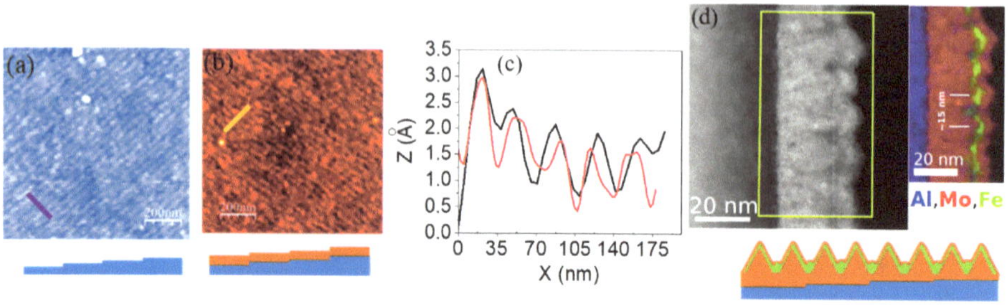

Figure 1. Microscopic images of the linear systems with the corresponding schematic representation in the lower part of the figure. (**a**) AFM image of the planar surface of Al_2O_3 substrate with atomic terraces of 35 nm average periodicity. (**b**) AFM image of the wetting layer of Mo surface following the structure of the Al_2O_3 substrate that was used for the deposition of the template presenting Mo linear trenches. (**c**) Topographic profiles realized over the lines marked in (**a**,**b**). (**d**) TEM image and the elemental spectroscopic analysis (inset) in a profile plane of a Fe sample prepared by deposition on Mo trenches. The sample is covered with a thin Mo layer against oxidation processes.

3.2. *Magnetic Characterization*

Usually, two types of magnetic measurements are performed for the characterization of the nanosystems: (i) hysteresis loops for the investigation of the magnetic reversal mechanisms and (ii) specific measurements of zero-field-cooling—filed-cooling (ZFC-FC) procedures for the investigation of the magnetic relaxation mechanisms. The measuring geometry for the vectorial investigation of the magnetization reversal depends on the orientation of the applied magnetic field and will be defined as follows according to the inset of Figure 2a: (i) longitudinal, with the magnetic field applied along Oz direction (along the wave), (ii) transversal, with the magnetic field applied along the Ox direction (perpendicular to the wave, but in the film plane) and (iii) perpendicular, with the magnetic field applied along to the Oy direction (perpendicular to the film plane).

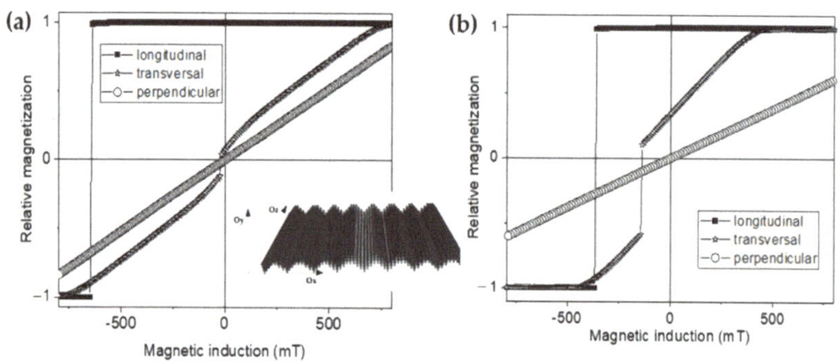

Figure 2. The magnetization reversal of an undulated Fe thin film with a thickness of 3 nm, deep of the valley of 5 nm, and lateral periodicity of 15 nm, in case of three specific directions of the applied magnetic field. The considered geometry with the associated reference system used in the computations is shown in the inset (**a**). The magnetization reversal of an undulated Fe thin film with a thickness of 3 nm, deep of the valley of 3 nm, and lateral periodicity of 15 nm, for the same three specific measurement geometries (**b**).

According to the present samples, three types of nanostructures should be critically revised: (i) assembly of magnetic nanowires, (ii) assembly of magnetic islands/clusters, and

(iii) undulated magnetic films. However, depending on the undulated film characteristics (lateral periodicity, deep and longitudinal uniformity) the last system can cover at the limit the first two cases, as well as the case of a quasi-uniform thin film. For example, undulated templates with deep well-formed trenches of high lateral periodicity and very low thickness of the magnetic film can lead to non-interacting filiform/quasi-cylindrical nanowires. Specific magnetization reversal mechanisms depend on the nanowire diameter [36,37]. If the lateral periodicity is decreased, the assembly of nanowires that interact by dipolar magnetic interactions is obtained [38]. If the longitudinal uniformity of the nanowires is lost, separate segments of nanowires of different aspect ratios (nanocluster-like) are obtained [39]. Depending on the separation gap and the transversal periodicity, such clusters can interact along the chain (unidimensional magnetic systems) as well as between chains (bi-dimensional magnetic systems). Moreover, depending on the shape and size of the magnetic clusters, they can become superparamagnetic above a certain temperature [40]. Depending on the wetting conditions and type of magnetic materials, a very broad distribution of magnetic clusters (both in size and aspect ratio) can be formed along the trenches. If the inter-chain magnetic interaction is mediated by the finest clusters which become superparamagnetic above a certain temperature, a unidimensional magnetic system is expected above this temperature and a bi-dimensional one, below. Magnetic relaxation and blocking temperatures of possible magnetic clusters can be observed in this case by typical ZFC-FC measurements.

In order to interpret the magnetic measurements of the prepared systems, firstly theoretical expectations on the magnetization reversal mechanism obtained by micromagnetic simulations will be briefly discussed in the case of some ideal systems derived from undulated magnetic thin films. The magnetization reversal of an undulated Fe thin film with a thickness of 3 nm, deep of the valley of 5 nm, and lateral periodicity of 15 nm is shown in Figure 2a. The simulations have been performed using the OOMMF (Object Oriented Micromagnetic Framework) public domain software, developed by the Applied and Computational Mathematics Division of National Institute of Standards and Technology (NIST) [41]. The considered geometry and the discretization space are presented in the inset. Half of the loops (from the highest positive applied magnetic field to the highest negative field) are shown. The three-magnetization reversal effects presented in Figure 2 correspond to the three above-mentioned directions of the applied magnetic field: (i) field applied along the wave (longitudinal geometry), (ii) field applied perpendicular to the wave but in the film plane (transversal geometry) and (iii) field applied perpendicular to the film plane (perpendicular geometry).

In longitudinal geometry, it is observed the typical jump of the relative magnetization (current magnetization relative to the spontaneous magnetization which can be experimentally approximated by saturation magnetization at low temperatures) from 1 to -1, specific to a magnetization reversal with the field applied along a magnetic easy axis. In this case, the easy axis is along the wave (Oz axis). On the other hand, the magnetization reversal in perpendicular geometry is linearly increasing with the applied field, according to a coherent spin rotation typical to a hard magnetic axis. Finally, in transversal geometry, the magnetization reversal approaches well a coherent spin rotation, except for a narrow interval of low fields (order of magnitude lower than the switching/jumping field in longitudinal geometry) where a fast reorientation of the spins take place. The situation is not similar to a magnetic field applied perpendicular to a narrow angular distribution of easy axes due to the rectangular shape of the loop in longitudinal geometry (field applied along the easy axis). Moreover, due to the lower value of the switching field in longitudinal geometry with respect to the saturation field in perpendicular geometry (that means with the film applied along an easy and hard axis of magnetization), the magnetization reversal mechanism in undulated thin films has to be represented by magnetic walls displacement anywhere a switching field is evidenced (e.g., in both longitudinal and transversal geometry) [36,40]. At variance to the film with a deep of the valley of 5 nm where the switching field in longitudinal geometry is about 6000 Oe, in the case of the film with a lateral periodicity of 15 nm and deep of the valley of 3 nm, the switching field in longitudinal geometry is about 4000 Oe whereas in transversal geometry the jump in magnetization is larger and at a switching field closer to the switching field in longitudinal geometry. At the same time, the coherent rotation in perpendicular geometry develops much harder (i.e., lower magnetic susceptibility) than in the former case. To note that in a planar geometry (0 nm deep of the valley), with in plane magnetic anisotropy, a full magnetization reversal from 1 to -1 appears at the same very low field of order of hundred Oe in both longitudinal and transversal geometries whereas the coherent reversal is the hardest. At this point, one may conclude on undulated thin films of Fe (and by extension of any soft ferromagnetic film) that the magnetic reversal mechanism depends not only on the direction of the magnetic field with respect to the structuring wave, but also on the wave characteristics with respect to the film thickness. In the case of a lateral periodicity of about 15 nm as in the present case, for nanometer-sized valley deeps larger than the film thickness, a strong in-plane uniaxial anisotropy (shape related) with the easy axis along the waves is evidenced by micromagnetic calculations. The magnetization reversal in longitudinal geometry takes place at a switching field higher than a few thousand Oe, via a fast domain wall displacement. In transversal geometry (with the field applied along the film but perpendicular to the wave), a coherent rotation takes place initially, followed by a final switching by domain wall displacement in lower negative fields of the order of hundred Oe. The coherent spin rotation which is in work in perpendicular geometry is saturated in fields higher than 10 kOe. By decreasing the valley deep down to the film thickness, the discrepancies between the magnetic reversal loops in longitudinal and transversal geometries become lower and lower, involving however almost similar switching fields of the order of kOe. The saturation of the coherent spin rotation in perpendicular geometry becomes harder and harder, involving applied magnetic fields of many tenths of kOe.

Besides, the magnetization reversal of magnetic clusters depends on both the size and the shape of the clusters. Ellipsoidal magnetic nanoparticles of enough low size and aspect ratio with magnetic monodomain structure can be superparamagnetic above the blocking temperature whereas below the blocking temperature the magnetization reversal respects the Stoner-Wohlfarth mechanism [41]. Rectangular hysteresis loops are obtained with the field applied along the easy axis, the switching field being dependent on the shape anisotropy energy and taking values of hundreds of Oe in the case of ellipsoidal Fe and FeCo nanoparticles of high aspect ratios. A coherent spin rotation takes place with the magnetic field applied perpendicular to the easy axis with the saturation field equating to the switching field. Given the trenched structure of the template, ellipsoidal magnetic nanostructures may be dominantly formed in case of unappropriated wetting of the deposited material. In fact, this limiting case can be obtained starting also from the unidirectional case of filiform configurations along the trenches. Accordingly, in the peculiar case of undulated templates with trenches of high lateral periodicity, non-interacting filiform nanowires (deposited only in the valley) can be obtained for the deposited effective thickness of the film lower than half of the deep of the valley. The magnetization reversal of the assembly respects the magnetization reversal of a nanowire, which is strongly dependent on its diameter. Coherent rotation behavior (Stoner-Wohlfarth like) is specific to nanowires of soft magnetic materials (Ni, Fe, and even Co) with diameters lower than 4–5 nm whereas for higher diameters other mechanisms are in work (transverse wall mode, curling mode, etc.). However, in all cases a distinction can be made in the reversal loops obtained with the applied field along the easy axis (longitudinal geometry in the present case) and perpendicular to the easy axis (either transversal geometry or perpendicular geometry in the present case). Rectangular loops are specific to longitudinal geometry and linear ones to transversal and perpendicular geometries (with saturation field equating to the switching field in case of only coherent rotation and a higher saturation field than the switching field in case of transverse wall mode). If dipolar interactions take place between the nanowires, deviations from both the rectangular and linear variation of the magnetization versus field may appear. Such much-rounded shapes of the loops can be also due to local discontinuities of the nanowires, leading to the distribution of magnetic segments, easily associated with a distribution of ellipsoidal nanoparticles of different sizes and aspect ratios.

Moreover, before the presentation of the experimental vectorial magnetic reversal curves depending on the direction of the applied magnetic field with respect to sample-related fixed axes, it is worth mentioning some important specificities for its interpretation. There are well-known pitfalls of SQUID measurements in the case of magnetic measurements on non-uniform magnetic configurations of extremely low magnetic moments [42,43]. It was proved that due to different filling factors of the SQUID pick-up coils, large variation in the measured signal (e.g., the total magnetic moment at saturation) can be observed depending on the orientation of the sample with respect to the coil geometry (such discrepancies are even more evident in case of nanostructures grown on diamagnetic substrates). Usually, the pick-up coils are concentric to the superconducting coil which generates the applied magnetic field, and hence the measured signal will depend on the orientation of the magnetic structure with respect to the field. Therefore, not the values of the saturation magnetization in different geometries will be critically discussed, but the overall vectorial behavior (the shape) of the hysteresis loops at different orientations of the applied field with respect to the nanostructured samples. It should be noted that the diamagnetic contribution of the substrate was properly subtracted in the case of the hysteresis loops presented in the next sections.

3.2.1. Magnetic Characterization of Fe Undulated Systems (with an Effective Thickness of 3 nm) Deposited on Mo Trenches

Magnetic properties of Fe nanostructures, with a thickness of 3 nm—as derived from the TEM image (Figure 1d), i.e., the magnetic anisotropy, the magnetic moment, and the easy axis of magnetization are determined from the observed magnetic behavior at various orientation of the applied field versus sample plane and a fixed in-plane direction as well as at different temperatures. Magnetic hysteresis loops acquired at 10 K and at different orientations of the applied field with respect to the structured undulated plane, as well as at different temperatures are presented in Figures 3a and 3b respectively.

Neglecting the values of the saturation magnetization (due to the pitfalls of such specific SQUID measurements) two main observations should be underlined from Figure 3a: (i) rounded hysteresis loops of similar shapes and coercive fields of about 280 Oe are obtained in transversal and longitudinal geometry at 10 K (no dominant easy axis specific to strongly undulated thin films) and (ii) a relative fast saturation within a not opened loop in perpendicular geometry. According to the above discussion, the only situation for such a behavior can be explained by a quasi-discontinuous undulated thin film formed by three-dimensional islands of Fe of nanometer size (about 5 nm), which strongly interact at low temperature. Hence, at 10 K, the magnetization reversal will be specific to a combination between the long-range magnetic structure of an undulated thin film with an equivalent thickness higher to the valley deep (whit negligible difference between magnetization reversal in transversal and longitudinal geometry) and an assembly of quasi-spheroidal nanoparticles which open the low coercive loop specific to the magnetic frozen state below the blocking temperature. As a consequence, due to nanoparticle-related components, the similar reversal loops in longitudinal and transversal geometries will present a relatively low coercive field as compared to a simple undulated thin film whereas. Also, the magnetic reversal in perpendicular geometry will present a higher magnetic susceptibility in low fields, followed by an apparent saturation due to the compensation of the diamagnetic susceptibility of the substrate with the positive susceptibility of the undulated film.

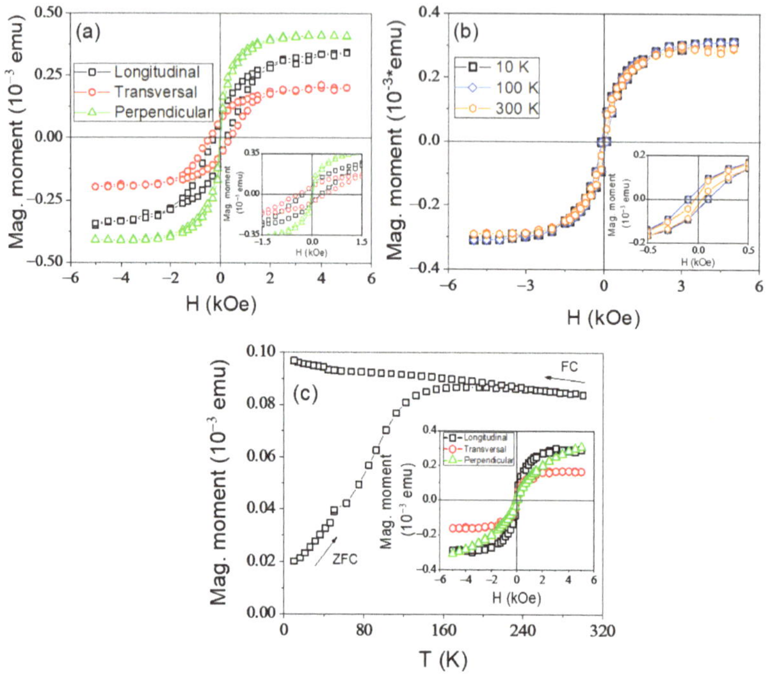

Figure 3. Hysteresis loops of the Fe nanostructured system collected at 10 K in longitudinal, transversal, and perpendicular geometry (**a**); hysteresis loops as a function of temperature in longitudinal geometry (**b**). Insets present an enlarged image of the hysteresis loops around the origin. ZFC-FC curves collected with 200 Oe applied field in longitudinal geometry. Inset shows the hysteresis loops collected at 300 K in different geometries (**c**).

By increasing the temperature, the finest nanoclusters (smallest Fe islands) become progressively superparamagnetic, canceling the interactions between the larger clusters. As a consequence, the long-range magnetic order specific to the undulated thin film structure disappears and the coercive field in the longitudinal magnetization reversal (specific to only an assembly of non-interacting magnetic nanoparticles) decreases progressively with the temperature while more and more nanoparticles become superparamagnetic (Figure 3b) This behavior is also nicely supported by the ZFC-FC curves presented in Figure 3c. The ZFC-FC procedure was carried out by measuring the magnetic signal with an applied magnetic field of 200 Oe while increasing temperature, after the previous cooling of the sample in a null magnetic field. As the temperature is increased, a progressive increase of the magnetic moment is observed due to a size distribution of the magnetic entities (and implicitly of the magnetic moment) as a consequence of their deblocking followed by their orientation towards the applied field. The range of temperatures at which the magnetic moment is increasing corresponds to the distribution of sizes of the magnetic entities and thus, of the blocking temperature. An average blocking temperature of about 136 K (in longitudinal geometry) was estimated at the maximum of zero-field cooling (ZFC) curves. From the variation of the coercive fields in Figure 3b with the $\sqrt{T}$ a final blocking temperature of about 150 K was determined. Despite the large error in the determination of this value due to the lack of experimental points, it is in good agreement with the blocking temperature estimated from the ZFC-FC curves. To note the very slow decrease of the curves above the branching point, proving a very broad size distribution of the Fe islands in the film. The magnetic moment measured at room temperature supports also the existence of magnetically frozen nanoparticles even at room temperature.

3.2.2. Magnetic Characterization of FeCo Undulated Systems (with an Effective Thickness of about 3 nm) Deposited on Mo Trenches

For comparison, a FeCo sample was prepared in the same experimental conditions as the Fe undulated system characterized above, keeping the same effective thickness of 3 nm of the FeCo magnetic layer. Magnetic reversal loops measured in different geometries and at different temperatures are presented in Figure 4a,b. There is not a notable difference between the magnetization reversal loops collected at 10 K with the field applied in the sample plane (longitudinal, transversal, and oblique geometries, the last one meaning the field oriented at 45 degrees with respect to longitudinal and transversal direction), similar to the previous case. The coercive field is in this case about 200 Oe (so, slightly lower than in the case of the Fe film), inferring on one hand an undulated long-range magnetic structure with an effective thickness higher than the valley deep, and on the other hand, rather the formation of oblate FeCo islands/nanoparticles (e.g., with lateral sizes larger than the highness). As a consequence, the magnetic susceptibility in perpendicular geometry is much lower in the case of the FeCo oblate nanoparticles/spheroids than in the case of the quasi-spheroidal Fe ones, and therefore, the overall contribution to the magnetic susceptibility at 10 K coming from the undulated FeCo thin film with long-range magnetic order, the system of FeCo oblate nanoparticles in the magnetic frozen regime and the diamagnetic substrate becomes negligible. This is an additional reason for the very low values of magnetization over the whole range of applied magnetic fields during the magnetic reversal in perpendicular geometry. On the other hand, the evolution of the coercive field with temperature in longitudinal geometry is quite similar to the one of the 3 nm Fe film. Following the same explanation, the finest FeCo oblate nanoparticles become superparamagnetic at higher temperatures (as proven also by the ZFC-FC measurements in Figure 4c), canceling out the long-range magnetic interaction of the non-homogeneous undulated FeCo thin film. The magnetic behavior is due to only the assembly of non-interacting FeCo nanoparticles which enter progressively the superparamagnetic state at a higher temperature.

The blocking temperature estimated from the ZFC curve measured in longitudinal geometry with an applied field of 200 Oe is about 157 K (Figure 4c). The insignificant decrease of magnetization above the branching temperature infers also FeCo nanoparticles of large size and shape distribution, many particles remaining in the blocked state also at room temperature. However, the larger blocking temperature and coercive field in longitudinal geometry in the case of the FeCo system as compared to the Fe system, give support for an in-plane anisotropy of the largest nanoparticles, higher in the case of FeCo systems. Nevertheless, this situation becomes possible only for quasi-ellipsoidal nanoparticles (prolate spheroids) oriented along the trenches. Hence, a higher in-plane uniaxial anisotropy of the system can be kept at 300 K (easy axis along the ellipsoidal nanostructures) in the FeCo systems. Given this behavior, two FeCo systems grown on trenches of higher periodicity will be considered, with the aim of diminishing the lateral dipolar interaction between the ellipsoidal nanoparticles which are magnetically blocked at room temperature.

3.2.3. Magnetic Characterization of FeCo Nanostructured System (with a Thickness of 3 nm) Deposited on Mo Trenches Prepared at Higher Temperature Than Those Presented in Section 3.2.2

Considering a further possibility to control the magnetic properties of these nanosystems, we prepared Mo templates using a temperature gradient in the range (500–200 °C), which induces a larger periodicity than that of the first two samples [22]. These templates were used for the deposition of FeCo films of two different effective thicknesses.

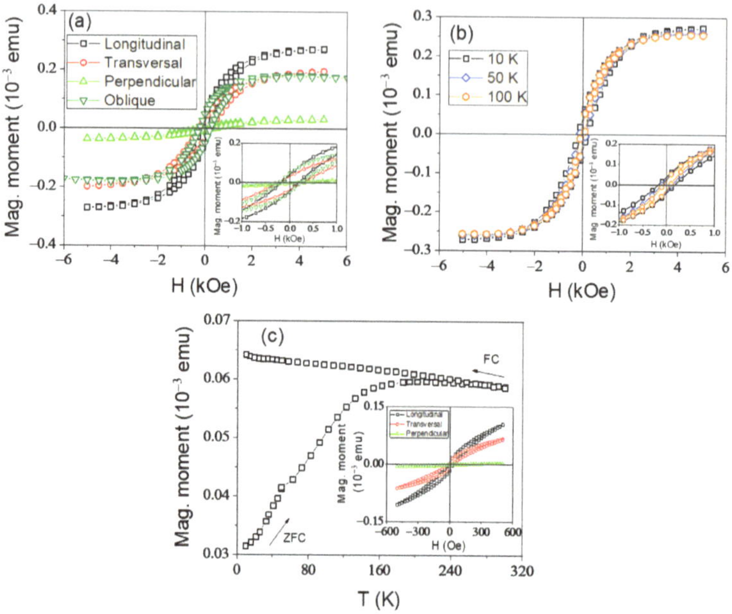

Figure 4. Hysteresis loops of FeCo sample collected at 10 K in longitudinal, transversal, oblique, and perpendicular geometry (**a**); hysteresis loops as a function of temperature in longitudinal geometry (**b**). Insets present an enlarged image of the hysteresis loops around the origin. ZFC-FC curves collected with 200 Oe applied field in longitudinal geometry. Inset shows the hysteresis loops collected at 300 K in different geometries (**c**).

Hysteresis loops were collected at 10 K in different geometries and at different temperatures and the ZFC-FC curves of the FeCo system with the same thickness (3 nm) and deposition conditions similar to the one presented in Section 3.2.2, are shown in Figure 5. However, in this case, the facetted template should have a higher periodicity of the trenches than the previous system. If concerning the magnetization reversal in longitudinal and transversal geometries the situation is similar to of the previous FeCo film (with almost the same shape and a coercive field of 200 Oe in both geometries), a significant difference appears in the perpendicular geometry (Figure 5a). This behavior might be induced by the Co composites and the magnetization reversal mechanism that is strongly dependent on the geometrical configuration of the finest FeCo nano-islands/nanoparticles which are blocked at low temperature and assure the long-range magnetic interaction of the undulated film and develops along a more rounded hysteresis loop, of high coercive field. The only explanation for this behavior, is the formation of oblate islands/spheroids grown on the wave side, in such a way that a field perpendicular to the undulated film will make a small but finite angle with the plane of the quasi-bidimensional island. According to the Stoner-Wohlfarth model, the magnetization reversal develops in this case gradually (with low magnetic susceptibility), but with a sensible coercive field. By increasing the temperature, the finest FeCo nanoparticles become superparamagnetic, cutting out the interaction among the largest prolate islands (nanoparticles assimilated to prolate spheroids with the long axis along trenches. Accordingly, the coercive field of the assembly of non-interacting prolate nanoparticles decreases with temperature, as the effect of more and more nanoparticles entering the superparamagnetic regime (Figure 5b), similar to the case of the previous FeCo film. However, the ZFC-FC curves, in this case, suggest a much-increased blocking temperature above 300 K (Figure 5c), meaning longer prolate magnetic nanostructures along the trenches.

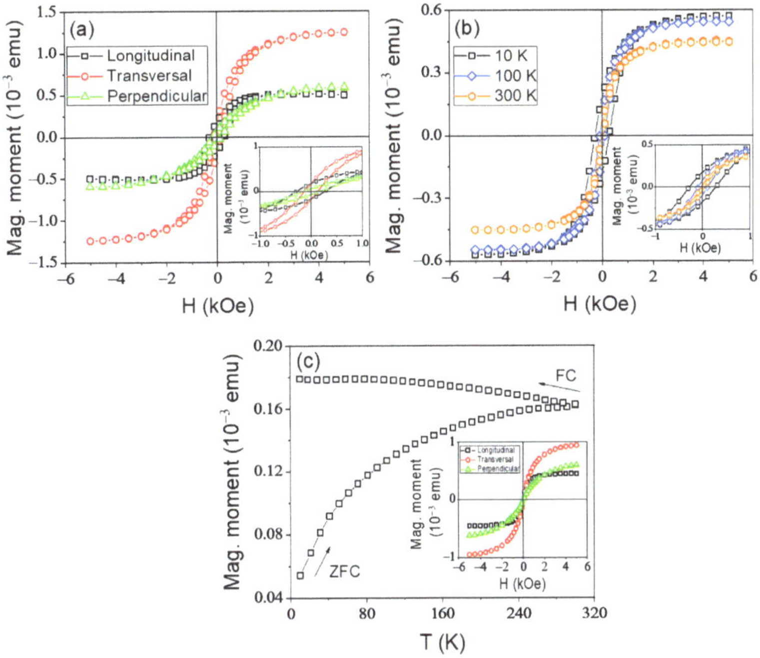

Figure 5. Hysteresis loops of a 3 nm FeCo film grown on more distanced trenches, collected at 10 K in longitudinal, transversal, and perpendicular geometry (**a**); hysteresis loops as a function of temperature in longitudinal geometry (**b**). Insets present an enlarged image of the hysteresis loops around the origin; ZFC-FC curve in longitudinal geometry. Inset shows the hysteresis loops collected at 300 K in different geometries (**c**).

3.2.4. Magnetic Characterization of a Thicker FeCo System than the One Presented in Section 3.2.3

In order to further study the influence of the film thickness on the magnetic properties of such nanostructures, a much thicker FeCo system of approximately 15 nm (5 times higher thickness than the systems presented above) was prepared on Mo linear trenches. The grooved Mo templates were deposited at a temperature gradient of (500–300 °C) similarly to the system previously presented. Hysteresis loops collected at 10 K in different geometries and at different temperatures as well as the specific ZFC-FC curves are illustrated in Figure 6. A noticeable difference can be observed in Figure 6a as compared to Figure 5a. That is the higher coercive field (800 Oe as compared to 200 Oe) reflected by the two loops of similar shape collected in longitudinal and transversal geometry. That means the undulated thin film contribution to the in-plane anisotropy of the system becomes dominant over the oblate nanoislands grown on the wave side, as will be easily expected from the deposition of a thicker film. On the contrary, in the perpendicular geometry where the spins of the undulated film respond very hardly to the field, the dominant increase of magnetization is given by the magnetization reversal of the finest oblate nanoislands which are magnetically frozen at low temperature and hence the shape of the loop is similar to the previous case. At increasing temperature, the finest oblate nanoparticles become superparamagnetic and cancel out the long-range magnetic interaction between larger clusters, leading to magnetic configurations specific to undulated thin films. Hence, the coercive field of the assembly of the prolate nanoclusters decreases at increasing temperatures due to the increased number of superparamagnetic nanoparticles/nanoislands (Figure 6b). The average blocking temperature, well above 300 K as deduced from the ZFC-FC curves presented in Figure 6c is even higher than in the previous case, proving the formation of prolate nanoislands of higher aspect ratio (or anisotropy constant).

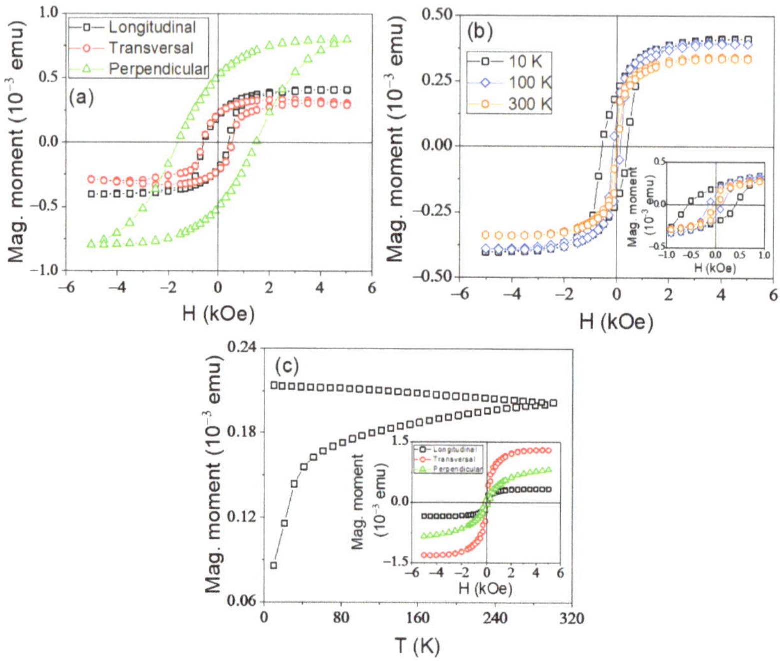

Figure 6. Hysteresis loops of 15 nm FeCo film grown on more distanced trenches, collected at 10 K in longitudinal, transversal, and perpendicular geometry (**a**); hysteresis loops as a function of temperature in longitudinal geometry (**b**). Insets present an enlarged image of the hysteresis loops around the origin. ZFC-FC curve in longitudinal geometry. Inset shows the hysteresis loops collected at 300 K in different geometries (**c**).

Due to the higher effective thickness of the inhomogeneous FeCo film, the in-plane magnetic properties of the FeCo system were also characterized at room temperature by magneto-optical measurements using a MOKE magnetometer.

Specific vectorial hysteresis loops (at different orientations of the field with respect to the trenches direction), as well as polar representations of the relative remnant magnetization (remnant to saturation magnetization ratio) and of the normalized coercive field (current coercive field reported to the maximal coercive field) obtained from the MOKE hysteresis loops collected at different azimuthal angles, are shown in Figure 7. The ample variation of the shape of the hysteresis loops with the azimuthal angle, translated also in the large oscillations of both the relative remnant magnetization and normalized coercive field, proves clearly the high uniaxial anisotropy of the system along the trenches achieved at room temperature in such systems.

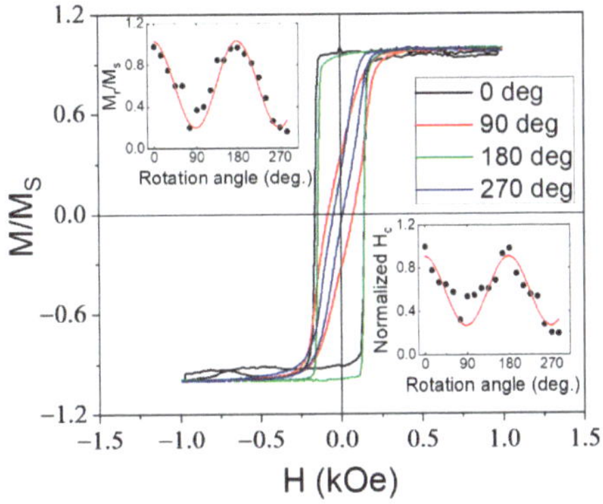

Figure 7. MOKE loops of FeCo structures with a thickness of 15 nm deposited on Mo with linear trenches, collected at different azimuthal angles with respect to the direction of trenches; Angular representation of the relative remanent magnetization and normalized coercive field are shown in insets.

4. Conclusions

Nanostructured undulated thin films of Fe and FeCo were prepared on trenched Mo templates. We characterize the vectorial magnetic properties by tuning the shape, size, thickness, and composition parameters of the system. Inhomogeneous undulated thin films consisting of low size oblate islands and larger prolate islands with the magnetic easy axis along the trenches were obtained in all cases. Both the long-range magnetic ordered undulated system and the nanoparticulate system contribute to the magnetization reversal at low temperatures. By increasing the temperature, the low-size oblate nanoparticles become superparamagnetic and cut off the long-range interactions among the larger prolate nanoparticles. Hence, at room temperature, the magnetization reversal is specific to an assembly of non-interacting oblate ellipsoidal nanoparticles which behave as nanowires with the easy axis of magnetization oriented along the trenches. The uniaxial anisotropy of the system at room temperature and the complex anisotropy at low temperature can be controlled by induced specificities of trenches, thickness, and composition of the magnetic material. The obtained results open new paths for the extension of the studies concerning the adjustment of magnetic properties on various linear magnetic systems and thin films.

Author Contributions: Conceptualization, B.B., G.S. and V.K.; methodology, investigation, and data curation A.E.S., B.B., G.S., A.C.K., C.L., L.T., N.I. and A.L.; writing—original draft preparation, A.E.S., B.B. and V.K.; writing—review and editing, B.B. and V.K.; visualization, A.E.S. and B.B.; supervision, B.B. and V.K.; project administration, B.B. and V.K.; funding acquisition, A.E.S., B.B. and V.K. All authors have read and agreed to the published version of the manuscript.

Funding: This work was supported by the Romanian Ministry of Research and Innovation through projects: Core Program PN19 (Contract No. 21N/2019), PN-III-P1-1.2-PCCDI-2017-0152 (contract no. 75PCCDI/2018), PN-III-P2-2.1-PED-2021-0378 (contract no. 575PED/2022) and PN-III-P1-1.1-PD-2019-1141 (contract no. PD 163/2020). The research leading to these results has also received funding from the NO-RO grants 2014–2021, under Project contract No 39/2021.

Institutional Review Board Statement: Not applicable.

Informed Consent Statement: Not applicable.

Data Availability Statement: Not applicable.

Acknowledgments: The authors thank Cristian Mihail Teodorescu for support coordinating the 75PCCDI/2018 project. The authors acknowledge Daniel Bürgler from Research Center Jülich (FZJ) for making possible the use of test substrates.

Conflicts of Interest: The authors declare no conflict of interest.

References

1. Guimarães, A.P. Magnetism of Thin Films and Multilayers. In *Principles of Nanomagnetism*, 1st ed.; Springer: Berlin/Heidelberg, Germany, 2009; Volume 7, pp. 112–117.
2. Dorantes-Dávila, J.; Pastor, G.M. Magnetic Anisotropy of One-Dimensional Nanostructures of Transition Metals. *Phys. Rev. Lett.* **1998**, *81*, 208–211. [CrossRef]
3. Staňo, M.; Fruchart, O. Magnetic nanowires and nanotubes. In *Handbook of Magnetic Materials*, 1st ed.; Elsevier: Amsterdam, The Netherlands, 2018; Volume 2, pp. 155–267.
4. Bhatti, S.; Sbiaa, R.; Hirohata, A.; Ohno, H.; Fukami, S.; Piramanayagam, S.N. Spintronics based random access memory: A review. *Mater. Today* **2017**, *20*, 530–548. [CrossRef]
5. Mansueto, M.; Chavent, A.; Auffret, S.; Joumard, I.; Nath, J.; Miron, I.M.; Ebels, U.; Sousa, R.C.; Buda-Prejbeanu, L.D.; Prejbeanu, I.L.; et al. Realizing an Isotropically Coercive Magnetic Layer for Memristive Applications by Analogy to Dry Friction. *Phys. Rev. Appl.* **2019**, *12*, 44029. [CrossRef]
6. Buddea, T.; Gatzen, H.H. Thin film SmCo magnets for use in electromagnetic microactuators. *J. Appl. Phys.* **2006**, *99*, 08N304. [CrossRef]
7. Mansueto, M.; Chavent, A.; Auffret, S.; Joumard, I.; Vila, L.; Sousa, R.C.; Buda-Prejbeanu, L.D.; Prejbeanu, I.L.; Dieny, B. Spintronic memristors for neuromorphic circuits based on the angular variation of tunnel magnetoresistance. *Nanoscale* **2021**, *13*, 11488–11496. [CrossRef]
8. Pan, L.; Wang, F.; Wang, W.; Chai, G.; Xue, D. In-plane Isotropic Microwave Performance of CoZr Trilayer in GHz Range. *Sci. Rep.* **2016**, *6*, 21327. [CrossRef]
9. Ali, Z.; Basaula, D.; Eid, K.F.; Khan, M. Anisotropic properties of oblique angle deposited permalloy thin films. *Thin Solid Films* **2021**, *735*, 138899. [CrossRef]
10. Li, J.; Zhan, Q.; Zhang, S.; Wei, J.; Wang, J.; Pan, M.; Xie, Y.; Yang, H.; Zhou, Z.; Xie, S.; et al. Magnetic anisotropy and high-frequency property of flexible FeCoTa films obliquely deposited on a wrinkled topography. *Sci. Rep.* **2017**, *7*, 2837. [CrossRef]
11. Cao, D.; Zhu, Z.; Feng, H.; Pan, L.; Cheng, X.; Wang, Z.; Wang, J.; Liu, Q. Applied magnetic field angle dependence of the static and dynamic magnetic properties in FeCo films during the deposition. *J. Magn. Magn. Mater.* **2016**, *416*, 208–212. [CrossRef]
12. Terada, H.; Ohya, S.; Anh, L.D.; Iwasa, Y.; Tanaka, M. Magnetic anisotropy control by applying an electric field to the side surface of ferromagnetic films. *Sci. Rep.* **2017**, *7*, 5618. [CrossRef]
13. Lau, Y.-C.; Sheng, P.; Mitani, S.; Chiba, D.; Hayashi, M. Electric field modulation of the non-linear areal magnetic anisotropy energy. *Appl. Phys. Lett.* **2017**, *110*, 022405. [CrossRef]
14. Sánchez, E.H.; Rodriguez-Rodriguez, G.; Aragón, R.; Arranz, M.A.; Rebollar, E.; Castillejo, M.; Colino, J.M. Anisotropy engineering of soft thin films in the undulated magnetic state. *J. Magn. Magn. Mater.* **2020**, *514*, 167149. [CrossRef]
15. Lu, W.; Lieber, C.M. Nanoelectronics from the bottom up. *Nat. Mat.* **2007**, *6*, 841–850. [CrossRef] [PubMed]
16. Pauzauskie, P.; Yang, P. Nanowire photonics. *Mater. Today* **2006**, *9*, 36–45. [CrossRef]
17. Arranz, M.A.; Sánchez, E.H.; Rebollar, E.; Castillejo, M.; Colino, J.M. Form and magnetic birefringence in undulated Permalloy/PET films. *Opt. Express* **2019**, *27*, 21285–21294. [CrossRef]
18. Patoisky, F.; Lieber, C. Nanowire nanosensors. *Mater. Today* **2005**, *8*, 20–28. [CrossRef]
19. Chen, R.; Lee, J.; Lee, W.; Li, D. Thermoelectrics of nanowires. *Chem. Rev.* **2019**, *119*, 9260–9302. [CrossRef]
20. Rousset, S.; Croset, B.; Girard, Y.; Prévot, G.; Repain, V.; Rohart, S. Self-organized epitaxial growth on spontaneously nano-patterned templates. *C. R. Phys.* **2005**, *6*, 33–46. [CrossRef]
21. Gambardella, P.; Dallmeyer, A.; Maiti, K.; Malagoli, M.C.; Eberhardt, W.; Kern, K.; Carbone, C. Ferromagnetism in one-dimensional monatomic metal chains. *Nature* **2002**, *416*, 301–304. [CrossRef]
22. Borca, B.; Fruchart, O.; Cheynis, F.; Hasegawa, M.; Meyer, C. Growth and magnetism of self-organized arrays of Fe (110) wires formed by deposition on kinetically grooved W (110). *Surf. Sci.* **2007**, *601*, 4358–4361. [CrossRef]
23. Borca, B.; Fruchart, O.; Kritsikis, E.; Cheynis, F.; Rousseau, A.; David, P.; Meyer, C.; Toussaint, J.C. Tunable magnetic properties of arrays of Fe (1 1 0) nanowires grown on kinetically grooved W(1 1 0) self-organized templates. *J. Magn. Magn. Mater.* **2010**, *322*, 257–264. [CrossRef]
24. Kuncser, A.; Kuncser, V. Magnetization reversal via a Stoner–Wohlfarth model with bi-dimensional angular distribution of easy axis. *J. Magn. Magn. Mater.* **2015**, *395*, 34–40. [CrossRef]
25. Borca, B.; Fruchart, O.; David, P.; Rousseau, A.; Meyer, C. Kinetic self-organization of trenched templates for the fabrication of versatile ferromagnetic nanowires. *Appl. Phys. Lett.* **2007**, *90*, 142507. [CrossRef]
26. Fruchart, O.; Nozieres, J.P.; Givord, D. Growth and interface magnetic anisotropy of epitaxial Mo/Fe/Mo (1 1 0) and W/Fe/W (1 1 0) ultrathin films. *J. Magn. Magn. Mater.* **1999**, *207*, 158–167. [CrossRef]
27. Prokop, J.; Kukunin, A.; Elmers, H.J. Magnetic anisotropies and coupling mechanisms in Fe/Mo (110) nanostripes. *Phys. Rev. Lett.* **2005**, *95*, 187202. [CrossRef]
28. Kishimoto, M.; Latiff, H.; Kita, E.; Yanagihara, H. Morphology and magnetic properties of FeCo particles synthesized with different compositions of Co and Fe through co-precipitation, flux treatment, and reduction. *J. Magn. Magn. Mater.* **2019**, *476*, 229–233. [CrossRef]

29. Beato-López, J.J.; Urdániz-Villanueva, J.G.; Pérez-Landazábal, J.I.; Gómez-Polo, C. Giant stress impedance magnetoelastic sensors employing soft magnetic amorphous ribbons. *Materials* **2020**, *13*, 2175. [CrossRef]
30. Tekgül, A.; Şahin, T.; Köçkar, H.; Alper, M. Structural, magnetic and GMR properties of FeCo (Cu)/Cu magnetic multilayers electrodeposited at high cathode potentials of the magnetic layer. *Optoelectron. Adv. Mater. Rapid Commun.* **2020**, *14*, 189–195.
31. Satywali, B.; Kravchuk, V.P.; Pan, L.; Raju, M.; He, S.; Ma, F.; Petrović, A.P.; Garst, M.; Panagopoulos, C. Microwave resonances of magnetic skyrmions in thin film multilayers. *Nat. Commun.* **2021**, *12*, 1909. [CrossRef]
32. Chan, K.Y.; Teo, B.S. Investigation into the influence of direct current (DC) power in the magnetron sputtering process on the copper crystallite size. *Microelectron. J.* **2007**, *38*, 60–62. [CrossRef]
33. Sasaki, T.; Koshizaki, N.; Beck, K.M. Comparison of Pt=TiO_2 nanocomposite films prepared by sputtering and pulsed laser deposition. *Appl. Phys. A* **1999**, *69*, S771–S774. [CrossRef]
34. Sun, L.; He, J.; Chen, Y.; Yue, F.; Yang, P.; Chu, J. Comparative study on Cu2ZnSnS4 thin films deposited by sputtering and pulsed laser deposition from a single quaternary sulfide target. *J. Cryst. Growth* **2012**, *361*, 147–151. [CrossRef]
35. Le, M.-T.; Sohn, Y.-U.; Lim, J.-W.; Choi, G.-S. Effect of Sputtering Power on the Nucleation and Growth of Cu Films Deposited by Magnetron Sputtering. *Mater. Trans.* **2010**, *51*, 116–120. [CrossRef]
36. Kuncser, A.; Antohe, S.; Kuncser, V. A general perspective on the magnetization reversal in cylindrical soft magnetic nanowires with dominant shape anisotropy. *J. Magn. Magn. Mater.* **2017**, *423*, 34–38. [CrossRef]
37. Costas, A.; Florica, C.; Matei, E.; Toimil-Molares, M.E.; Stavarache, I.; Kuncser, A.; Kuncser, V.; Enculescu, I. Magnetism and magnetoresistance of single Ni–Cu alloy nanowires. *Beilstein J. Nanotechnol.* **2018**, *9*, 2345–2355. [CrossRef]
38. Matei, E.; Enculescu, I.; Toimil-Molares, M.E.; Leca, A.; Ghica, C.; Kuncser, V. Magnetic configurations of Ni–Cu alloy nanowires obtained by the template method. *J. Nanopart. Res.* **2013**, *15*, 1863. [CrossRef]
39. Locovei, C.; Radu, C.; Kuncser, A.; Iacob, N.; Schinteie, G.; Stanciu, A.; Iftimie, S.; Kuncser, V. Relationship between the Formation of Magnetic Clusters and Hexagonal Phase of Gold Matrix in AuxFe1− x Nanophase Thin Films. *Nanomaterials* **2022**, *12*, 1176. [CrossRef]
40. Kuncser, V.; Palade, P.; Kuncser, A.; Greculeasa, S.; Schinteie, G.; Kuncser, V. *Size Effects in Nanostructures*; Kuncser, V., Miu, L., Eds.; Springer Series in Materials Science; Springer: Berlin/Heidelberg, Germany, 2014; p. 205.
41. Donahue, M.J.; Porter, D.G. *OOMMF User's Guide, Version 1.0, Interagency Report, NISTIR 6376*; National Institute of Standards and Technology: Gaithersburg, MD, USA, 1999.
42. Stamenov, P.; Coey, J.M.D. Sample size, position, and structure effects on magnetization measurements using second-order gradiometer pickup coils. *Rev. Sci. Instr.* **2006**, *77*, 015106. [CrossRef]
43. Buchner, M.; Hofler, K.; Henne, B.; Ney, V.; Ney, A. Basic principles, limits of detection, and pitfalls of highly sensitive SQUID magnetometry for nanomagnetism and spintronics. *J. Appl. Phys.* **2018**, *124*, 161101. [CrossRef]

coatings

Article

Effect of a Discontinuous Ag Layer on Optical and Electrical Properties of ZnO/Ag/ZnOStructures

Petko Vitanov [1], Tatyana Ivanova [1,*], Hristosko Dikov [1], Penka Terziyska [2], Maxim Ganchev [1], Nikolay Petkov [3], Yordan Georgiev [4,5] and Asen Asenov [6]

1 Central Laboratory of Solar Energy and New Energy Sources, Bulgarian Academy of Sciences, 72 Tzarigradsko Chaussee Blvd, 1784 Sofia, Bulgaria
2 Institute of Solid State Physics, Bulgarian Academy of Sciences, 72 Tzarigradsko Chaussee Blvd, 1784 Sofia, Bulgaria
3 Physical Sciences, Munster Technological University, Rosa Avenue, Bishopstown, Cork, Ireland and Tyndall National Institute, Lee Maltings, Cork, T12 R5CP, Ireland
4 Institute of Ion Beam Physics and Materials Research, Helmholtz-Zentrum Dresden-Rossendorf (HZDR), BautznerLandstrasse 400, 01328 Dresden, Germany
5 Institute of Electronics, Bulgarian Academy of Sciences, 72 Tsarigradsko Chaussee, 1784 Sofia, Bulgaria
6 School of Engineering, The University of Glasgow, Glasgow G12 0QQ, Scotland, UK
* Correspondence: tativanovame@yahoo.co.uk

Abstract: ZnO/Ag/ZnO nanolaminate structures were deposited by consecutive RF sputtering at room temperature.The optical transparency, sheet resistance, and figure of merit are determined in relation to the deposition time of Ag and to the film thickness of the ZnO top layer. An improved transmittance has been found in the visible spectral range of the ZnO/Ag/ZnO structure compared to ZnO multilayers without Ag. High transmittance of 98% at 550 nm, sheet resistance of 8 Ω/sq, and figure of merit (FOM) of 111.01 $\times$ 10^{-3} $\Omega^{-1}$are achieved for an optimized ZnO/Ag/ZnO nanolaminate structure. It is suggested that the good optical and electrical properties are due to the deposition of the discontinuous Ag layer. The electrical metallic type conductivity is caused by planar located silver metal granules. The deposition of a discrete layer of Ag nano-granules is confirmed by atomic force microscopy (AFM) and cross-section high-resolution transmission electron microscopy (HRTEM) observations.

Keywords: transparent conductive oxide (TCO); magnetron sputtering; ZnO/Ag/ZnO; transparent conducting nanolaminate structures; discontinuous Ag layer; oxide/metal/oxide

1. Introduction

In solar cells and optoelectronic devices, an ideal transparent top electrode must effectively collect electrical carriers while exhibiting minimal optical and electrical losses. Independent from the application, all transparent electrode materials should combine the following properties: transparency in the spectral range of the device, high lateral conductivity (low sheet resistance), good ohmic contact with adjacent layers, and low impact on the underlying layers during their deposition. Recently, the main objective of developing new materials for transparent conductive oxide (TCO) films is to achieve lower resistivity and higher transmittance in the spectral visible range and to replace the conventional ITO conductors. Transparent conducting oxides (TCOs), namely SnO_2, ITO, doped ZnO, etc., are widely investigated in order to be applied as electrodes in a variety of optoelectronic devices such as solar cells and light-emitting diodes [1]. Among transparent electrodes, metal oxide/metal/metal oxide (OMO) structures can achieve improved optical and electrical performances comparable to single TCO layers and very thin metallic films [2]. The development of ITO-alternative low-cost, earth-abundant, and mechanically stable TCOs exhibiting excellent transparency and electrical properties is very important. OMO structures as multilayer TCOs possess excellent optical and electrical properties that can be achieved at much lower thickness even after combining the individual layer thicknesses [3]. The top and bottom oxide layers provide high optical transparency along with protection for the metal layer from degrading [4]. Thus, a ZnO-based multilayer TCO can be cheaper compared to ITO as the complete thickness of the structure ($\leq$100 nm) is much lower than that of ITO (~400 nm) to obtain similar electrical and optical properties [5].

The optical and electrical properties of oxide/metal/oxide structures are well documented [5–12], demonstrating that the dielectric/metal/dielectric thin films with different metals such as Ag, Au, and Cu have improved conductivity and high optical transmission [5,8–12].

Herein we make use of electrical conductivity in granular (discontinuous) type films [13]. The granules are metallic particles with sizes ranging usually from a few to hundreds of nanometers, embedded into an insulating matrix. The nanolaminate OMO structure is formed using one or two different dielectric layers. The electrical conductivity can be tuned by controlling the lateral dispersion of the metal granules within the nanolaminate. The most important part of the fabrication of nanolaminate OMO structures with specified optical and electrical properties is the deposition of a discontinuous metal layer. The advantages of this approach are reproducibility, effective process control, and a selective formation of conductive areas in a dielectric structure. Nanolaminate structures with dielectric layers such as TiO_2, MoO_x, and Al_2O_3 [14], and various metals such as Ag, Cr, Ni, and Pt [15] have been studied. These studies were conducted for various applications of OMO structures [16–18] such as indium-free TCO, optoelectronic devices, energy and flexible electronics, and flexible photovoltaic cells.

In this study, we report on the formation of transparent and conductive ZnO/Ag/ZnO nanolaminate structures obtained only by magnetron sputtering at room temperature. Using atomic force microscopy (AFM) and high-resolution transmission electron microscopy (HRTEM) measurements, we evidence that the deposition of Ag on the ZnO layer forms a discontinuous layer. These islands-type structures in conjunction with the optimized thickness of the top zinc oxide layer increase the structure transparency, achieving a high value of the figure of merit: $-111 \times 10^{-3}\ \Omega^{-1}$. An additional advantage of this structure is that it retains its properties after annealing up to 400°C making it a promising candidate as a highly efficient transparent electrode for optoelectronic devices.

2. Materials and Methods

2.1. Nanolaminate Structure Deposition

ZnO/Ag/ZnO nanolaminate structures were prepared by high-frequency magnetron sputtering technique on glass and Si substrate. The equipment was an RF magnetron sputtering system model CFS-4ES Tokuda system (Tokuda, Seisakusho Co., Ltd., Ginan-cho, Japan) with a 3 sputtering source P-GUN75 (3 inches). The substrate holder was a disk with a diameter of 200 mm and had a water cooling system.The used targets are Ag and ZnO with purities of 99.9% (Kurt J. Lesker; Company LTD., Hastings, UK), respectively. Argon was used as a sputtering gas. The three targets (75 mm in diameter each) were positioned at eccentric positions in front of the substrate holder.The holder was water cooling; therefore a little radiation heat was generated. The targets and the substrate holder were vertically arranged (mode-side sputtering) at a distance of 8 cm. The substrate holder can be rotated at a speed of 10–80 rpm. The three-target configuration allows us to deposit successively different layers without opening the chamber. Sputtering was performed at room temperature without any post-process treatment.The base chamber pressure before the deposition was approximately 1×10^{-6} mbar and the sputtering process was performed at a chamber pressure of approximately 6×10^{-3} mbar for ZnO and 1×10^{-3} mbar for Ag. ZnO and Ag were deposited using an RF power of 200 W. During the sputtering process, the substrate holder was rotated at speed of 20 rpm for ZnO and 60 rpm for Ag. The ZnO/Ag/ZnO multilayer structures were consecutively deposited onto glass substrates (Borofloat glass) and Si wafers (p-type CZ, 3″ polished wafers with resistivity 10 Ω.cm).

Before being loaded into the sputtering chamber, the glass substrates were cleaned in an ultrasonic bath with deionized water for 15 min and finally dried in a N_2 stream.

In our experiments, the deposition times of undercoat and overcoat ZnO layers were 3 min and 6 min, while the deposition time of Ag was set to 5 and 10 s. When Ag is sputtered for a longer time, the metal grains change to high-density distribution. As the result, the conduction ability of fabricated samples is improved. Structures without silver nano-granules were prepared as reference samples under the same conditions. The description of the prepared samples is given in Table 1.

Table 1. The sample preparation and the obtained film thickness of the nanolaminate structure.

№	Sample	Sputtering Time			Layer Thickness (nm)		
		ZnO (Bottom)	Ag	ZnO (Top)	ZnO (Bottom)	Ag	ZnO (Top)
1	ZnO/Ag/ZnO	3 min	7 s	3 min	29	Nanoclusters	29
2	ZnO/Ag/ZnO	3 min	7 s	6 min	29		63
3	ZnO/Ag/ZnO	3 min	10 s	3 min	29		29
4	ZnO/ZnO	3 min	-	3 min	29		29
5	ZnO/ZnO	3 min	-	6 min	29		63

The thickness of ZnO layers was determined by ellipsometric measurements on test samples on a silicon substrate at the different sputtering times of the layers. More complicated is the problem of determining the thickness of the Ag layer. We hypothesize and prove an island structure of Ag granules. In this case with metal islands, where the free space between them is large, the role of the surface substrate dominates. For this reason, we consider that the real thickness cannot be determined from ellipsometric measurements. For this reason, some researchers use the duration of the sputtering process [19] or various Ag deposition rates [20].

2.2. Characterization Methods

The optical properties of the nanolaminate OMO structures are studied by means of UV-VIS-NIR Shimadzu spectrophotometer UV 3300 (Shimadzu Corporation, Kyoto, Japan). Spectral ellipsometry (SE)was applied to optically characterized nanolaminate OMO structures. A spectroscopic ellipsometer J. A. Woollam M-2000 (J.A. Woollam Co. Instruments, Lincoln, NE, USA) was used to acquire film thickness and optical constants in the spectral region of 200–1000 nm. The SE measurements were collected at angles of incidence equal to 45°, 55°, 65°, and 75°, respectively. The sheet resistance was measured using the four-point probe method using a VEECO instrument, model FPP-100, USA.

The surface morphology was investigated by AFM measurements that were carried out using a Multimode V (Bruker, ex. Veeco, Santa Barbara, CA, USA). Imaging was performed in tapping mode and height, with a scan rate of 2 Hz and 512-line image resolution. SEM and HRTEM analyses were performed to evaluate the surface morphology of the Ag layer. TEM analysis was performed with a JEOL 2100, 200 kV (JEOL Ltd., Tokyo, Japan), double tilt holder, beam aligned at the {110} zone axis of the carrier wafer.The cross-section is made with Helios Nanolab Dual Beam FIB (CIC nanoGUNE, San Sebastian, Spain): final thinning at 93 pA 30 kV, final polish 5 kV 56 pA, in situ lift out.

3. Results and Discussions

3.1. Ellipsometric Measurements

ZnO structures of different thicknesses deposited on silicon and glass substrates were investigated by applying spectroscopic ellipsometry (SE). SE spectra were measured at room temperature in a spectral range from 193 nm to 1000 nm. The angle of incidence varies between 45° and 75° (10° steps). SE measurements produce two parameters; the amplitude ratio (ψ_{Exp}) and phase difference (Δ_{Exp}) between the incident and the reflected polarized beams. The extraction of the optical constants n and k from the measured quantities ψ_{Exp}

and Δ_{Exp} is done using an adequate optical model and fitting procedure. Two types of structures were measured: sample 4 and sample 5 (see Table 1). Sample 4 was obtained after two consecutive ZnO depositions of three minutes each with a pause of 30 s between them. Sample 5, with two ZnO layers, was formed by deposition of the first layer for 3 min and the second one for 6 min.

The refractive index (n) and extinction dispersion (k) of the sputtered ZnO structure are presented in Figure 1. The extinction above 400 nm is very low, indicating the high transparency of the film. The refractive index is varied from 2.04 to 1.86 in the spectral range of 400 to 1000 nm.These values are in agreement with the literature data. The refractive index of sputtered ZnO films is reported to be in the range of 1.96 to 2.06 at 500 nm, depending on the film thickness and technological conditions [21].

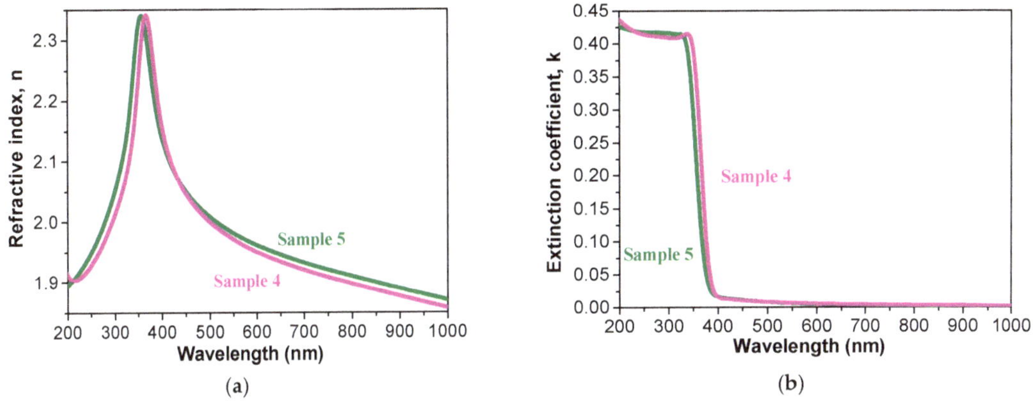

Figure 1. Dispersion of (**a**) refractive index and (**b**) extinction coefficients of magnetron sputtered ZnO/ZnO structures, sample 4 and sample 5.

The total thickness of the ellipsometric measurements for sample 4 is 58 nm and for sample 5 is 92 nm. The average deposition rate is 9.9 nm/min.

Figure 2 shows the SE parameters ψ_{Exp}, and Δ_{Exp} measured for sample 4 at a different angle of incidence for samples 1, 3, and 4. Samples 1 and 3 are nanolaminate film structures with Ag nano-granules i.e., ZnO (28 nm)/Ag/ZnO (28 nm), the difference is the density of Ag nano-granules (deposition time of Ag). Sample 4 is only the ZnO layer. These results are similar to the results presented in [22].

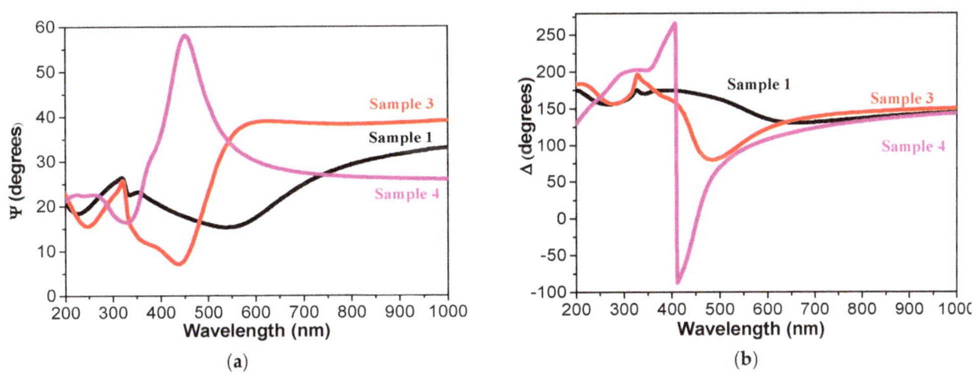

Figure 2. Ellipsometric angles (Ψ, Δ) of (**a**) sample 1 (ZnO (29 nm)/Ag (7 sec)/ZnO (29 nm)), sample 3 (ZnO (29 nm)/Ag (10 sec)/ZnO (29 nm)), and (**b**) sample 4 (ZnO (29 nm)/ZnO (29 nm)) on glass substrate. Ψand Δ are measured at an angle of incidence of 45°.

The spectral behavior of ψ_{Exp} and Δ_{Exp} depends on the deposition time of silver. This can be used as a technological criterion for the presence of silver particles.

3.2. Optical Properties and Sheet Resistance

Multilayered ZnO/Ag/ZnO structures were deposited layer by layer on glass substrates using an RF magnetron sputtering system. The sputtering time and thickness of the layers are listed in Table 1.

In previous articles [14–16], the relationship between optical transmission and sheet resistance of nanolaminate structures of the ZnO/Ag/ZnO type, depending on the thickness of the Ag layer (or deposition time) has been reported. The effect of the thickness of the top layer (ZnO above Ag nano-granules) is also studied in this research. Figure 3 shows the transmittance spectra of the samples described in Table 1.

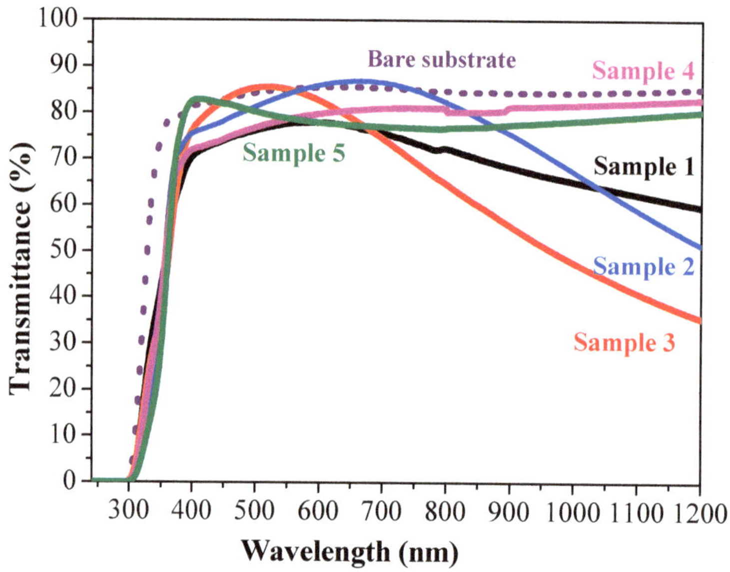

Figure 3. Transmittance spectra for ZnO/ZnO and nanolaminate structure ZnO/Ag/ZnO. The bare substrate is given for comparison.

Table 2 presents the results for the sheet resistance, average transmittance, and figure of merit of the samples from Table 1. In many previous studies of oxide/metal/oxide nanolaminate structures, it has been reported that sheet resistance and optical transmission depend on the thickness of the metal layer [6,8,9]. The thickness of the metal layer varies from 5 nm to 20 nm. Our research shows that increasing the sputtering time increases the density and size of the metal granules, which is the reason for the increase in electrical conductivity. These results are reported for TiO_2/Ag/TiO_2 nanolaminate structures [15]. A similar dependence for the studied samples ZnO/Ag/ZnO is presented in Table 2 for sample 1 (7 s sputtering time) and sample 3 (10 s). It must be noted that the transmittance is measured against air and the reported transparency values are for the whole system glass substrate + nanolaminate structure. Table 2 presents the average transmittance value in the spectral range of 400–800 nm considering and subtracting the bare glass substrate spectrum. It is seen that the ZnO/Ag/ZnO nanolaminate structures are very transparent reaching almost 98%.

Table 2. Sheet resistance, R_{sheet}, of nanolaminate structures, average transmittance, $T_{average}$, and FOM values, estimated for nanolaminate structures. The average transmittance is determined for the spectral range of 400–800 nm and extracting the substrate transmittance.

Structure	R_{sheet} (Ω/sq)	$T_{average}$ (%)	FOM × 10^{-3} ($Ω^{-1}$)
Sample 1 ZnO (29 nm)/Ag (7 s)/ZnO (29 nm)	23	90.84	16.64
Sample 2 ZnO (29 nm)/Ag (7 s)/ZnO (63 nm)	8	98.82	111.01
Sample 3 ZnO (29 nm)/Ag (10 s)/ZnO (29 nm)	13	94.01	41.48

The figure of merit (FOM), proposed by Haacke [23,24], is an effective way to evaluate the electrical and optical properties of transparent conductors and can be determined by:

$$FOM = \frac{T_{av}^{10}}{R_{sheet}}, \tag{1}$$

where T_{av} is the average transmittance in the spectral range 400–800 nm and R_{sheet} is the sheet resistance measured by a four-point probe. The results of FOM are summarized in Table 2.

In Table 3, a comparison of reported sheet resistance and figure of merit of OMO structures, deposited using different methods is given. It must be noted that for commercial ITO, the values are T = 90.2%, R_{sheet} = 30 Ω/sq, and $FOM = 12 \times 10^{-3}\ Ω^{-1}$. The FOM values obtained in this work are close to or even higher than previously reported values (Table 3). Thus, OMO nanolaminate structures with superior optical and electrical properties and figures of merit have been fabricated using consecutive magnetron sputtering at room temperature. The fabricated OMO structures with lower sheet resistance and lower thickness can prove advantageous for reducing the substrate cost in the devices [3,25].

Table 3. Overview of different OMO structures, including their deposition method, average transmittance, sheet resistance, and figure of merit (FOM).

Structures	Deposition Method	R_{sheet} (Ω/sq)	$T_{average}$ (%)	FOM × 10^{-3} ($Ω^{-1}$)	Reference
ZnO/Cu/ZnO	RF/DC sputtering	14.04	68 (550 nm)	63.7	[26]
ZnO/Ag/ZnO	RF/DC sputtering	3.01	90	236	[6]
ZnO/Ag/ZnO	RF sputtering PET substrate	4.98	92–95	104.5	[27]
ZnO/Ag/ZnO	Reactive sputtering	82	5.3	24	[28]
MGZO/Ag/MGZO	reactive plasma deposition	10.0	94.7	58.0	[29]
ZnO/Ag/ZnO	RF sputtering	5.4	87.0	22.4	[3]
AZO/Ag/AZO		5.3	92.0	45.5	[3]
AZO/Au/AZO	RF sputtering	14.31	82.1 (550 nm)	9.69	[30]
AZO/Cu/AZO	RF sputtering	143.4	70	1.97	[31]
ITO/Ag/ITO	MOCVD	3.8	91.3	106.1	[32]
ITO/Ag/ITO	RF sputtering	264.3	92.0	1.65	[33]
ITO/Ag/ITO	Impulse magnetron sputtering	7.29	97.0	101.16	[34]
ITO/Cu/ITO		10.43	74.0	4.83	[34]
$TiO_2/Ag/TiO_2$	RF sputtering	3.5	89	69.6	[33]
$TiO_2/Ag/TiO_2$	RF sputtering	3.31	96.8	69.6	[9]
$TiO_2/Ag/TiO_2$	sputtering	13	88.7 (550 nm)	23.2	[35]
$SnO_2/Ag/SnO_2$	RF sputtering	9.0	94.8	60.0	[11]
$Ta_2O_5/Ag/Ta_2O_5$	RF sputtering	2.53	91.64	157.04	[36]

The nanolaminate structures with an equal film thickness as ZnO (sample 1 and sample 3) possess sheet resistance values that decrease when increasing the deposition time of Ag. The comparison between sample 1 and sample 3 reveals that the sheet resistance has been reduced by a factor of two while the silver deposition time was increased from 3 to 6 min. Notably, sample 2, having almost double the thickness of the top ZnO layer in sample 1, showed sheet resistance that is significantly decreased. The effect of the Ag incorporation of the Ag layer and the thicker top layer is also observed in the transmittance spectra (see Figures 4 and 5).

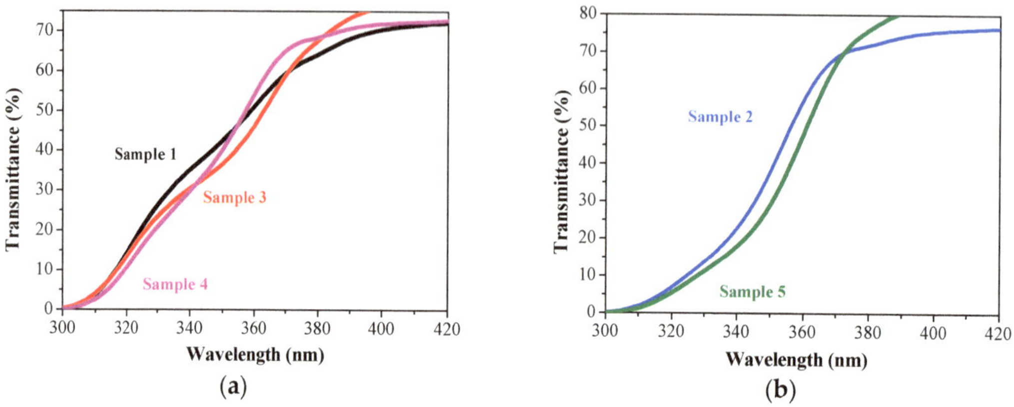

Figure 4. (**a**) presents the absorption edge of samples with the top and bottom ZnO layers having the same thickness of 29 nm, but the Ag sputtering time has been changed and (**b**) is the comparison between sample 2, ZnO (29 nm)/Ag (7 s)/ZnO (63 nm)and sample 5, ZnO (29 nm)/ZnO (63 nm).

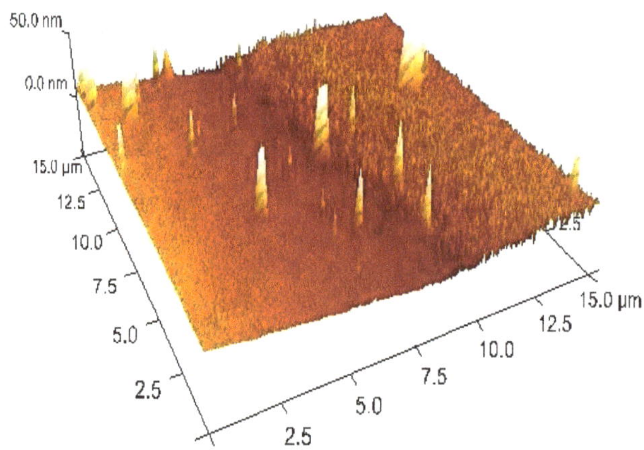

Figure 5. A3D AFM image of sputtered Ag on Si through the mask, the border mask/Ag is clearly seen.

The nanolaminate structure with Ag (sample 2) is more transparent in the visible spectral ranges 450–700 nm when compared to sample 5, with a similar thickness of the ZnO layers but no Ag particle deposition. Similar optical behavior is found for samples 1 and 3 when compared to sample 4 without Ag. Sample 1 has a transparency of 74%–76%, lower than sample 4 (74%–79%). In the near IR region, the transparency of nanolaminate structures with Ag decreases, which is typical of OMO structures. Sample 3 reveals the strongest decrease in the wavelength above 650 nm up to 1200 nm (NIR spectral range), induced most probably by the higher density of the silver particles.

The transmittance spectra of samples 2 and 3 show a broad enhanced transmission in the visible range with maximum transmittances at wavelengths of about 650 nm and 500 nm, respectively. In contrast, the transmittance spectra for sample 1 showed no enhancement of the visible light transmittance. Structurally, sample 1 has received the same sputtering sequence as sample 2, only the thickness of the top ZnO layer has double the thickness (see Figure 3). Such an enhancement of the visible light transmittance by metal (Ag) nano-granule incorporation has already been observed for the ZnO/Ag/ZnO films [37]. The authors of [37] claim that this effect can be attributed to the coupling between the incident light and surface plasmon polaritons (SPPs) of the Ag mid-layer with a layer-plus-islands structure. The coupling or resonance between incident light and SPPs can lead to many interesting physical phenomena, such as extraordinary light transmission [38] and surface-enhanced Raman scattering [39]. In [22], it is also shown that Ag clusters not only have evidently lowest possible absorbance but also might gain from plasmonic features such as re-emission of the light coupled by the SPs at the random silver-island grating or enhanced transmittance.The optical transmission enhancement of ZnO films with incorporated Ag was also investigated by others [40] and they stated that light transmission is enhanced by SPR.

Figure 4 manifests the change of the absorption edge, depending on the Ag layer (different deposition times) and ZnO thickness. The absorption edge is shifted towards the longer wavelengths due to the Ag layer. It is also dependent on the Ag sputtering time. A similar optical behavior of the absorption edge has been previously reported for structures of ZnO/Ag/glass [37,41].

The optical and electrical characterization reveals that Ag granules deposited between two ZnO films lead to improved transparency, lower sheet resistance, and a high figure of merit. These results are encouraging for the application of these nanolaminate structures as highly effective TCOs.

3.3. Morphology of the Ag Layer and Nanolaminate Structure

As the presence of silver particles contributes significantly to the improved conductivity of the nanolaminate structures, we have investigated the morphology of the silver layer and its surface topography. Figure 5 shows an AFM image of the silicon surface whereby the silver is deposited through a mask.

Figure 6a, b show the 3D and plain AFM images of sputtered Ag on Si substrate after 7 s and 15 s sputtering time. It is clear that the continuous Ag layer was not formed after magnetron sputtering according to the morphology images in Figures 7 and 8. The Ag nanoclusters are randomly distributed with distinct shapes and sizes, forming isolated crystal grains. The diameter of the Ag particles is about 30–50 nm. As the sputtering time of the Ag layer is increased, the size (diameter) of the Ag particles is also increased, but their height did not change obviously. From the AFM measurement, the RMS roughness, the arithmetic average roughness Ra and the maximum roughness depth of samples were determined. For the shorter deposition time (7 s), the Ag layer has the following surface parameters: RMS is 3.04 nm, Ra is 2.03 nm, and R_{max} is 26.8 nm. Interestingly, the increase of the deposition time to 15 sec results in lower surface parameters: RMS is 2.32 nm, R_a is 1.73 nm, and R_{max} is 21.2 nm. Similar results were also reported for the deposition of copper nanoclusters [8].

The film morphology of silver was studied by the HRTEM method. For this purpose, the substrate used was an oxidized silicon wafer, coated with silicon nitride (200 nm) for consecutive deposition of ZnO and Ag, whereby the latter was sputtered for 7 s. The deposition was made through a mask to define areas with and without silver particles within the same sample for analysis. Finally, a top blanket layer of ZnO was deposited. The nanolaminate structure was investigated with cross-section TEM and the images are shown in Figures 7–9, whereby lower magnification (larger imaging area) images are resented in Figure 7 outlining the difference between regions with and without Ag nano-granules. Higher magnification images of these two regions are presented in Figures 8 and 9, respectively.

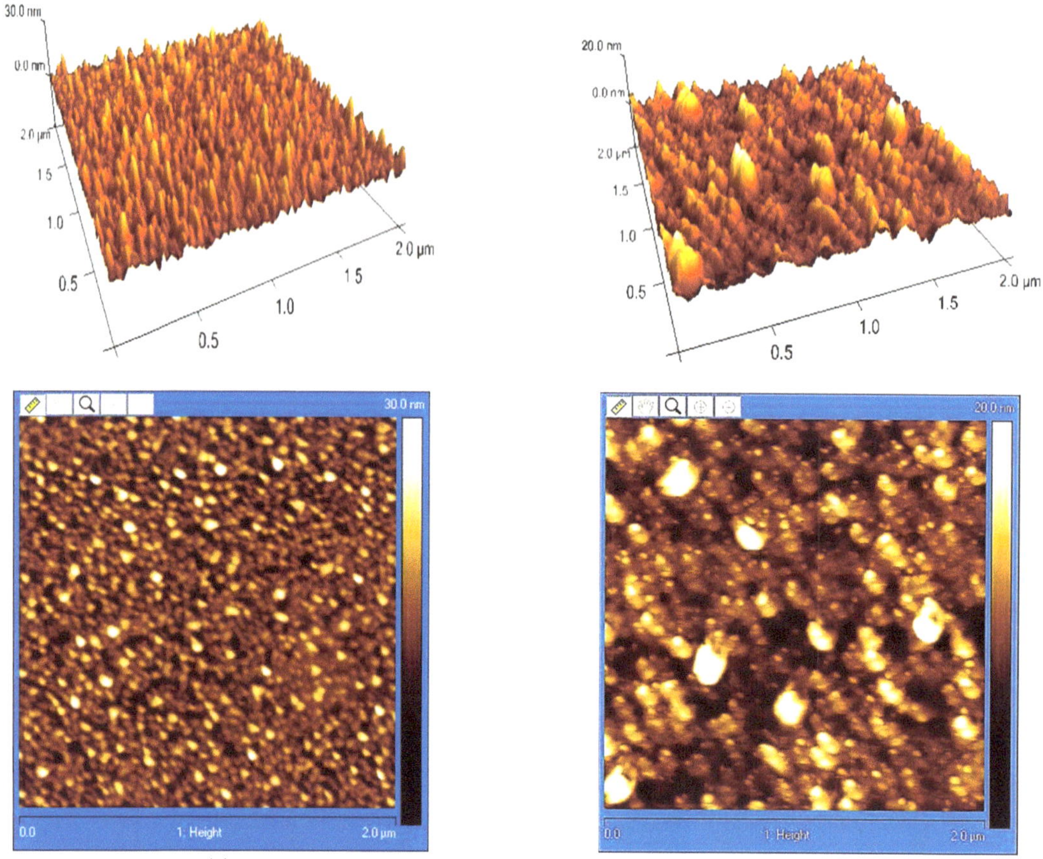

(a) (b)

Figure 6. Surface topography of Ag layer after (**a**) 7 s and (**b**) 15 s sputtering time.

The polycrystalline structure of the consecutively sputtered ZnO layers is clearly distinguished (as can be seen in Figure 8).

In comparison, the region with Ag NPs shows that the sputtering for 7 sec results in forming of a discontinuous film with Ag nanoparticles with a random in-plane orientation as well as a size of about 100 nm. These 3D nanoclusters have penetrated into the ZnO film and even into the underlying Si_3N_4 layer (Figure 9). The dark and bright bands in the Ag nanoclusters are twinning or multiple stacking faults. This is typical for Ag nano-granules when it is formed at low temperature.

Figure 10 shows an SEM image of nanolaminate structure ZnO (29 nm)/Ag (7 sec)/ZnO (29 nm). The Ag islands are clearly seen. This result agrees with AFM and TEM analysis.

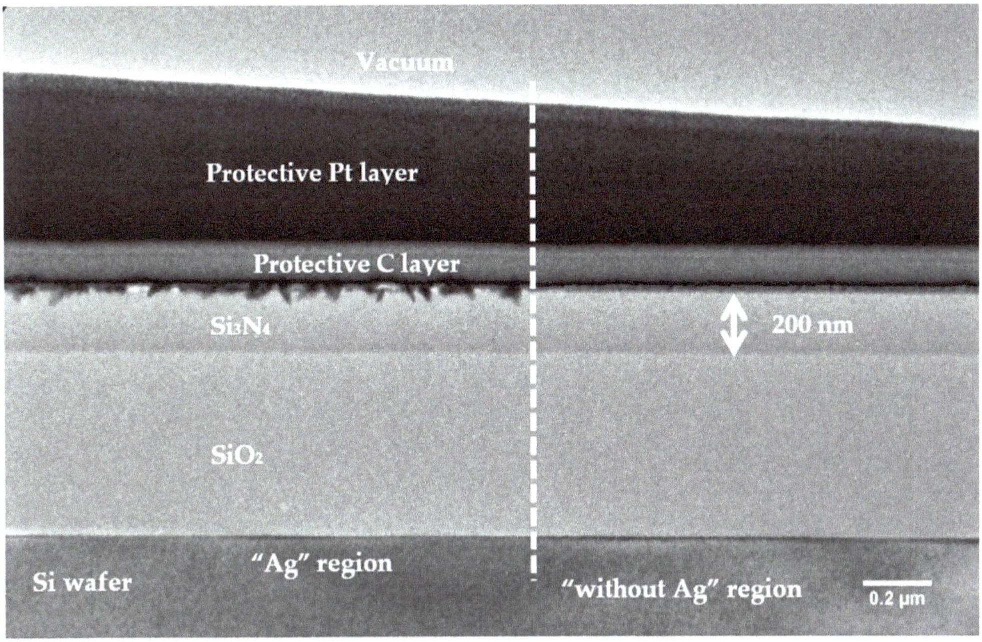

Figure 7. TEM cross-section micrograph of the nanolaminate structure with two areas: the "Ag" region is with Ag nano-granules and the "without Ag" region is without Ag deposition.

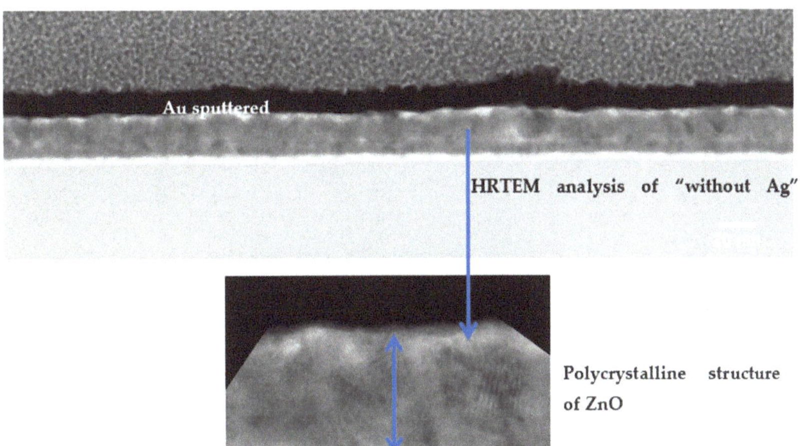

Figure 8. TEM cross-section image of the "without Ag" region without Ag granules and the polycrystalline structure of the ZnO layer.

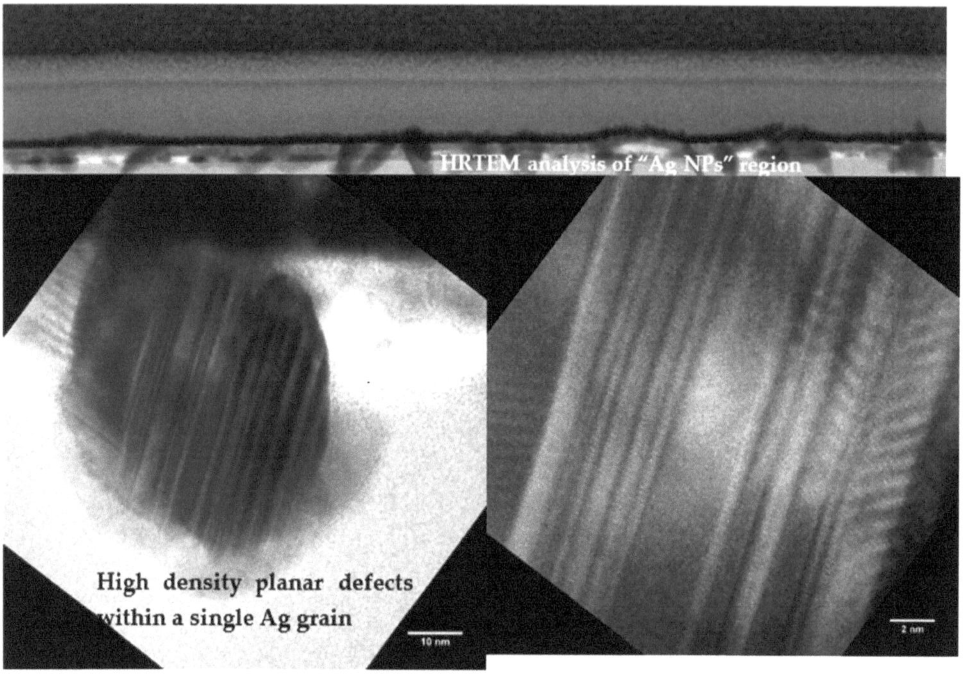

Figure 9. TEM cross-section of the "Ag granules" region and image of a single Ag granule with planar defects.

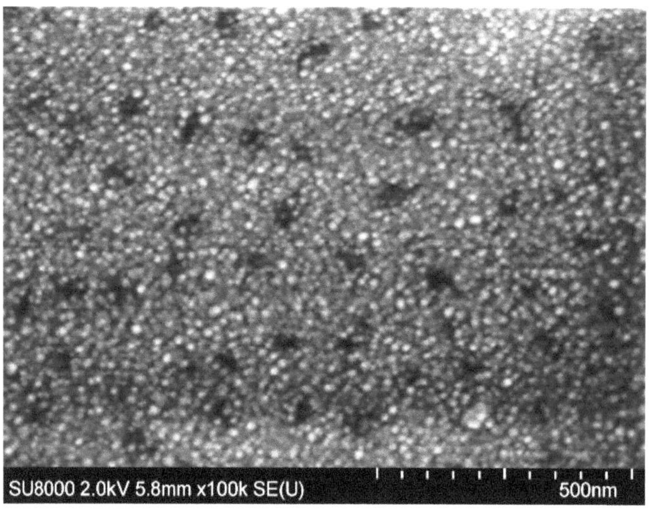

Figure 10. SEM image of the nanolaminate structure ZnO (29 nm)/Ag (7 s)/ZnO (29 nm).

The most important part of the technological process of nanolaminate (OMO) structure with specified optical and electrical properties is the deposition of the discontinuous metal layer. The optical properties and electrical conductivity are modified with planar-located metal granules.

Usually, TCOs have a semiconductor behavior as a negative temperature coefficient of receptivity (TCR). The nanolaminate structure with metal granulate demonstrates metallic conductivity because it has a positive TCR [15]. Therefore, the carriers are delocalized due to thermal activation, and conductivity is dominated by phonon scattering.

Our hypothesis is based on the usage of electronic conductivity in granular (discontinuous) type materials [13]. The key parameter that determines most of the physical properties of the granular array is the average tunneling conductance G between neighboring grains. It is convenient to introduce the dimensionless conductance [13], (corresponding to one spin component):

$$g = \frac{G}{(2e^2/h)}, \quad (2)$$

which is measured in units of the quantum conductance (e^2/h). The samples with $g > 1$ exhibit metallic transport properties, while those with $g < 1$ show insulating behavior.

This hypothesis is confirmed by using the OMO structure as a strain resistor [16,41,42]. The exponential dependence of the tunneling probability on the intergranular distance suggests the suitability of the material as a high-sensitivity strain sensor. When the OMO nanolaminate structure is strained in tension, the average distance between the particles increases. The tunneling probability decreases, and the resistivity of the microresistor increases [41].

3.4. Effect of Temperature Annealing

The impact of the thermal treatment on the optical and electrical properties of sample 2—ZnO (29 nm)/Ag (7 sec)/ZnO (63 nm)—was studied. When sample 2 was exposed to low-temperature annealing at 200 °C/30 min, the transparency and the sheet resistance were almost unchanged, however, when treated at 400 °C/15 min, the transmittance is reduced by 10 to 15% to an average transmittance of 84.6% in the spectral range of 400–800 nm (see Figure 11). In parallel, the measured sheet resistance is changed from 8 to 20 Ω/sq and correspondingly, the FOM was significantly decreased to $10.57 \times 10^{-3}\ \Omega^{-1}$.

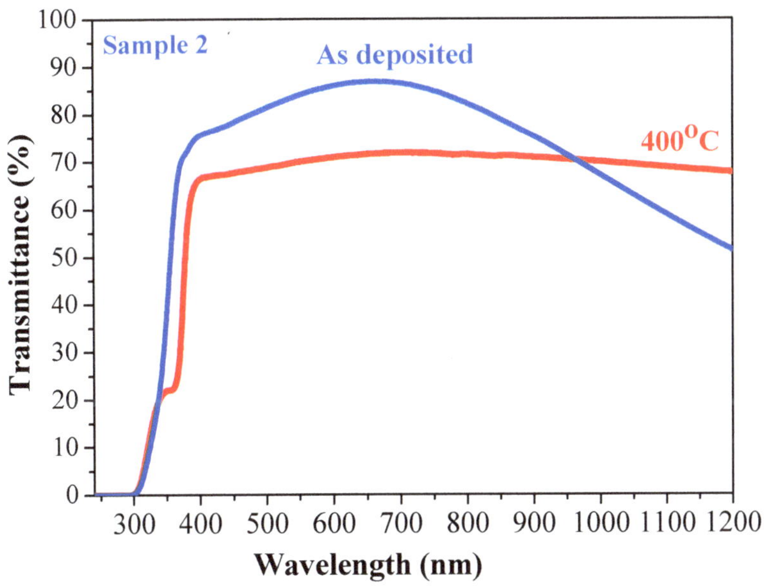

Figure 11. Transmittance spectra of the as-deposited and the annealed at 400 °C/15 min. Sample 2: ZnO (29 nm)/Ag (7 s)/ZnO (63 nm).

The sheet resistance is relatively low and the nanolaminate structure is still transparent. These results are encouraging as they suppose a kind of stability during thermal treatments.

This is an advantage for applying these structures as optical contacts in devices where thermal treatments are required.

4. Conclusions

In conclusion, we have demonstrated highly efficient transparent ZnO/Ag/ZnO structures deposited at room temperature which are especially suitable for the application as the TCO layer.The selected technological conditions for magnetron sputtering of silver led to the formation of a discontinuous layer of Ag nanoclusters and these clusters define the optical and electrical properties of ZnO/Ag/ZnO nanolaminate structures. The technological process for sputtering Ag is reproducible and reliable in terms of thickness and structure. This has been proven on the basis of 10 years of experimental experience. This study also presents two indirect methods for detecting the presence of silver nanoclusters:spectroscopic ellipsometry and TEM analysis. In order to achieve superior optical transparency and low sheet resistance, the crucial role has not only the sputtering process of Ag but the optimal film thickness of the ZnO top layer as well. Varying ZnO layer thickness, the figure of merit (FOM) changes from 16.64×10^{-3} to $111.01 \times 10^{-3} \Omega^{-1}$. The optimized ZnO/Ag/ZnO structure achieves a maximum transmittance of 98% (550 nm) and sheet resistance of 8 Ω/sq. After low-temperature annealing at 200°C/30 min, the transparency and the sheet resistance were almost unchanged. In order to achieve superior optical transparency and low sheet resistance, the crucial role has not only the sputtering process of Ag but the optimal film thickness of the ZnO top layer as well. The optimized oxide/metal/oxide structure achieves a maximum transmittance of 98% (550 nm) and sheet resistance of 8 Ω/sq.

The advantage of the proposed multisource magnetron sputtering deposition of nanolaminate structures is that the formation is performed at room temperature and at a relatively high deposition rate of nearly 9.9 nm/min. This makes the proposed approach for the fabrication of TCO coatings easily applicable for lab-scale and commercial devices. This technological process can be applied to flexible substrates as well.

Author Contributions: P.V.: conceptualization, writing—original draft preparation, supervision; T.I.: investigation, writing—original draft preparation, writing—review and editing; H.D.: investigation, Validation; P.T.: investigation; M.G.: investigation; N.P.: investigation, writing—review and editing; Y.G.: investigation, writing—review and editing; A.A.: supervision. All authors have read and agreed to the published version of the manuscript.

Funding: The authors kindly acknowledge financial support from project № BG05M2OP001-1.002-0014 "Center of competence HITMOBIL—Technologies and systems for generation, storage and consumption of clean energy", funded by Operational Programme "Science and Education For Smart Growth" 2014–2020, co-funded by the EU from European Regional Development Fund.

Institutional Review Board Statement: Not applicable.

Informed Consent Statement: Not applicable.

Data Availability Statement: The data are not publicly available.

Acknowledgments: The authors kindly acknowledge financial support from project № BG05M2OP001-1.002-0014 "Center of competence HITMOBIL—Technologies and systems for generation, storage and consumption of clean energy", funded by Operational Programme "Science and Education For Smart Growth" 2014–2020, co-funded by the EU from European Regional Development Fund.

Conflicts of Interest: The authors declare no conflict of interest.

References

1. Sharma, V. Transparent conducting electrodes based on zinc oxide. In *Metal Oxides, Nanostructured Zinc Oxide*; Awasthi, K., Ed.; Elsevier: Amsterdam, The Netherlands, 2021; pp. 291–318.
2. Singh, R.; Gupta, M.; Mukherjee, S.K. Effect of Ag layer thickness on optical and electrical properties of ion-beam-sputtered TiO_2/Ag/TiO_2 multilayer thin film. *J.Mater. Sci Mater.Electron.* **2022**, *33*, 6942–6953. [CrossRef]
3. Barman, B.; Kumar, S.; Dutta, K.S. Fabrication of highly conducting ZnO/Ag/ZnO and AZO/Ag/AZO transparent conducting oxide layers using RF magnetron sputtering at room temperature. *Mater. Sci. Semicond. Process.* **2021**, *129*, 105801. [CrossRef]
4. Nezhad, E.H.; Haratizadeh, H.; Kari, B.M. Influence of Ag mid-layer in the optical and thermal properties of ZnO/Ag/ZnO thin films on the glass used in Buildings as insulating glass unit (IGU). *Ceram. Int.* **2019**, *45*, 9950–9954. [CrossRef]
5. Miao, D.; Jiang, S.; Shang, S.; Chen, Z. Highly transparent and infrared reflective AZO/Ag/AZO multilayer film prepared on PET substrate by RF magnetron sputtering. *Vacuum* **2014**, *106*, 1–4. [CrossRef]
6. Sahu, D.R.; Huang, J.L. High quality transparent conductive ZnO/Ag/ZnO multilayer films deposited at room temperature. *Thin Solid Films* **2006**, *515*, 876–879. [CrossRef]
7. Sahu, D.R.; Lin, S.Y.; Huang, J.L. ZnO/Ag/ZnO multilayer films for the application of a very low resistance transparent electrode. *Appl. Surf. Sci.* **2006**, *252*, 7509–7514. [CrossRef]
8. Song, S.; Yang, T.; Lv, M.; Li, Y.; Xin, Y.; Jiang, L.; Wu, Z.; Han, S. Effect of Cu layer thickness on the structural, optical and electrical properties of AZO/Cu/AZO tri-layer films. *Vacuum* **2010**, *85*, 39–44. [CrossRef]
9. Kim, J.H.; Moon, Y.J.; Kim, S.K.; Yoo, Y.Z.; Seong, T.Y. Al-doped ZnO/Ag/Al-doped ZnO multilayer films with a high figure of merit. *Ceram. Int.* **2015**, *41*, 14805–14810. [CrossRef]
10. Jeon, K.; Youn, H.; Kim, S.; Shin, S.; Yang, M. Fabrication and characterization of WO_3/Ag/WO_3 multilayer transparent anode with solution-processed WO_3 for polymer light-emitting diodes. *Nanoscale Res. Lett.* **2012**, *7*, 253. [CrossRef]
11. Yu, S.; Zhang, W.; Li, L.; Xu, D.; Dong, H.; Jin, Y. Optimization of SnO_2/Ag/SnO_2 tri-layer films as transparent composite electrode with high figure of merit. *Thin Solid Films* **2014**, *552*, 150–154. [CrossRef]
12. Kim, J.H.; Kim, D.H.; Seong, T.Y. Realization of highly transparent and low resistance TiO_2/Ag/TiO_2 conducting electrode for optoelectronic devices. *Ceram. Int.* **2015**, *41*, 3064–3068. [CrossRef]
13. Beloborodov, I.S.; Lopatin, A.V.; Vinokur, V.M.; Efetov, K.B. Granular electronic system. *Rev. Mod. Phys.* **2007**, *79*, 469–518. [CrossRef]
14. Dikov, H.; Ivanova, T.; Vitanov, P. Oxide/metal/oxide nanolaminate structures for application of transparent electrodes. *J. Phys. Conf. Ser.* **2016**, *764*, 012021. [CrossRef]
15. Dikov, H.; Vitanov, P.; Ivanova, T.; Stavrov, V. Optical and electrical properties of nanolaminate dielectric structures. *J. Phys. Conf. Ser.* **2016**, *700*, 012054. [CrossRef]
16. Dikov, C.; Vitanov, P.; Ivanova, T.; Stavrov, V.; Tomerov, E.; Stavreva, G.; Bineva, I. Optical and electrical properties of TiO_2/Pt/TiO_2 nanolaminate structures. *J. Phys. Conf. Ser.* **2018**, *992*, 012033. [CrossRef]
17. Su, Y.C.; Chiou, C.C.; Marinova, V.; Lin, S.H.; Dikov, H.; Vitanov, P.; Hsu, K.Y. Liquid crystal electro-optic modulator based on transparent conductive TiO_2/Ag/TiO_2 multilayers. *Opt. Quant. Electron.* **2018**, *50*, 242. [CrossRef]
18. Chiou, C.C.; Hsu, F.H.; Petrov, S.; Marinova, V.; Dikov, H.; Vitanov, P.; Lin, S.H. Flexible light valves using polymer-dispersed liquid crystals and TiO_2/Ag/TiO_2 multilayers. *Opt. Express* **2019**, *21*, 1691. [CrossRef]
19. Yu, X.; Zhang, D.; Wang, P.; Murakami, R.; Ding, B.; Song, X. The optical and electrical properties of ZnO/Ag/ZnO films on flexible substrate. *Int. J. Modern Phys. Conf. Ser.* **2012**, *6*, 55–56. [CrossRef]
20. Zhao, Z.; Alford, T.L. The optimal TiO_2/Ag/TiO_2 electrode for organic solar cell application with high device-specific Haacke figure of merit. *Sol. Energy Mater. Sol. Cells* **2016**, *157*, 599–603. [CrossRef]
21. Al-Kuhaili, M.F.; Durrani, S.M.A.; El-Said, A.S.; Heller, R. Enhancement of the refractive index of sputtered zinc oxide thin films through doping with Fe_2O_3. *J. Alloys Compd.* **2017**, *690*, 453–460. [CrossRef]
22. Sytchkova, A.; Grilli, M.L.; Rinaldi, A.; Vedraine, S.; Torchio, P.; Piegari, A.; Flory, F. Radio frequency sputtered Al:ZnO-Ag transparent conductor: A plasmonic nanostructure with enhanced optical and electrical properties. *J. Appl. Phys.* **2013**, *114*, 094509. [CrossRef]
23. Haacke, G. New figure of merit for transparent conductors. *J. Appl. Phys.* **1976**, *47*, 4086. [CrossRef]
24. Bou, A.; Torchio, P.; Vedraine, S.; Barakel, D.; Lucas, B.; Bernède, J.C.; Thoulon, P.Y.; Ricci, M. Numerical optimization of multilayer electrodes without indium for use in organic solar cells. *Sol. Energy Mater. Sol. Cells* **2014**, *125*, 310–317. [CrossRef]
25. Kim, G.; Lim, J.W.; Lee, J.; Heo, S.J. Flexible multilayered transparent electrodes with less than 50 nm thickness using nitrogen-doped silver layers for flexible heaters, Mater. *Res. Bull.* **2022**, *149*, 111703. [CrossRef]
26. Rabizadeh, M.; Ehsani, M.H. Effect of heat treatment on optical, electrical and thermal properties of ZnO/Cu/ZnO thin films for energy-saving application. *Ceram. Int.* **2022**, *48*, 16108–16113. [CrossRef]
27. Kim, J.H.; Lee, J.H.; Kim, S.W.; Yoo, Y.Z.; Seong, T.Y. Highly flexible ZnO/Ag/ZnO conducting electrode for organic photonic devices. *Ceram. Int.* **2015**, *41*, 7146–7150. [CrossRef]
28. El Hajj, A.; Lucas, B.; Chakaroun, M.; Antony, R.; Ratier, B.; Aldissi, M. Optimization of ZnO/Ag/ZnO multilayer electrodes obtained by Ion Beam Sputtering for optoelectronic devices. *Thin Solid Films* **2012**, *520*, 4666–4668. [CrossRef]
29. Zhang, Y.; Liu, Z.; Ji, C.; Chen, X.; Hou, G.; Li, Y.; Zhang, X. Low-Temperature Oxide/Metal/Oxide Multilayer Films as Highly Transparent Conductive Electrodes for Optoelectronic Devices. *ACS Appl. Energy Mater.* **2021**, *4*, 6553–6561. [CrossRef]

30. Chu, C.H.; Wu, H.W.; Huang, J.L. AZO/Au/AZO tri-layer thin films for the very low resistivity transparent electrode applications. *Mater. Sci. Eng. B* **2014**, *186*, 117–121. [CrossRef]
31. Mendil, D.; Challali, F.; Touam, T.; Bockelée, V.; Ouhenia, S.; Souici, A.; Djouadi, D.; Chelouche, A. Preparation of RF sputtered AZO/Cu/AZO multilayer films and the investigation of Cu thickness and substrate effects on their microstructural and optoelectronic properties. *J. Alloys Compd.* **2021**, *860*, 158470. [CrossRef]
32. Lee, S.Y.; Park, Y.S.; Seong, T.Y. Optimized ITO/Ag/ITO multilayers as a current spreading layer to enhance the light output of ultraviolet light-emitting diodes. *J. Alloys Compd.* **2019**, *776*, 960–964. [CrossRef]
33. Im, H.S.; Kim, S.K.; Lee, T.J.; Seong, T.Y. Combined effects of oxygen pressures and RF powers on the electrical characteristics of ITO-based multilayer transparent electrodes. *Vacuum* **2019**, *169*, 108871. [CrossRef]
34. Li, H.; Gao, Y.J.; Yuan, S.H.; Wuu, D.S.; Wu, W.Y.; Zhang, S. Improvement in the figure of merit of ITO-Metal-ITO sandwiched films on poly substrate by high-power impulse magnetron sputtering. *Coatings* **2021**, *11*, 144. [CrossRef]
35. Wu, C.C. High flexible touch screen panel fabricated with silver-inserted transparent ITO triple-layer structures. *RSC Adv.* **2018**, *8*, 11862–11870. [CrossRef] [PubMed]
36. Song, T.S.; Cho, J.W.; Kim, J.H.; Kim, S.K.; Seong, T.Y. High ultraviolet transparent conducting electrodes formed using tantalum oxide/Ag multilayer. *Ceram. Int.* **2022**, *48*, 3536–3543. [CrossRef]
37. Wang, P.; Zhang, D.; Kim, D.H.; Qiu, Z.; Gao, L.; Murakami, R.; Song, X. Enhancement of light transmission by coupling to surface plasmon polaritons of a layer-plus-islands silver layer. *J. Appl. Phys.* **2009**, *106*, 103104. [CrossRef]
38. Shafiq, R.; Khan, A.D.; Al-Harbi, F.F.; Ali, F.; Armghan, A.; Asif, M.; Rehman, A.U.; Ali, F.; Arpanaei, E.M.; Alibakhshikenari, M. Optical Transmission Plasmonic Color Filter with Wider Color Gamut Based on X-Shaped Nanostructure. *Photonics* **2022**, *9*, 209. [CrossRef]
39. Huang, Y.; Fang, Y.; Zhang, Z.; Zhu, L.; Sun, M. Nanowire-supported plasmonic waveguide for remote excitation of surface-enhanced Raman scattering. *Light Sci. Appl.* **2014**, *3*, e199. [CrossRef]
40. Zhang, D.; Yabe, H.; Akita, E.; Wang, P.; Murakami, R.; Song, X. Effect of silver evolution on conductivity and transmittance of ZnO/Ag thin films. *J. Appl. Phys.* **2011**, *109*, 104318.
41. Stavrov, V.; Stavreva, G.; Tomerov, E.; Dikov, H.; Vitanov, P. Self-sensing cantilevers with nano-laminated Dielectric-Metal-Dielectric resistors. In Proceedings of the 2017 40th International Spring Seminar on Electronics Technology (ISSE), Sofia, Bulgaria, 10–14 May 2017; pp. 1–6.
42. Dukic, M.; Winhold, M.; Schwalb, C.; Adams, J.D.; Stavrov, V.; Huth, M.; Fantner, G.E. Direct-write nanoscale printing of nanogranular tunnelling strain sensors for sub-micrometre cantilevers. *Nat. Commun.* **2016**, *7*, 12487. [CrossRef]

coatings

MDPI

Review

Recent Advances in Surface Functionalization of Magnetic Nanoparticles

Cezar Comanescu [1,2]

[1] National Institute of Materials Physics, Atomistilor 405A, 077125 Magurele, Romania; cezar.comanescu@infim.ro
[2] Faculty of Physics, University of Bucharest, 405 Atomistilor St., 077125 Magurele, Romania

Abstract: In recent years, significant progress has been made in the surface functionalization of magnetic nanoparticles (MNPs), revolutionizing their utility in multimodal imaging, drug delivery, and catalysis. This progression, spanning over the last decade, has unfolded in discernible phases, each marked by distinct advancements and paradigm shifts. In the nascent stage, emphasis was placed on foundational techniques, such as ligand exchange and organic coatings, establishing the groundwork for subsequent innovations. This review navigates through the cutting-edge developments in tailoring MNP surfaces, illuminating their pivotal role in advancing these diverse applications. The exploration encompasses an array of innovative strategies such as organic coatings, inorganic encapsulation, ligand engineering, self-assembly, and bioconjugation, elucidating how each approach impacts or augments MNP performance. Notably, surface-functionalized MNPs exhibit increased efficacy in multimodal imaging, demonstrating improved MRI contrast and targeted imaging. The current review underscores the transformative impact of surface modifications on drug delivery systems, enabling controlled release, targeted therapy, and enhanced biocompatibility. With a comprehensive analysis of characterization techniques and future prospects, this review surveys the dynamic landscape of MNP surface functionalization over the past three years (2021–2023). By dissecting the underlying principles and applications, the review provides not only a retrospective analysis but also a forward-looking perspective on the potential of surface-engineered MNPs in shaping the future of science, technology, and medicine.

Keywords: magnetic nanoparticles; surface functionalization/engineering; theranostics; drug delivery; biomedical applications; targeted therapy; biomolecular conjugation; multifunctional nanoparticles

1. Introduction

Magnetic nanoparticles (MNPs) have emerged as versatile entities with profound implications across various scientific frontiers. Their unique physicochemical properties, including superparamagnetism and high surface area-to-volume ratios, have positioned them as compelling candidates in fields ranging from biomedicine [1,2] to catalysis [3,4]. However, the strategic engineering of their surfaces by means of surface modifying agents (such as amine, diimide, carboxyl, aldehyde, hydroxyl, etc.) has unlocked a new dimension of functionalities, allowing further modification by molecule attachment and thus driving recent advancements and expanding their potential [5].

In the realm of multimodal imaging, the surface functionalization of MNPs has become a cornerstone in enhancing diagnostic accuracy [6–8]. By judiciously modifying the surface chemistry, researchers have endeavored to optimize their interaction with biological entities, improve biocompatibility, and enhance their imaging contrast properties. This has led to the development of MNPs with tailored surface coatings, thereby enabling the precise targeting of specific biomarkers and cell populations. The introduction of functional moieties onto the MNP surface has not only amplified their potential as magnetic resonance imaging (MRI) contrast agents but has also opened avenues for multimodal imaging techniques, merging the power of MRI with other imaging modalities such as fluorescence or positron emission tomography (PET).

In drug delivery, the surface functionalization of MNPs has been pivotal in overcoming the formidable challenges associated with efficient and targeted drug transport [9]. These modified surfaces allow for the conjugation of therapeutic agents, enabling controlled release profiles and enhancing the pharmacokinetics of drugs. Furthermore, surface functionalization offers the means to encapsulate drugs within protective shells, shielding them from premature degradation or clearance, while facilitating site-specific release at the intended destination. This approach has not only improved drug efficacy but has also mitigated off-target effects, bringing us closer to the long-envisioned realm of personalized medicine with a keen attention to the fate of the MNPs inside the human body [10]. The catalytic landscape has equally been reshaped by the ingenuity of MNP surface functionalization. Tailoring the surfaces of MNPs with catalytic moieties has yielded heterogeneous catalysts with unparalleled activity and selectivity. This has proven particularly advantageous in complex and intricate catalytic transformations, enabling efficient conversions with reduced side reactions. The advent of well-defined surface architectures has further enabled precise control over catalytic sites and reactions, fostering a synergy between catalysis and nanotechnology that promises to revolutionize chemical synthesis (Figure 1).

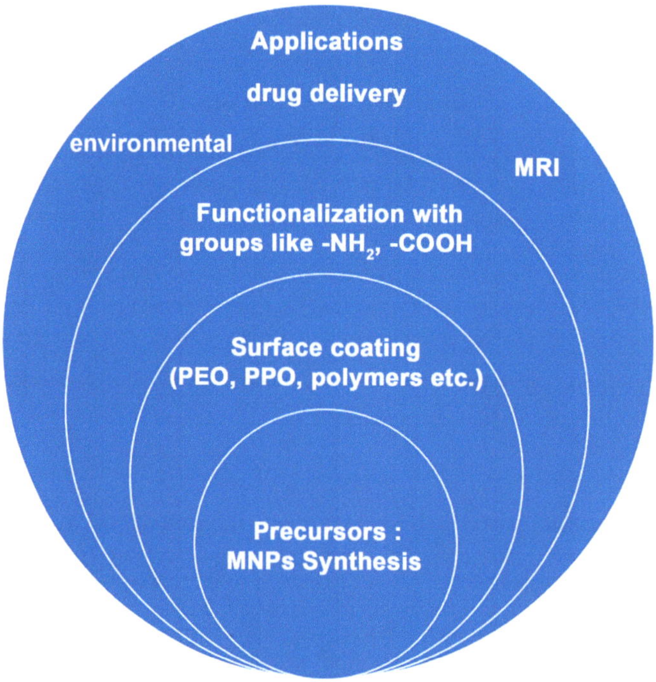

Figure 1. General schematic diagram of MNPs' synthesis and functionalization.

This review navigates the recent strides in MNP surface functionalization, unveiling the intricacies of various strategies, their impact on enhancing imaging capabilities, optimizing drug delivery, and catalytic prowess. The review underscores the interplay between surface chemistry and function, providing insights into the key mechanisms underlying these enhancements. Additionally, characterization techniques illuminate the modified surface engineering, shedding light on their structure and behavior. Conex applications such as cutaneous wound treatment have been shown to be responsive to functionalized hydrogels based on magnetic core (like Fe_3O_4, $MnFe_2O_4$ and other ferrites) [11]. Despite the strides made in MNP surface functionalization, several critical challenges persist. Achieving precise control over surface modifications, understanding the intricate interplay between surface alterations and functional outcomes, and ensuring long-term stability remain areas of active investigation. Furthermore, as MNPs traverse the path from laboratory innovation to clinical application, it is imperative to address issues of biocompatibility, toxicity, and clinical translatability. These challenges underscore the pressing need for a comprehensive review that consolidates recent advances in MNP surface functionalization and provides a roadmap for future research. The interplay between surface chemistry and function has a clear role in enhancing drug loading and release, improving biocompatibility and specific targeting and binding, MRI contrast, stimuli-responsive behavior, and reduced aggregation for enhanced stability. For instance, when functionalized with mesoporous silica, MNPs benefit from a high surface area for drug loading, with high surface areas and pore volumes affording high drug-loading capacity. Furthermore, stimuli-responsive coatings like pH-sensitive polymers control drug release, ensuring targeted delivery. On the other hand, coating MNPs with biocompatible polymers like polyethylene glycol (PEG) creates a hydrophilic layer that reduces nonspecific protein adsorption, thus preventing recognition by the immune system and leading to prolonged circulation times and enhanced biocompatibility. Functionalization with targeting ligands, such as folic acid or antibodies, provides specificity. The functional group's affinity for receptors on target cells promotes selective binding, ensuring that the nanoparticles effectively reach and interact with the desired biological targets. Stimuli-responsive coatings, like pH-sensitive polymers, alter their conformation and properties in response to changes in environmental conditions. For instance, in an acidic tumor microenvironment, pH-responsive coatings swell, leading to controlled drug release. An aspect which is key to successful utilization in biologic systems is the prolonged stability of MNP formulations; surface coatings with steric hindrance properties, like PEG, create a barrier that prevents particles from aggregating. The repulsion between PEG chains on adjacent nanoparticles ensures colloidal stability, particularly in complex biological environments.

This review aims to address these gaps by presenting a thorough analysis of the recent strides in the surface functionalization of MNPs. Specifically, the multifaceted strategies employed to engineer MNPs at the nanoscale are addressed, with a particular focus on organic and inorganic coatings, ligand exchange mechanisms, and self-assembly approaches. Through a comprehensive exploration of recent research, this review encapsulates the burgeoning field of MNP surface functionalization, offering a panoramic view of its recent and multifaceted applications in the biomedical field. The subsequent sections delve into specific surface functionalization strategies and their implications in multimodal imaging [12] and drug delivery. The evolving challenges and future prospects are also discussed, underscoring the transformative potential of surface-engineered MNPs in addressing some of the most pressing scientific and technological frontiers of our time. In summary, this review endeavors to synthesize recent advancements in the surface functionalization of MNPs, providing a holistic perspective on their potential to revolutionize scientific and medical landscapes. Through analysis of strategies, outcomes, and applications, the aim is not only to consolidate existing knowledge but also to identify new avenues for research and innovation in this intriguing field.

2. Synthesis of Magnetic Nanoparticles

The synthesis of magnetic nanoparticles (MNPs) represents a pivotal starting point in tailoring their properties for diverse applications. The field has witnessed a surge in innovative approaches to engineer MNPs with precise control over size, shape, crystallinity, and magnetic properties. While the MNP shape is not a main cause of oxidative stress that leads to apoptosis, it plays a key role in cellular uptake. Controlling size and morphology (shape, aspect ratio) can be achieved for most iron oxides (including FeOOH, Fe_2O_3, and Fe_3O_4), given their prime magnetic feature that makes them suitable, among others, for magnetorheological fluids [13]. One strategy in this sense is to use size-controlling agents, which effectively also act as surface coating, like polyethyleneimine (PEI) [14]. A concise overview of the prominent synthesis methods and their influence on surface functionalization will consider the chemical and physical methods described below.

The chemical synthesis route remains a cornerstone in producing MNPs with tunable characteristics. Co-precipitation, a widely adopted method, involves the controlled precipitation of metal salts in the presence of reducing agents or surfactants; for instance, a mixture of cation precursors containing Fe^{2+} and Fe^{3+} in a stoichiometric ratio can be precipitated with hydroxide (NH_4OH, NaOH) under a protective atmosphere to yield magnetite Fe_3O_4. This approach yields monodisperse MNPs with controllable sizes, making it a popular choice for subsequent functionalization. Similarly, thermal decomposition (polyol method) involves the decomposition of metal precursors at elevated temperatures, facilitating the formation of MNPs with narrow size distributions and high crystallinity, oftentimes allowing additional tuning of shape (cubic, hexagonal, etc.) and size (very small NPs of a few nm and very narrow PDI polydispersity index can be achieved through this route). These chemically synthesized MNPs offer versatile platforms for surface functionalization due to their well-defined surfaces and high crystallinity.

Physical methods, such as laser ablation and sputtering, have emerged as viable alternatives for producing MNPs tuned for specific applications. Laser ablation involves irradiating a target material with high-energy laser pulses, thereby inducing ablation and condensation of nanoparticles. This technique allows for precise control over size and composition, enabling tailored surface modifications. For instance, Franzel et al. reported the synthesis of superparamagnetic MNPs consisting of Fe_3O_4 and Fe_3C upon laser ablation of an Fe foil in ethanol [15]. Superparamagnetism refers to the property exhibited by certain nanoparticles, particularly magnetic nanoparticles, that do not have a permanent magnetic moment but that can respond strongly to an external magnetic field, and it represents an important characteristic for applications like targeted drug delivery and magnetic resonance imaging (MRI). Further modification of the synthesis by altering reaction media (water, organic solvents) affords variation in the composition of MNPs, including core–shell structures of type iron–iron oxide, carbon coating, etc. Sputtering, on the other hand, relies on the ejection of target material atoms by energetic ion bombardment. This yields MNPs with minimal contamination suitable for subsequent surface engineering, like, for instance, FeCo NPs of very high saturation magnetization (226 emu/g) or multifunctional MNPs coated with PEG polyethylene glycol for improved solubility and enhanced biocompatibility [16]. Magnetization, as a core principle, refers to the property of a material to become magnetized in the presence of an external magnetic field. In the case of MNPs, their small size leads to unique magnetic behaviors. When subjected to an external magnetic field, the magnetic moments of individual nanoparticles align with the field, resulting in an overall magnetic polarization. This phenomenon is known as superparamagnetism. The level of magnetization is influenced by factors such as the size of the nanoparticles (sheer size, aspect ratio, shape, volume), their composition, and the strength of the applied magnetic field. Understanding and manipulating magnetization properties is essential in tailoring the behavior of magnetic nanoparticles for specific applications. The ability to control magnetization allows for precise targeting of nanoparticles to specific anatomical sites in magnetic drug delivery. Biocompatibility refers to the ability of a material or substance to function safely within a biological system without causing harm or adverse reactions; in this context, MNPs should not elicit harmful responses from the body's tissues or immune system.

Biogenic synthesis has garnered attention for its eco-friendly approach and facile surface functionalization potential. Utilizing microorganisms, plants, or their extracts, this method harnesses the biological entities' inherent ability to reduce metal ions and form MNPs. The resulting MNPs often exhibit unique surface functionalities due to the biomolecules involved in their synthesis, thus opening avenues for surface modifications [17]. Carvallo et al. reported the use of magnetotactic bacteria to synthesize magnetosomes coated with 1,2-dioleoyl-*sn*-glycero-3-phosphocholine (DOPC) or citric acid for use in magnetic hyperthermia applications [9,17]; the MNPs coated with citric acid showed a higher SAR-specific absorption rate and were thus better suited for biomedical applications [17]. When IONs were synthesized by magnetotactic bacteria and were further utilized to synthesize magnetosomes coated with citric acid or 1,2-dioleoyl-sn-glycero-3-phosphocholine (DOPC), the magnetosomes showed reduced magnetostatic interactions compared to those in neat magnetosomes [17].

The influence of synthesis parameters on MNP surfaces cannot be understated. The influence of the shape and size of MNPs on their MFH effect has been investigated recently, concluding that specific synthesis parameters should be met in order to produce the highest possible SAR and hence the desired effect in MFH; for instance, ellipsoidal NPs with the highest SAR were found to be those of 10 nm (equatorial size) and an aspect ratio of 2 [18]. The choice of precursor materials, solvents, and reaction conditions significantly affects the surface chemistry. Organic ligands or capping agents introduced during synthesis can impart initial surface functionalities, setting the stage for subsequent modifications [19]. Temperature, reaction time, and precursor concentration also dictate MNP properties, including surface energy and reactivity (Table 1). In a typical synthesis, Fe(II) and Fe(III) precursors in a 1:2 molar ratio are added to a mixture of oleic acid (OA) and oleylamine (OAm) in a round-bottom flask and then heated up to 300 °C (depending on the solvent of choice, benzyl ether, dioctyl ether etc.), after which a dark brown solution is produced. After cooling and washing (typically with ethanol, C_2H_5OH), stirring and/or sonication with additional OA can successfully lead to the organic coating of MNPs, which can be resuspended in organic solvents (hexane or higher alkanes). Additional heating may be required, after which cooling and separation (usually by a permanent magnet) lead to the final MNPs that can be dried or stored for further use. It is essential to carry out the co-precipitation reaction under a protective inert gas atmosphere in order to avoid the further oxidation of magnetite to hematite. Various modifications of the synthetic procedure are found throughout the literature, but the core principles remain the same. Notably, detailed procedures provide a reproducible method for the synthesis of magnetic nanoparticles, ensuring consistent results for subsequent applications in various fields.

The synthesis of MNPs serves as the foundation for their subsequent surface functionalization. Chemical, physical, and biogenic synthesis methods offer diverse avenues for tailoring MNP characteristics, paving the way for precise and effective surface engineering. Understanding the impact of synthesis parameters on surface properties is paramount for devising successful surface functionalization strategies in magnetic resonance imaging (MRI), drug delivery, and catalysis. A concise and brief overview of the synthesis methods typically employed in MNPs' synthesis is summarized in Table 1.

The surface properties of magnetic nanoparticles (MNPs) are intricately linked to the synthesis parameters employed during their fabrication. These parameters include reaction temperature, precursor concentrations, surfactant types, and reaction times. The careful manipulation of these factors can yield MNPs with tailored surface characteristics, influencing their behavior in diverse applications. For instance, studies by Lu et al. [20] and Majidi et al. [21] systematically investigated the impact of reaction temperature on MNP surface functionalization while also serving as comprehensive reviews on synthetic methods of generating MNPs. The results demonstrated a notable increase in the density of surface functional groups as the temperature was elevated from 100 °C to above 200 °C. Moreover, reaction times have been shown to impact the size distribution and surface roughness of MNPs [20].

Surfactant choice and concentration also exert significant control over MNP surface properties; for instance, the use of oleic acid (OA) as a surfactant resulted in a higher degree of surface coverage compared to other surfactants. The nature of the shell (organic/inorganic, magnetic or non-magnetic) has an influence on its magnetic properties, as the ligands can modify the anisotropy and magnetic moment of the metal atoms located at the surface of the particles [20].

Table 1. Synthetic overview of synthesis methods for MNPs and their main characteristics.

Synthesis Method	Description	Characterization Tools
Co-Precipitation	In this classic method, metal salts ($FeCl_3$, $FeCl_2$) are dissolved in a solvent (water, ethanol) followed by the addition of reducing agents ($NaBH_4$, NH_3, NH_2NH_2). The reduction of metal ions yields MNPs.	Transmission electron microscopy (TEM) and X-ray diffraction (XRD) elucidate the core structure of MNPs.
Thermal Decomposition	Organic metal precursors (iron pentacarbonyl) are decomposed at elevated temperatures (200–300 °C) in organic solvents (oleic acid, oleyl amine) to generate MNPs.	TEM and XRD reveal their size and crystallinity.
Microemulsion	Water-in-oil microemulsions, containing surfactants (CTAB) and co-surfactants (butanol, hexanol), are employed to control MNP nucleation and growth. Iron salts ($FeSO_4$) in the aqueous phase react with reducing agents ($NaBH_4$, hydrazine, etc.) to form MNPs.	Dynamic light scattering (DLS) and UV-Vis spectroscopy monitor the reaction progress.
Sol-Gel	Silica-coated MNPs are synthesized via hydrolysis and condensation of silane precursors (TEOS) in the presence of MNPs. The resulting silica shell stabilizes the MNPs and provides functional groups for subsequent modifications.	FTIR spectroscopy can attest for shell formation.
Microfluidics	Continuous-flow microreactors facilitate controlled nucleation and growth of MNPs by mixing iron precursors ($Fe(acac)_3$) with reducing agents (hydrazine) under controlled flow conditions.	In-line spectroscopy monitors the reaction kinetics.
Hydrothermal	Iron precursors ($FeCl_3$, $Fe(acac)_3$) are hydrothermally treated at elevated temperatures (150–300 °C) and pressures (10–100 atm) in a solvent (water, ethylene glycol, benzyl ether) to yield crystalline MNPs.	Synthesis Scanning electron microscopy (SEM) and XRD confirm particle morphology and crystallinity.
Electrochemical Synthesis	Electrodeposition involves the reduction of iron ions onto an electrode surface. Precursors (iron salts) are dissolved in an electrolyte solution, and an electric current is applied.	Electrochemical techniques monitor the deposition process, while SEM reveals surface morphology.
Sonolysis	Ultrasound irradiation of metal salt solutions ($FeCl_3$) generates reactive species that reduce metal ions to MNPs.	The process is monitored by UV-Vis spectroscopy, while TEM elucidates their morphology and size.
Radiolysis	Irradiation of metal salt solutions ($FeCl_3$) with ionizing radiation (gamma rays, electron beams) induces the reduction of metal ions to MNPs.	Size and composition are determined by TEM and inductively coupled plasma mass spectrometry (ICP-MS).
Biogenic Synthesis	Microorganisms (bacteria, fungi) or plant extracts reduce metal ions (Fe^{3+}) to MNPs.	Energy-dispersive X-ray spectroscopy (EDS) and Fourier-transform infrared spectroscopy (FTIR) confirm the presence of biomolecules on the MNP surface.

Furthermore, concentration gradients of precursor materials play a pivotal role in dictating MNP surface composition, and oftentimes non-stoichiometric ratio is necessary in order to obtain pure Fe_3O_4 (magnetite). These examples highlight the critical role of quantitative data in elucidating the influence of synthesis parameters on MNP surfaces. By employing advanced characterization techniques and systematically varying synthesis conditions, researchers can precisely tailor MNP surface properties, opening avenues for enhanced performance in applications ranging from drug delivery to catalysis.

Regarding biogenic synthesis of magnetic nanoparticles, there are several potential concerns associated with their further use in a clinical setting. These concerns primarily revolve around safety, scalability, and regulatory approval. Biogenic synthesis relies on living organisms to produce nanoparticles, which introduces variability in terms of particle size, shape, and surface properties, and hence may not always meet the stringent standards required for clinical applications. Additionally, this type of synthesis can introduce impurities or contaminants that could pose risks in a clinical context, particularly if they include toxic substances or allergens. The use of biogenic synthesis methods may require extensive regulatory approval processes to ensure safety and efficacy; therefore, standardizing the synthesis process to meet regulatory standards can be challenging due to the biological variability inherent in living organisms. Scalability and reproducibility remain current challenges in order to meet clinical demand. The availability of specific organisms and their growth conditions may limit the production capacity, and replicating the exact conditions across different batches can be complex and oftentimes unpredictable. In some cases, the organisms used in biogenic synthesis may be genetically modified or engineered, which raises concerns about potential risks to the host organisms as well as the potential release of genetically modified organisms into the environment. Even though biogenic synthesis methods aim to produce NPs using biological entities, questions about the biocompatibility and potential toxicity of the resulting nanoparticles may arise. It is crucial to thoroughly evaluate the safety profile of biogenically synthesized nanoparticles for clinical use. Ethical consideration about the use of living organisms for nanoparticle synthesis also raises some issues related to the potential impact on ecosystems, especially if they are genetically modified or rare species.

Lastly, biogenic synthesis may lead to nanoparticles with different stability profiles compared to their chemically synthesized counterparts. Understanding the long-term stability and storage conditions of biogenically synthesized nanoparticles is crucial for their clinical viability. In conclusion, while the biogenic synthesis of magnetic nanoparticles holds promise for various applications, including clinical ones, there are notable concerns regarding safety, scalability, regulatory approval, and ethical considerations. Thorough evaluation, standardization, and rigorous testing are essential steps in addressing these concerns and advancing biogenic synthesis methods towards clinical applications [20,21].

These synthesis methods collectively offer a toolkit for tailoring MNPs with distinct characteristics, laying the groundwork for subsequent surface functionalization. Characterization techniques such as TEM, XRD, FTIR, and spectroscopic methods provide critical insights into their structural and surface properties, guiding the choice of surface modification strategies. The synthesis route serves as a pivotal bridge between MNPs' core composition and their surface chemistry, enabling a precise design for multimodal imaging, drug delivery, and catalytic applications.

The novel functionalization strategies discussed in this paper represent significant advancements over existing methods in terms of both efficiency and effectiveness, tackling issues such as specificity, safety, biocompatibility, and clinical translation. In terms of efficiency, these novel strategies enable highly precise targeting of specific cells or tissues. This surpasses conventional methods, which often rely on passive accumulation through the enhanced permeability and retention (EPR) effect. The new functionalization techniques offer unprecedented control over drug release kinetics, surpassing traditional encapsulation methods that may have limited control over release rates. Surface modifications in the novel strategies also enhance catalytic activity through tailored active sites, surpassing generic functionalization which may not optimize the catalytic potential.

In terms of effectiveness, the novel strategies greatly enhance therapeutic precision, enabling drugs to be delivered precisely where they are needed. This is a significant improvement over traditional methods where off-target effects can be a concern. The new functionalization techniques often lead to enhanced stability and longevity of the functionalized nanoparticles, surpassing traditional coatings that may be less robust while also affording the incorporation of multiple functionalities onto a single nanoparticle, allowing for synergistic effects [8,22,23]. Theranostics refers to combined therapy and diagnostics and represents a class of technologies that combine therapeutic and diagnostic capabilities in a single system. Notably, MNPs used in theranostics can both deliver a drug to a specific site and also be imaged to monitor the drug's distribution and therapeutic effects. This strategy outperforms existing methods that may be limited to a single function, and it is key to offering enhanced multifunctionality. In this context, MNPs are ideal candidates in theranostic platforms. For instance, MRI-guided NPs have been utilized in combined photodynamic therapy (PDT) and photothermal therapy (PTT), thus achieving chemical exchange on the tumor and, respectively, localized thermal damage at the tumor level [8]. Additional strides have been made in order to overcome biologic barriers such as the blood–brain barrier (BBB), and these consist of various functionalization of penetrating NPs with CPP (cell-penetrating peptides) such as hydrophilic (cationic; TAT, penetratin, R8), amphipathic (SynB, RGD, etc.) or hydrophobic (nonpolar, C105Y, PFV, Pep-7) [22]. An ongoing trend is to utilize MNPs in cell membrane-based biomimetic nanosystems for personalized disease theranostics including oncology, bacterial infections, brain diseases, and inflammatory diseases [23]. Theranostics exemplifies how the integration of multiple functionalities in MNPs can lead to highly effective multifunctional platforms. By combining targeted drug delivery with real-time imaging, researchers can develop personalized and optimized treatment strategies for cancer patients, minimizing side effects and maximizing therapeutic outcomes. This approach holds significant promise for the future of precision medicine in oncology [22,23].

Important advancements have been made in terms of specificity, with stimuli-responsive coatings in the novel strategies responding to specific cues in the microenvironment, thereby offering highly specific imaging capabilities. "Stimuli-responsive" describes materials, coatings, or systems that can change their properties or behavior in response to specific external stimuli; in relation to MNPs, this could refer to coatings that change their structure or release properties in response to factors like pH, temperature, or light, among other factors. Traditional contrast agents may lack this level of specificity. Meanwhile, recent strides often incorporate coatings or ligands that enhance biocompatibility, reducing potential toxicity concerns, which in turn surpasses older methods that may not have addressed biocompatibility to the same extent. Finally, the many examples summarized in this review point to their great promise in clinical applications, with many MNP systems undergoing clinical trials, surpassing existing methods that may still be in the preclinical stage.

In summary, the novel functionalization strategies discussed in this paper demonstrate superior efficiency and effectiveness compared to existing methods. They offer a range of advantages, including enhanced precision, stability, multifunctionality, and improved biocompatibility, and are poised to redefine various fields, from medicine to catalysis and beyond.

3. Characterization Techniques

Characterizing the intricate surface modifications of magnetic nanoparticles (MNPs) is paramount to comprehending their behavior and tailoring them for diverse applications [24]. This section describes the wide array of characterization techniques typically utilized to unveil the subtle intricacies of surface functionalization (Table 2).

These techniques collectively decipher the intricate landscape of surface functionalization, illuminating the impact of modifications on MNPs' physicochemical properties. Rigorous application of these characterization tools and more advanced connected methods can unveil the nuances of surface engineering, enabling the precise tailoring of MNPs for specific applications in multimodal imaging, drug delivery, catalysis, and beyond [25].

Table 2. Typical characterization tools used to describe functionalized MNPs.

Characterization Methods	Description
Transmission Electron Microscopy (TEM)	TEM offers high-resolution imaging, revealing MNP morphology, size, and core–shell structures. Contrast variations highlight surface coatings and confirm successful functionalization. Energy-dispersive X-ray spectroscopy (EDS) coupled with TEM maps elemental distribution across MNPs.
X-ray Photoelectron Spectroscopy (XPS)	XPS provides elemental composition information and oxidation states of surface-functionalized MNPs. Binding energy shifts indicate surface chemical interactions, validating ligand attachment or shell formation.
Fourier-Transform Infrared Spectroscopy (FTIR)	FTIR identifies functional groups on MNPs' surfaces through vibrational spectra. Shifts in or the appearance of peaks confirm ligand exchange or coating formation. Attenuated total reflection (ATR) FTIR enables the analysis of solid MNPs.
Dynamic Light Scattering (DLS)	DLS evaluates the hydrodynamic size and dispersity of surface-functionalized MNPs in solution. Changes in size or dispersity after functionalization reflect coating stability and influence on hydrodynamic behavior.
Zeta Potential Analysis	Zeta potential quantifies surface charge of MNPs, revealing electrostatic interactions between functional coatings and the surrounding medium. Zeta potential changes indicate successful charge modification.
Nuclear Magnetic Resonance (NMR)	Solution-state NMR elucidates surface functional groups and molecular dynamics. Ligand exchange or bioconjugation is confirmed by chemical shift changes or the appearance of new peaks.
UV-Visible Spectroscopy:	UV-Vis spectroscopy reveals ligand-specific absorbance, confirming functionalization. Shifts in absorption bands indicate changes in the electronic environment due to surface engineering.
Scanning Electron Microscopy (SEM)	SEM provides topographical information on MNP surfaces. Microstructural changes due to functionalization, such as shell formation or aggregation, are discerned.
Magnetic Measurements	Superconducting quantum interference device (SQUID) magnetometry quantifies magnetic behavior, confirming core–shell architecture and magnetic moments of surface-functionalized MNPs.
X-ray Diffraction (XRD)	XRD verifies MNP crystallinity and phase changes due to functionalization. Shifts in or broadening of diffraction peaks indicate surface modifications or encapsulation.

4. Surface Functionalization Strategies

Surface functionalization of magnetic nanoparticles (MNPs) constitutes a transformative approach, enabling tailored modifications that unlock their potential in diverse applications. This section expands on the array of surface functionalization strategies, elucidating their principles, advantages, and outcomes in MR imaging, drug delivery, and catalysis as some of the main applications of surface-functionalized MNPs. Below is a table summarizing the most widespread methods used to functionalize MNPs, according to their type of coating (organic, inorganic), specific precursors, and materials used, and the suitable characterization methods (Table 3).

Table 3. Summary of functionalization strategies for MNPs.

Type of Functionalization	Precursors/Materials and Short Description of Method	Observations	Characterization Techniques
Organic Coatings	Organic ligands, such as citrate, polyethylene glycol (PEG), or polymers, are grafted onto MNPs' surfaces through covalent or non-covalent interactions.	This imparts colloidal stability, biocompatibility, and modulates surface charge. The enhanced dispersibility of MNPs in biological media facilitates their utilization in drug delivery and targeted imaging.	Dynamic light scattering (DLS) and zeta potential measurements verify the stability and charge modification.
Inorganic Shells	MNPs are encapsulated within inorganic materials like silica, gold, or metal oxides. These shells enhance stability, protect MNPs from degradation, and enable bioconjugation. For instance, core–shell Fe_3O_4/Pt MNPs have been obtained via silylation/polymerization [26].	Silica-coated MNPs, for instance, offer a platform for versatile surface functionalization and controlled drug release.	Transmission electron microscopy (TEM) and X-ray photoelectron spectroscopy (XPS) confirm shell formation.
Ligand Exchange	The exchange of native surface ligands with functional molecules (amines, thiols) allows for precise control over MNP properties.	This strategy enables targeted drug delivery through the conjugation of targeting ligands or the attachment of therapeutic payloads.	Fourier-transform infrared spectroscopy (FTIR) confirms successful ligand exchange.
Self-Assembly	MNP surface functionalization can exploit the self-assembly of molecules (DNA, peptides) onto the MNP surface.	DNA-functionalized MNPs, for instance, enable programmable interactions, leading to controlled aggregation or dispersal.	Gel electrophoresis and fluorescence assays validate self-assembly.
Bioconjugation	MNPs are conjugated with biomolecules (antibodies, peptides) through affinity interactions, yielding specific targeting capabilities.	Antibody-conjugated MNPs offer exquisite cellular or molecular targeting in imaging and drug delivery.	Enzyme-linked immunosorbent assays (ELISA) confirm successful bioconjugation.
Click Chemistry	Bioorthogonal reactions (click chemistry) facilitate specific and robust surface functionalization.	Azide-terminated MNPs react with alkyne-modified molecules, yielding stable and versatile conjugates.	Copper-catalyzed or copper-free click reactions validate successful coupling.
Layer-by-Layer Assembly	Sequential deposition of polyelectrolyte layers onto MNPs yields multifunctional coatings.	This approach enables the controlled release and modulated surface charges.	Quartz crystal microbalance (QCM) and UV-Vis spectroscopy monitor layer-by-layer assembly.
Responsive Polymers	Stimuli-responsive polymers (pH, temperature) can be grafted onto MNP surfaces, facilitating controlled drug release in specific microenvironments [27].	Suitable for controlled drug release	The swelling behavior of polymers is probed using dynamic light scattering (DLS) and turbidimetry.
Host-Guest Systems	Cyclodextrins or cucurbiturils form host–guest complexes with molecules on MNP surfaces, enhancing stability and enabling controlled release.	Suitable for controlled drug release	Nuclear magnetic resonance (NMR) and UV-Vis spectroscopy

The surface functionalization of magnetic nanoparticles (MNPs) is a pivotal step that imparts them with specific functionalities, thereby enabling their use in diverse applications.

Several strategies have emerged, each with distinct advantages and limitations. Inorganic shells, for instance, exhibit enhanced stability and offer tunable properties by varying the composition and thickness of the inorganic shell. However, their synthesis can involve multi-step processes and requires precise control over reaction conditions, which can be more complex compared to other strategies, and there is a potential for core–shell mismatch, which can make seamless integration challenging. The ligand exchange variant allows for versatility due to the wide range of ligands available and can be carried out under mild conditions, thus minimizing the potential damage to the magnetic core. On the downside, the stability can be limited, as ligands might detach over time, and thus specific control over shell thickness can be challenging.

These strategies underscore the transformative impact of surface functionalization, enabling MNPs to transcend their innate capabilities. Characterization techniques like TEM, FTIR, and spectroscopy play a pivotal role in verifying successful functionalization, guiding the design of MNPs tailored to specific applications. As we delve further into this review, we explore the role of these strategies in enhancing multimodal imaging, drug delivery, and catalysis, illuminating the dynamic interplay between surface engineering and diverse applications.

5. Surface Functionalization for Drug Delivery

The strategic surface functionalization of magnetic nanoparticles (MNPs) has revolutionized drug delivery, enabling precise and efficient therapeutic interventions. This section delves into the multifaceted enhancements brought about by surface engineering, including controlled drug release, targeted delivery, and overcoming biological and histohematic barriers [22]. Responsive and targeted drug delivery based on MNPs (magnetite Fe_3O_4 mainly) coated with polymers and biopolymers [28], biomolecules or macromolecules (5-fluorouracil (5-FU), oxaliplatin, and irinotecan), can be successfully used for the treatment of various types of cancer, such as in colorectal cancer therapy [29]. A brief timeline overview starting from the SiO_2 coating of magnetite in 1985 is depicted in Figure 2 [29]. Inorganic shells such as SiO_2 (silica) still remain one of the preferred surface functionalization routes, especially for MNPs like γ-Fe_3O_4 [30]. Recent advances in nanoscience greatly benefit from the introduction of NPs and magnetic iron oxide nanoparticles in particular, owing to their unique set of properties and many advantageous biomedical applications [31], extending their use to stem cell biotechnology [32].

Various polymers have been employed for coating inorganic NPs (such as a Pt shell [26]), and MNPs in particular, such as hydrophilic polymers (like functional polyaspartamide [33]) that help them achieve water-solubility. Moreover, the stability of magnetite NPs obtained by Massart synthesis (co-precipitation of Fe(II) and Fe(III) precursors) was studied by Klekotka et al. [34], including their surface-functionalized counterparts, specifically SiO_2-covered Fe_3O_4 (SiO_2@Fe_3O_4) or magnetite grown on pre-formed Fe_3O_4 seeds (Fe_3O_4@Fe_3O_4) in polyol synthesis using oleyl amine/oleic acid as stabilizers and surface protection agents [34]. An overview of the main research direction currently being taken by the research community is given in Figure 3.

Surface modifications offer unprecedented control over drug release kinetics. Drug-loaded MNPs encapsulated within stimuli-responsive polymers (pH, temperature) exhibit on-demand drug release triggered by specific microenvironment cues [27]. The rate and extent of drug release are characterized using in vitro release assays. Surface-functionalized MNPs offer unparalleled targeting precision. The conjugation of targeting ligands (antibodies, peptides) guides MNPs to specific cell receptors or tissues, minimizing off-target effects. In vitro cell binding assays and in vivo biodistribution studies validate targeted delivery efficiency. Biodistribution refers to how a substance, like a nanoparticle, is distributed throughout the body, and this encompasses where the substance moves, how much of it accumulates in different tissues or organs, and how long it stays there.

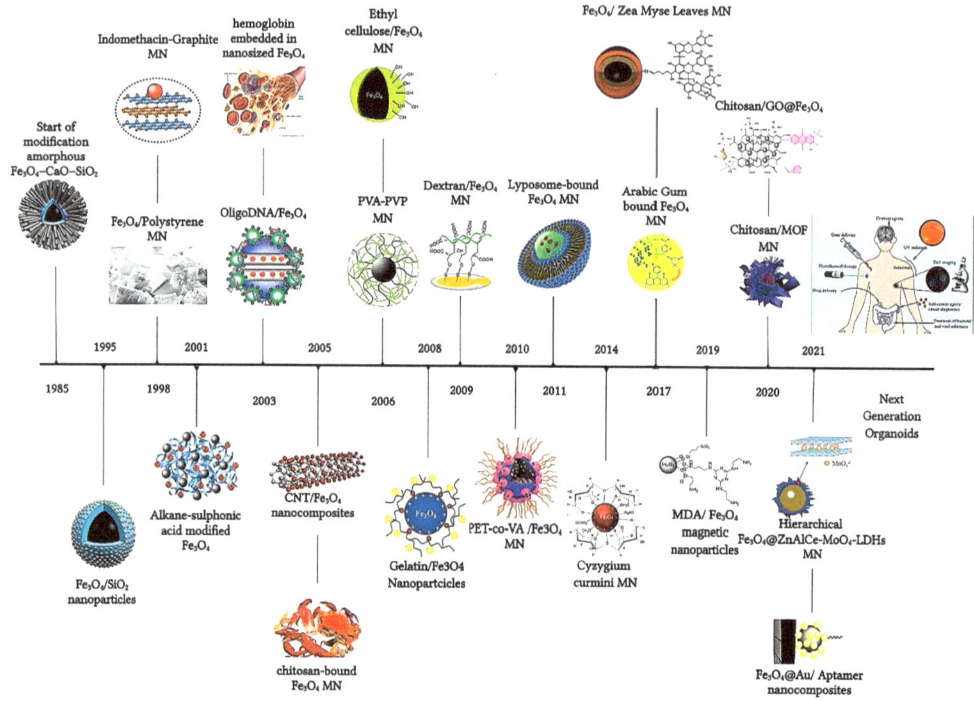

Figure 2. The timeline of magnetic nanoparticles in therapeutic and imaging application. Reprinted from reference [29] under a Creative Commons Attribution 4.0 International License.

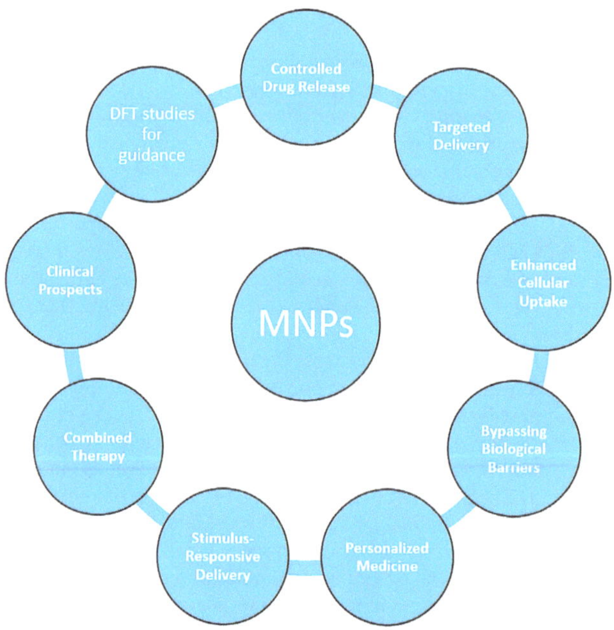

Figure 3. Representative research direction in functionalized magnetic nanoparticles.

Tissue engineering and regeneration has been shown to be possible by Chitosan (CS)-based NP formulations, including CS/ferromagnetic scaffolds, owing to the efficient interaction of -NH_2 groups (CS) with the MNPs' polymeric coating [35]. Such effective CS-coating of the magnetite magnetic core also prevents further oxidation of magnetite Fe_3O_4 to hematite Fe_2O_3. Biomedical applications of MPNs and SPIONs in particular have been recently reviewed by Jiang et al. related to their synthesis, in vivo protein detection, magnetic heating/MFH [36], and in vivo imaging and drug delivery [1]. Treatment of degenerative diseases can potentially be modulated by Wnt signaling and has inspired Hu et al. to immobilize Wnt fragment peptides on MNPs with promising results on human embryonic kidney (HEK293) Luc-TCF/LEF reporter cell line [37].

Surface engineering influences cellular uptake mechanisms, thereby enhancing drug internalization. Ligand-decorated MNPs undergo receptor-mediated endocytosis, improving drug delivery to intracellular compartments. Confocal microscopy and flow cytometry quantify enhanced cellular uptake. Surface-engineered MNPs enable intracellular drug delivery. Cell-penetrating peptides facilitate direct entry into cells, bypassing endocytotic pathways. Intracellular imaging and drug efficacy assays validate efficient intracellular drug delivery. These surface modifications enable MNPs to traverse biological barriers [22]. Stealth coatings (PEG) reduce opsonization, prolonging blood circulation, and enable passive accumulation in tumor tissues via the enhanced permeability and retention (EPR) effect. Pharmacokinetic studies and tumor accumulation assays validate EPR-driven delivery.

Surface functionalization permits the simultaneous loading of multiple drugs onto MNPs. Conjugation of distinct drug molecules or payloads enables combination therapy, targeting multiple pathways simultaneously. In vitro cytotoxicity assays elucidate synergistic therapeutic effects. Surface-functionalized MNPs facilitate personalized medicine approaches. Patient-specific targeting ligands enable tailored drug delivery, minimizing adverse effects and enhancing therapeutic outcomes. Genomic and proteomic profiling guide the design of personalized MNPs, while biomimetic approaches [14] enhance the outcome in various diseases such as brain cancer and inflammatory and bacterial infections [23]. Biomimetic NPs can also be used as cost- and time-effective magnetic biosensors, and they can concentrate Gram-positive and Gram-negative microorganisms for easier bacterial detection to concentrations as low as 10 CFU/mL using qPCR [14].

MNPs equipped with responsive coatings release drugs in response to external stimuli (magnetic field, light) [27,38]. Magnetic hyperthermia-triggered drug release exploits localized heating induced by alternating magnetic fields. Temperature measurements validate controlled release [5,9,17,39,40]. Dual-functional MNPs merge drug delivery with therapeutic modalities. The conjugation of chemotherapeutic drugs with photosensitizers or gene therapy agents [41] offers synergistic treatment approaches. Cell viability assays and in vivo tumor regression studies demonstrate combination therapy efficacy.

Surface-functionalized MNPs have entered clinical trials, demonstrating promise in cancer therapy and beyond. The translation of these advanced platforms from preclinical studies to clinical practice requires rigorous safety assessments, optimization of targeting strategies, and scalable production techniques. While base nanoparticles pose several potential shortcomings and toxicity-related issues, functionalized NPs alleviate many of the main downsides preventing the use of MNP-based nanoplatforms: reactive oxygen species (ROS) generation, targeted delivery, and biodegradation. For instance, conjugated polymer nanoparticles (CPNs) modified with dopants (like iron oxide) behave as theranostic agents with prospects for multifunctionality in imaging and treatment [42]. Arias-Ramos et al. synthesized by nanoprecipitation conjugated polymer nanoparticles CPNs based on 2 nm-thick oleic acid-caped MNPs ($NiFe_2O_4$, Fe_3O_4 of ~5 nm) and fluorescent conjugated polymer (CP) poly(9,9-dioctylfluorene-alt-benzothiadiazole, F8BT) or polystyrene grafted with ethylene oxide functionalized with carboxyl groups (PS-EG-COOH), with good results towards glioblastoma (GBM, a common tumor of the central nervous system) using light delivery to the brain tissue by means of fiber optics [42].

Simulations using quantum chemistry (DFT) have also been performed, for instance on the system of tirapazamine (TPZ, anticancer drug) and the magnetic nanoparticle (MNP) $Fe_6(OH)_{18}(H_2O)_6$, where the interaction between the MNP and TPZ was shown to be facilitated by intraring N-atom, $-NH_2$, and -NO groups present in the TPZ molecule, concluding that interaction energetics via the first two are more accessible than via the -NO moiety [43]. There were two envisioned pathways for the covalent binding of TPX onto MNPs via their surface -OH hydroxyl groups (Figure 4).

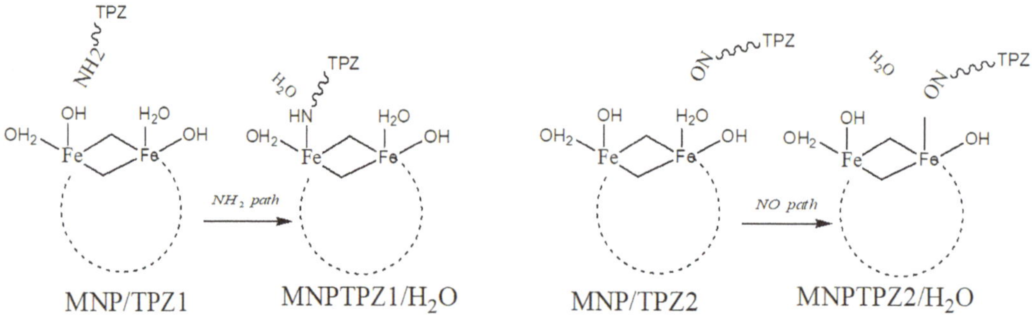

Figure 4. TPZ–MNPs binding via $-NH_2$ or -NO mechanisms.

Other functionalization strategies focused on the polymer coating of magnetite NPs by atom transfer radical polymerization (ATRP) in order to yield magnetic functionalized supports of the general formula Fe_3O_4@MSN-PDMAEMA-FA of ~180 nm to be used for Dox (doxorubicin) drug loading Dox@Fe_3O_4@MSN-PDMAEMA(-FA). These drug delivery systems were tested against breast cancer cells (MCF-7) and resistant cancer cells (MCF-7 ADR) [44]. Beagan et al. used magnetic mesoporous silica nanoparticles (MMSNs) coated with pH-responsive polymer 2-Diethyl amino ethyl methacrylate (DEAEMA) grafted by surface-initiated ARGET atom transfer radical polymerization (ATRP); these surface-functionalized MMSNs were further modified by anionic groups' functionalities that would then covalently bind FA (folic acid) as a targeting agent [44]. The functionalization strategy used is shown in Figure 5, and afforded long-circulation time for the smart drug delivery system.

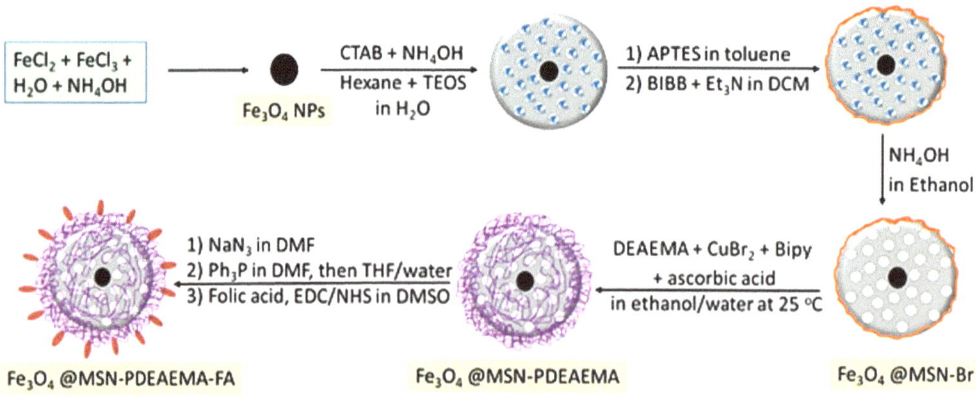

Figure 5. Synthesis procedure for poly(2-(diethylamino) ethylmethacrylate) brushes capped with folic acid grafted on magnetic mesoporous silica nanoparticle (MMSNs) surfaces via SI-ARGET ATRP [44].

The functionalization strategy outlined in Figure 5, involving the synthesis of poly(2-(diethylamino) ethylmethacrylate) (PDEAEMA) brushes capped with folic acid grafted on magnetic mesoporous silica nanoparticles (MMSNs) surfaces via SI-ARGET ATRP, offers several distinct advantages related to tailored surface functionality, grafting density, enhanced biocompatibility, specific targeting via folic acid, combined magnetic and mesoporous features, and potential for imaging applications. The use of surface-initiated atom transfer radical polymerization (SI-ARGET ATRP) enables precise control over the grafting process, allowing for customization of the surface with PDEAEMA brushes and folic acid, thereby providing tailored surface functionality. SI-ARGET ATRP is known for its ability to achieve high grafting densities, ensuring a densely packed layer of PDEAEMA brushes on the MMSNs surface. The incorporation of PDEAEMA brushes enhances the biocompatibility of the MMSNs. PDEAEMA is a pH-responsive polymer that becomes protonated under acidic conditions, mimicking the slightly acidic environment of cancer cells, and this property can aid in the selective targeting of cancer cells. More specifically, in an acidic environment, such as within the endosomes or lysosomes of cancer cells, the PDEAEMA brushes undergo protonation, leading to swelling and the subsequent release of encapsulated drugs. Through further conjugation via folic acid, this strategy targets cancer cells due to its high affinity for folate receptors, which are overexpressed on the surface of many cancer cells. The presence of folic acid on the MMSNs' surface ensures specific binding and uptake by cancer cells, maximizing therapeutic efficacy [44]. The MMSNs possess both magnetic properties and a mesoporous structure, thereby allowing for magnetic guidance to target sites, and provide a high surface area for drug loading, thus enabling multifunctional drug delivery platforms. Furthermore, the magnetic properties of MMSNs offer the potential for imaging applications, such as magnetic resonance imaging (MRI). This dual functionality allows for theranostic applications, where therapy and imaging are integrated into a single platform, making this approach highly promising for targeted drug delivery in cancer therapy.

Magnetic nanoplatforms are currently enhanced by loading with enzymes, affording easier recovery and reutilization as well as improved stability and catalytic activity [45]. Belleti et al. synthesized hybrid Au/Fe_3O_4 NPs of ~15 nm mean diameter using L-cysteine (Cys) as the polymer capping agent or dithiol-terminated polyethylene glycol ($PEG(SH)_2$), yielding NPs of type $PEG(SH)_2$Au/F_3O_4NPs or CysAu/F_3O_4NPs [46]. These nanoparticle systems were further conjugated with luciferase enzymes able to catalyze bioluminescent reactions (*Pyrearinus termitilluminans* green-emitting click beetle luciferase, PyLuc and *Phrixotrix hirtus* red-emitting railroad worm luciferase, RELuc). A brief overview of the current landscape in MNP coating and functionalization is given in Table 4.

Table 4. Functionalization of various types of MNPs with main applicability in drug delivery.

NP Type	Coating Agent	Active Agent	Characterization Methods	Activity	Obs.	Ref.
$NiFe_2O_4$, Fe_3O_4 of ~5 nm	Oleic acid coating (2 nm)	Conjugated polymer poly(9,9-dioctylfluorene-alt-benzothiadiazole, F8BT) or polystyrene grafted with ethylene oxide functionalized with carboxyl groups (PS-EG-COOH)	UV-VIS, DLS, TEM, MRI evaluation, cytotoxicity in vitro (U-87 MG and T98G cells, MTT and Live/Dead cell viability assays), fluorescence imaging	Glioblastoma (GBM); MR imaging (MRI)	Theranostic agents with prospects for multifunctionality in imaging and treatment; preclinical MRI studies	[42]
magnetic nanoparticle $Fe_6(OH)_{18}(H_2O)_6$	–	Tirapazamine (TPZ,)	DFT study	Anticancer (not evaluated)	Drug binding via intraring N-atom, $-NH_2$	[43]
Fe_3O_4@MSN-PDMAEMA-FA (~180 nm)	MSN-PDMAEMA-FA	DOX (doxorubicin)	SEM, TEM, FTIR, surface area, TGA, XPS, UV, DLS	Anticancer (MCF-7, and MCF-7 ADR cells)	Excellent biocompatibility, minimally toxicity	[44]
Au/Fe_3O_4 NPs (15 nm diameter)	L-cysteine (Cys); dithiol-terminated polyethylene glycol ($(PEG(SH)_2)$),	Luciferase (enzyme)	FTIR, FESEM, luminescence, bioluminescence	Enzymatic activity, luminescence activity	48% activity preservation in case of CysAuNPMag	[46]

Table 4. *Cont.*

NP Type	Coating Agent	Active Agent	Characterization Methods	Activity	Obs.	Ref.
Superparamagnetic Iron Oxide Nanoparticles (SPIONs)	Citrate coating	SPIONs@citrate (52–58 nm hydrodynamic Z size) loaded into human T cells (27 or 80 μg/mL)	TEM, Magnetic susceptibility, hydrodynamic Z-average size, zeta potential	Anticancer; infiltration of SPIONs into primary human CD3+ T cells (1,4 pg Fe/cell))	Magnetic delivery of immune cells (dynamic regime)—potential future application	[47]
maghemite (γ-Fe_2O_3), magnetite (Fe_3O_4) 10.9 ± 1.6 nm	Oolyol synthesis from $Fe(acac)_3$ and diethylene glycol (DEG)	IONPs@DEG obtained with continuous growth method	TEM, XPS, XRD, magnetization	MRI; T_1-imaging, T_2-imaging (high relaxivities r_2:163.4 $mM^{-1}s^{-1}$, r_1:135.0 $mM^{-1}s^{-1}$)	Water-dispersible NPs obtained when T = 190°, 220° and 235 °C (higher T leads to agglomeration); high wt% C (XPS)	[48]
biological magnetite nanoparticles (BMs), by magnetotactic bacteria *Magnetovibrio blakemorei* strain MV-1^T	Glutaraldehyde GA, poly-L-lysine PLL (linking reagents)	Amphotericin B, AmB (to yield BM–PLL–AmB and BM–PLL–GA–AmB conjugates)	TEM, FTIR	Controlled drug release, magnetic hyperthermia (in PBS media)	Magnetosomes with high encapsulation efficiencies and drug loadings (0.1‰ PLL: 52.7%, and 25.3 mg per 100 mg; while 0.1‰ PLL–GA 12.5%: 45.0%, 21.6 mg per 100 mg)	[49]
Magnetic nanoparticles (MNPs) of Fe_3O_4	Chitosan (CS)	Telmisartan (TEL), yielding MNP–CS–TEL TEL is an angiotensin II receptor blocker (ARB), treating high blood pressure, heart failure, diabetic kidney disease, and cancer	FTIR, TGA, XRD, FE-SEM (field emission scanning electron microscope), TEM, VSM (vibrating sample magnetometer), BET surface area analyzer	Anticancer drug therapy, as carriers of Telmisartan (TEL); tested against PC-3 human prostate cancer	Chitosan CS coating validated by FTIR and TGA data. Grafting TEL, a poorly soluble drug, on the surface-coated of MNPs (MNP-CS) by amide bond between amino groups of chitosan CS and carboxylic groups of TEL	[50]
Magnetic nanoparticles (maghemite NPs) on anodic alumina nanotubes, to give magnetic anodic alumina nanotubes (MAANTs)	Silanization by means of (3-aminopropyl) triethoxysilane (3-APTES, 99%)	Protein padding of albumin-fluorescein isothiocyanate conjugate (FITC-BSA)	Environmental scanning electron microscopy (ESEM), energy dispersive X-ray (EDX); FTIR (ATR), ζ-potential, dynamic light scattering (DLS); TEM, fluorescence	Drug delivery and biosensing applications	Proteolytic hydrolysis (amide bond breaking) in presence of cathepsin B- protease (growth and initial stages of tumor metastasis), releasing fluorescent fragments of the protein	[51]
Cobalt Nanoparticles	Carbon-coating, yields CCo nanoparticles	Functionalization with Sulfonated Arene Derivatives via aqueous in situ diazotization reaction	FTIR, SEM, elemental analysis	Catalysis (anionic ROP of glycidol), or recyclable anticoagulant	Covalent linkage of an in situ generated diazonium on the graphene-like surface	[52]
Fe_3O_4 nanoparticles (MNPs)	Mesoporous silica (SBA-15), To yield PEI grafted Fe_3O_4@SiO_2@SBA-15 labeled FA	Doxorubicin (DOX)	FTIR, TGA, XRD, VSM, SEM, EDX, UV-Vis	Targeted delivery to MCF-7 cell line (breast cancer)	pH-sensitive mesoporous magnetic and biocompatible nanocarrier, high internalization	[53]
Fe_3O_4	SiO_2-NH_2 coating, alkyne surface functionalization to yield magnetic nanoparticle Fe_3O_4@SiO_2-NH_2-alkyne	Azide-functionalized *E. Coli*	XRD, FTIR, TEM, SEM, immobilization yield (Y), activity recovery (E),	Biomedicine and catalysis –conversion of glycerol into DHA; Tested against recombinant *E. coli* harboring glycerol dehydrogenase	Click chemistry affords covalent bonding between azide and alkyne groups; immobilization yield 83%, activity recovery 94%	[54]
Au/Fe	Glutathione (GSH)-capped on hybrid gold-magnetic-iron-oxide NPs (Au-Mag-GSH)	-	Colloidal stability, DLS: Z-potential, HR-TEM, (S)TEM, SEM, EDX, FTIR, Photoluminescence (PL) excitation and emission spectra, cytotoxicity	Biomedical applications	NPs were found to not be toxic at typically used concentrations (1.5 μg/mL)	[55]

Table 4. *Cont.*

NP Type	Coating Agent	Active Agent	Characterization Methods	Activity	Obs.	Ref.
Superparamagnetic iron oxide nano-rods (IONRs) based on magnetite	Branched polyethyleneimine (BPEI) to yield PEI-coated Fe_3O_4 nano-rods; cyclohexane layer prevents NPs oxidation during synthesis	Carnosine dipeptide (β-alanine and L-histidine)	SEM-EDS, TEM, XRD, FT-IR, TGA, GC-MS, DLS, zeta potential, magnetic properties (by whole body 1.5 T MRI system)	Cancer treatment: glioblastoma brain tumors (GBM); inhibition of post-surgery metastasis; MRI monitoring	Superparamagnetic nano-rods tested against U87 human glioblastoma astrocytoma cell line. Carnosine was fully released by mild hyperthermia (40 °C)	[56]
$Ag_{(1-x)}Ni_xFe_2O_4$	Polyethylene glycol (PEG)	Curcumin	XRD, FTIR, SEM, TEM, VSM, UV-Vis	Drug delivery	Synthesis of Ag-doped Ni ferrite nanoparticle; PEG was used as a solvent during synthesis; curcumin loading was pH-dependent	[57]
$NiFe_2O_4$ (~5 nm)	No pre-functionalization required	Serum albumin (BSA)	DLS, FT-IR, TEM, SAED	Immobilized metal affinity chromatography of proteins (IMAC)	BSA binding fitted Langmuir isotherm; high capacity 916 mg BSA/g dried NPs	[58]
SPIONs	Polymers: Poly(2-ethyl-2-oxazoline) (PEtOZ); Poly(2-ethyl-2-oxazoline-co-2-isopropyl-2-oxazoline) (PEtIOZ)	Opsonins and Albumin	TEM, TGA/DSC, DLS, Isothermal Titration Calorimetry, ProtParam tool (Computation of Protein Properties)	Biomedical applications	Protein corona formation on poly(2-alkyl-2-oxazoline)-grafted SPIONs depends on protein size, flexibility, and charge	[59]
MNPs based on commercial 75%–80% (*w*/*w*) Fe_3O_4 (diameter 100 nm)	Streptavidin-functionalized, encapsulation with hydroxyethyl starch	Oligonucleotide-functionalized (stability test 92% after 3 months at 4 °C)	Fluorescence, AC Susceptibility	Circle-to-circle amplification (C2CA)' diagnostic	Newcastle disease virus and Salmonella as target sequences	[60]
γ-Fe_2O_3/CeO_2 Maghemite(seeds)/cerium oxide MNPs	PEG/neridronate with PEG of 2000 or 5000 Da, producing γ-Fe_2O_3/CeO_2@-PEG_{2k} and γ-Fe_2O_3/CeO_2@PEG_{5k}	–	TEM, SAED, EDX, EELS, DLS, relaxometry, fluorescence	Antioxidant, MRI tracing (high r_2 relaxivity); theranostic platform	Improved colloidal stability in PBS, enhanced biocompatibility; CeO_2—radical scavenger	[61]
iron oxide Fe_3O_4 MNPs	Oleic acid (OA), silica (TEOS), cationic polymer *P* poly[*N*-isopropylacrylamide-co-(3-acrylamidopropyl) trimethylammonium chloride], P(NIPAm-co-AMPTMA)	Vancomycin (Van)	TEM, XRD, FTIR, TGA, SEM, DLS, magnetization curves	Antibacterial (*Shigella boydii, Bacillus cereus, Staphylococcus aureus* and *Escherichia coli*)	MNPs synthesized by co-precipitation/microemulsion method; Fe_3O_4/SiO_2/P(NIPAM-coAMPTMA). Vancomycin (Van) creates stronger H–bonding between Van and C-Terminal L-lysyl-D-alanyl-D-alanine of bacteria	[62]
Superparamagnetic Iron Oxide Nanoparticle SPIONs (Fe_3O_4, 40 nm size)	HAD/OA ratio changes MNPs shape	–	TEM, XRD, Electrophoretic mobility measurements, magnetization, SAR/hyperthermia	Hyperthermia (MH, MFH)	SPIONs with spherical, cuboidal (SAR_{max}) or rod-like shape; efficient MFH (rt to 45 °C, in 60 s, 20 kA/m, 136–205 kHz)	[63]
iron oxide nanoparticle (ION)	–	5-aminolevulinic acid (ALA)	Computational study/quantum chemistry	Anticancer therapy	Configurations optimized at optimized at B3LYP/6-31G(d,p) in aq. solution; H-bonding plays a central role	[64]
MNPs manganese ferrite ($MnFe_2O_4$), spinel structure	Citrate-stabilized (14.4 ± 2.6 nm), lipid-coated (8.9 ± 2.1 nm)	Doxorubicin (Dox)	Fluorescence, Förster resonance energy transfer (FRET), STEM, XRD, Raman, SQUID, rheology, UV-Vis, hyperthermia	Drug delivery/release; theranostic	Dehydropeptide-based supramolecular magnetogels; improved drug release of lipid-coated vs. citrate-coated MNPs	[65]

Table 4. *Cont.*

NP Type	Coating Agent	Active Agent	Characterization Methods	Activity	Obs.	Ref.
γ'-Fe_4N (prepared by gas nitridation from commercial γ-Fe_2O_3, 20 nm)	The first report of surface-modified iron nitrides; α''-$Fe_{16}N_xZ_{2-x}$, and α'-$Fe_8N_xZ_{1-x}$, by wet ball milling	–	PPMS (Ms, Hc), XRD, TEM, DLS, FTIR, seta-potential	Envisioned biomedical applications (DNA, protein or drug delivery)	γ'-Fe_4N have 3 times higher saturation magnetizations than IONPs (H_C = 310 Oe, M_S (15 kOe) = 182.7 emu/g; M_R = 45 emu/g)	[66]
Fe_3O_4	Chitosan (Cs), with/without silica; Cs-f-SiO_2@Fe_3O_4, Cs-f-Fe_3O_4	Silymarin (SIL)	Cytotoxicity tests (MCF-7, MTT assay)	Drug delivery, anticancer, antioxidant,	99–120 mg SIL/g functionalized MNPs (Folin–Ciocalteu method)	[67]
SPION	APTES Modification (SPION@APTES)	Toxin: dianthin-epidermal growth factor (DiaEGF) or endosomal escape enhancers (EEE), glycosylated triterpenoids SO1861	Enzymatic Activity, TEM, DLS, DSC, In Vitro Cytotoxicity, Relaxivity	Targeted tumor therapy, drug delivery	2000-fold enhancement in tumor cell cytotoxicity, 6.7-fold gain in specificity; steric stabilization inhibits agglomeration	[68]
MNPs	Functionalized graphene oxide (acylated, G-COCl), polylactic acid, polyvinyl alcohol, polyethylene glycol, and nilotinib (second layer), sodium alginate, polyethylene glycol, poly (lactic-co-glycolic acid), polylactic acid and nilotinib gel (third layer)	Nilotinib (Tasigna™, medication for chronic myelogenous leukemia)	UV-Vis, FTIR, FT-NMR, VSM, SEM, TEM, TGA	Drug delivery	Nilotinib 400 mg, super paramagnetic particles 0.01 g and total MFGO mass (0.1 g) were kept constant in all samples. Faster drug release at acidic pH 3 (24 h) vs. slightly basic pH 7.4 (48 h).	[69]
SPIONs, ^{166}Ho doped iron oxide	Au layer coating	Monoclonal antibody trastuzumab (Tmab)	TEM, TGA, cytotoxicity studies	Multimodal cancer therapy	[^{166}Ho] Fe_3O_4@Au NPs (150 nm) conjugated with Tmab targets HER2+ receptors. Cytotoxic effect toward SKOV-3 ovarian cancer cells.	[70]
MNPs (Fe_2O_3, ~15 nm, 29 emu/g)	Silica coating, a multistep synthesis, activated NP couples a triethylene glycol spaced glycosyl imidazole; silyl propyl-H-imidazole functionalization, glycosylation and deacetylation to NpFeSiImSugar NPs.	–	TLC (thin-layer chromatography), ATR FTIR, TG-DTA, TEM, SEM, EDX, XRD, VSM, BET, ^{1}H and ^{13}C NMR	Nucleic acid (NA) extraction	hydrogen bonding between the surface bonded carbohydrate and nucleic acid targets (NpFeSiIm-Sugar/DNA complex) to ensure nucleic acid selectivity and avoid protein contamination. high DNA particle loading ratio of 30–45 wt% (MNP/DNA ratio)	[71]
SPIONs	Caffeic acid (Caf-SPIONs); citrate-stabilized (Cit-SPION)	Bovine serum albumin (BSA)	Hydrodynamic size (Z-Average), polydispersity index (PDI), zeta potential at pH 7.3, volumetric susceptibility, and iron content Atomic Emission Spectroscopy (AES); HPLC-UV; fluorescence; magnetic field simulations (COMSOL)	Anticancer	Caf-BSA-SPIONs; tested against A375M melanoma cells, fibroblasts;	[72]

Table 4. *Cont.*

NP Type	Coating Agent	Active Agent	Characterization Methods	Activity	Obs.	Ref.
SPION ("SPIO")	Glycyrrhizin-chitosan coating (SPIO@Chitosan-GL)	Glycyrrhizin (anti-inflammatory, anti-ulcer, anti-allergic, antioxidant, anti-tumor, anti-diabetic, hepatoprotective)	MRI, immunofluorescence	Anti–inflammatory	Monitoring of pancreatic islets and mesenchymal stem cell (MSC) spheroids; inhibition of inflammatory damage-associated molecular pattern (DAMP) protein in mice	[73]
Fe_3O_4 MNP	Silica-coating, Polyamide 6 (PA6) by in situ polymerization, to yield self-healable magnetic nanoparticle polymer (SHMNP) composite	–	TEM, XRD, XRD (small/wide angle, SAXS/WAXS), DSC, magnetic properties (SQUID, 100 K, 400 K), ZFC-FC	Self-healable polymer nanocomposites	Multiferroic-polyamide 6 (PA6) nanocomposite	[74]
hematite (α-Fe_2O_3)	Polylactic acid (PLA)	–	FTIR, TGA, DSC, VSM	Biomedical applications (3D printing, cardiovascular stents)	Stimuli-responsive PLA/α-Fe_2O_3 nanocomposites	[75]
MNPs	MNP-CA-PEI nanoparticles 290.74 ± 63.84 nm: citric acid (CA)-modified MNP cross-linked with polyethyleneimine (PEI) (carbonyldiimidazole as the crosslinker)	-(GFP plasmid; MNP/nucleic acid polyplexes)	DLS, TGA, SQUID, FTIR, zeta-potential, surface characterization (Langmuir, Freundlich), fluorescence	Multifunctional magnetic nanocarriers (siRNA, shRNA), gene delivery	Nucleic acid delivery by caveolae-mediated endocytosis; adsorption isotherm follows pseudo-first order kinetics; HEK 293 cells were used.	[76]
Fe_3O_4–Ag	In situ reduction of Ag with gallic acid (reducing agent), silica shell	–	FE-SEM, FTIR	Antibacterial, antitumor	Nanoflower-like or nanodumbbell (NP/gallic acid ratio of 10:1) multifunctional nanocomposites	[77]
SPIONPs (10 nm, toluene), cubosomes. CD44 and CD221)	Hyaluronic acid (HA, ligand for CD44) and antibodies (Abs) against CD221 coupled to cubosomes via electrostatic attraction and thiol-Michael reaction	Helenalin	Cryo-EM, SAXS,	Anticancer	Tested on rhabdomyosarcoma cells (RMS) and control (fibroblast) cells	[78]
$CoFe_2O_4$@$BaTiO_3$ (CFO@BTO), by solvothermal synthesis (38 nm, M_S = 47.4 emu/g)	Surface functionalization with amphiphilic polymer, polyisobutylene-alt-maleic anhydride (PMA)	Doxorubicin (DOX) and methotrexate (MTX)	XRD, HRSEM, HRTEM, SAED, MH measurements, ZFC–FC, DLS < zeta-potential, drug release kinetics	Drug delivery; cancer treatment, suitable for chemo-resistant cancers	Core–shell magnetoelectric NPs obtained as nanorods; 98% drug release in 20 min (MF = 4 mT). Tested on HepG2 and HT144 cells and 3D spheroid models ($p < 0.05$).	[79]
IONPs	Thiolated β-cyclodextrin (β-CD-SH), through Fe–S bonding (TβCD-IONPs)	Doxorubicin (DOX), to yield DOX-TβCD-IONPs	XPS, drug release (modelled by Higuchi model)	(Targeted) cancer treatment	Ellipsoidal shape, ~14 nm. Cellular response dependent on IONPs' functionalization. Tested against breast cancer cell line MCF-07.	[80]
MNP (Fe_3O_4)	Citrate coating	–	XRD, TEM, magnetic measurements (VSM)	Biomedical applications; drug delivery, imaging diagnostic (especially in liver, where iron accumulates)	In vivo models showed increased iron content in liver. No viability issues against different cell lines (HaCaT and HepG2).	[81]
Luminescent, LMNPs Fe_3O_4@$BaMoO_4$:Eu^{3+} (LMNPs)	3-aminopropyl-triethoxysilane (APTES). β-cyclodextrin (β-CD)	Triazole derivatives	XRD, FTIR, PL, TEM, VSM	Drug carrier	Hybrid nanoparticle system LMNPs@APES-CD had high drug loading of 61.69 mg/g	[82]

Table 4. *Cont.*

NP Type	Coating Agent	Active Agent	Characterization Methods	Activity	Obs.	Ref.
Fe_3O_4	Polyethyleneimine, gold, silica, and graphene derivatives	dsDNA	STEM, ICP-MS/MS, UV-Vis, TGA, FTIR, charge (ζ-potential). ICP-MS/MS	DNA isolation	Fe_3O_4@PEI MNPs adsorbed DNA efficiently	[83]
Fe_3O_4	–COOH coating	rtPA	SEM, TEM, XRD, EDX, VSM, FTIR	Thrombolytic nanomedicine	Efficient thrombolysis (rabbit carotid artery occlusion model)	[84]
Magnetic beads (MBs), $Zn_xFe_{3-x}O_4$ nanoparticles (ZnFeNPs), 13 ± 3 nm, M_S = 81 emu/g	Polymeric matrix of poly(lactic-co-glycolic) acid (PLGA), generating polymeric MBs that were covered with polyethyleneimine (PEI) (MB@PEI), obtaining particles of 96 ± 16 nm	Affinity protein neutravidin (NAV), by glutaraldehyde crosslinking	TEM, HRTEM, DLS, XRD, FTIR, ICP-MS	Biosensing (detection of Tau protein)	MB@NAV has high MS (higher than commercial NAV formulation). Alzheimer's disease biomarker (Tau protein) could be detected using MB@NAV at very low 63 mg/mL	[85]
Fe_3O_4	Silica, double-chain surfactant, ionic liquids (ILs)	Epirubicin hydrochloride (EPI)	XRD, FT-IR TG, TEM. liquid chromatography–fluorescence detection (LC-FL)	Analytical extraction methods	didodecyldimethy-lammonium bromide was the most effective surfactant for adsorption tests on MNPs	[86]
$Ni_xFe_{3-x}O_4$ NPs (x = 0.0, 0.2, 0.4, 0.6, 0.8, and 1.0)	Aminosilane coating	–	XRD, Raman, EDX, FTIR, DLS	Cancer treatment (MCF7 and HeLa cell lines); drug delivery, hyperthermia	Fe^{2+} substitution by Ni^{2+}, inverse spinel structure; hydrodynamic radii 10 nm (DLS). Coating decreases agglomeration and cell viability	[87]
IONPs	Oleic acid (OA) and tetraethylene glycol (TEG)	Dexamethasone (Dexa)	NMR, HR-MS, TEM, DLS, HPLC,	Biological applications, drug delivery	Novel encapsulation system based on triazole-derived micelle precursor	[88]
MNP (Fe_3O_4)	Hyperbranched polyglycerol and carboxymethyl cellulose, to Fe_3O_4@PG and Fe_3O_4@PG/CMC-PEG@DOX	Doxorubicin (DOX)	IR, NMR, TG, VSM, XRD, DLS, HR-TEM and UV–Vis	Drug delivery, anticancer, contrast agent in magnetic resonance imaging MRI	biocompatibility toward normal cells (HEK-293), high toxicity against cancerous cells (HeLa).	[89]
SPIONs (Fe_3O_4)	Copolyester, poly (globalide-*co*-ε-caprolactone) (PGlCL), modified with amino acid cysteine (Cys) via a thiol-ene reaction (PGlCLCys); folic acid (FA)	Methotrexate (MTX); conjugation achieved via -NH_2 group of cysteine	^1H NMR, Gel Permeation Chromatography (GPC), HPLC, FTIR, DSC, TEM/SAED, XRD, magnetic properties (VSM), TGA, DLS	Enzymatic release, antitumor nanoplatform (tested on tumor cells MDA-MB 231)	SPIONS (SPION@PGlCLCys) enable further conjugation with active biomolecules; drug loaded SPION@PGlCLCys_MTX obtained by carbodiimide-mediated coupling (amide bond)	[90]
Fe, MnFe, CoFe	Oleic acid, allyl amine (synthesis), polypropylene sulphide PPS-coating to produce PPS-MNPs (80 ± 15 nm)	DOX and CUR loaded PPS–MNPs	DLS, TEM, FTIR, XRD, VSM, TGA, Calorimetric magnetic fluid hyperthermia, UV-Vis (drug encapsulation efficiency), drug loading (HPLC)	Hyperthermia, Cancer treatment (tested on human epithelial cells HEK293)	MNPs of 8 nm, 12 nm and 16 nm prepared by seed-mediated method from M-acetylacetonates; high breast cancer cell death (95%)	[91]
MNP (Fe_3O_4)	3-amino propyl triethoxy silane (APTES) coating to MNPs@SiO_2/CMT (carboxymethyl tragacanth)	Doxorubicin (Dox)	FT-IR, SEM, EDX, TEM, XRD, VSM, TGA, and zeta potential, fluorescence microscopy	Anti-cancer	Effect studied on MCF7 human breast cancer cells	[92]

Table 4. *Cont.*

NP Type	Coating Agent	Active Agent	Characterization Methods	Activity	Obs.	Ref.
Fe_3O_4	PEI-modified	Dopamine (DA), self-polymerized	FTIR, AFM, SEM, agarose gel electrophoresis, fluorescence microscopy, flow cytometry	Gene vector (DNA delivery)	Fe_3O_4@PDA@PEI modified NPs are stable hydrophilic NPs (50–150 nm)	[93]
MNPs (iron oxide, 18 nm diameter)	Mesoporous silica shell (93 nm diameter)	*trans*-resveratrol (post-silanization)	DLS, zeta potential, NMR, IR, magnetism (SQUID, ZFC-FC), fluorescence resonance energy transfer (FRET)	Biomedical (protein–ligand, immunoassay design)	Superparamagnetic behavior useful for magnetic bioseparation; high zeta-potential kept the mesoporous NPs in alkaline solution	[94]
magnetic nanocomposites (MNCs), nanocapsules (NCs) based on Fe_3O_4	Oleic-acid-modified, Nylon-6 coated	Doxorubicin (Dox)	DLS, ζ-potential, FTIR, TEM, drug loading/release (UV-Vis), cytotoxicity studies	Anticancer; drug delivery (against A549 and HEK 293FT cell lines)	High DOX loading: 732 μg/mg (DOX/MNC) and 943 μg/mg (DOX/NC); pH-sensitive drug release	[95]
Fe_3O_4 nanoclusters	Stabilizers: 3,4-dihydroxybenzhydrazide (DHBH) and poly [3,4-dihydroxybenzhydrazide] (PDHBH)	–	TEM, XPS, FTIR, VSM (magnetization)	Anticancer, hyperthermia (human normal dermal fibroblasts-BJ, colon adenocarcinoma-CACO2, and melanoma-A375)	In situ solvothermal process of MNPs-magnetite nanoclusters, MNC of 50 emu/g and 60 emu/g	[96]
Cobalt-Iron Ferrite Nanoparticles $Co_xFe_{1-x}Fe_2O_4$ (x = 0.0, 0.2,0.4, 0.6, 0.8, and 1.0)	Surface modification with amino-silane (AEPTMS)	–	SEM, EDX, XRD, FTIR	Cytocompatibility, biomedical applications	MTT assay on fibroblast cells (cytotoxicity tests); Co/Fe ratio influences cytotoxicity	[97]
Fe_3O_4	L-Cysteine (L-Cys)-coating	Doxorubicin (Dox)	EDS, SAED, XRD, FTIR, TEM. XPS, Mössbauer spectroscopy, SQUID	Drug delivery, anticancer (anti-melanoma)	Fe_3O_4-L-Cys-Dox NPs showed anti-melanoma activity on mouse (B16F10) and human (A375) metastatic melanoma	[98]

Various MNP systems have been designed to overcome embolism caused by these nanoparticles, which requires coating with biocompatible and non-cytotoxic polymers, such as poly (globalideco-ε-caprolactone) (PGlCL), modified with the amino acid cysteine (Cys) via a thiol-ene reaction (PGlCLCys) [90]. The reaction scheme utilized to produce a coating of SPIONS (SPION@PGlCLCys) and further conjugation with folic acid (FA) or the anti-cancer drug methotrexate (MTX) is described in Figure 6. Cysteine (Cys) was chosen specifically due to its good compatibility with SPIONs, which it binds to through carboxylic and thiol groups, while the selection of biomolecules (FA, MTX) was aimed at anticancer treatment, as experiments showed a 45% release of MTX within 72 h under enzymatic-triggered release and a reasonable reduction in tumor cell (breast carcinoma MDA-MB 231) viability of 20%. The tumoral cell viability remains high, although the MTX loading wss quite low at 3.20 μg MTX/mg IONP [90].

The use of surface-functionalized magnetic nanoparticles (MNPs) in medical therapies shows great promise, but it is important to be aware of potential side effects and complications related to biocompatibility and toxicity, accumulation and biodistribution, retention and clearance, inflammatory/allergic response, potential interference, agglomeration, under/overdosing, unintended effect on nearby tissues/organs, rare element toxicity (present in some surface coatings), incomplete drug release, clinical translation, and regulatory approval.

pled with in vitro and in vivo validation studies, underscore the transformative potential of surface-engineered MNPs in revolutionizing drug delivery paradigms, offering novel avenues for personalized and targeted therapies [100].

6. Challenges and Future Perspectives

While remarkable strides have been made in the surface functionalization of magnetic nanoparticles (MNPs), several challenges and promising avenues lie ahead. This section delves into the intricacies of these challenges and outlines future directions that hold the potential to reshape the field. Some of the most stringent aspects that require further improvements have been highlighted in Figure 8, and they include: biocompatibility and toxicity, long-term stability, seamless and time-effective clinical translation, qualitative and quantitative control, multifunctionality, a deeper understanding of the structure–function relationship, enhanced in vivo behavior and targeting precision, multimodal integration, and exploring other areas of interest beyond biomedicine.

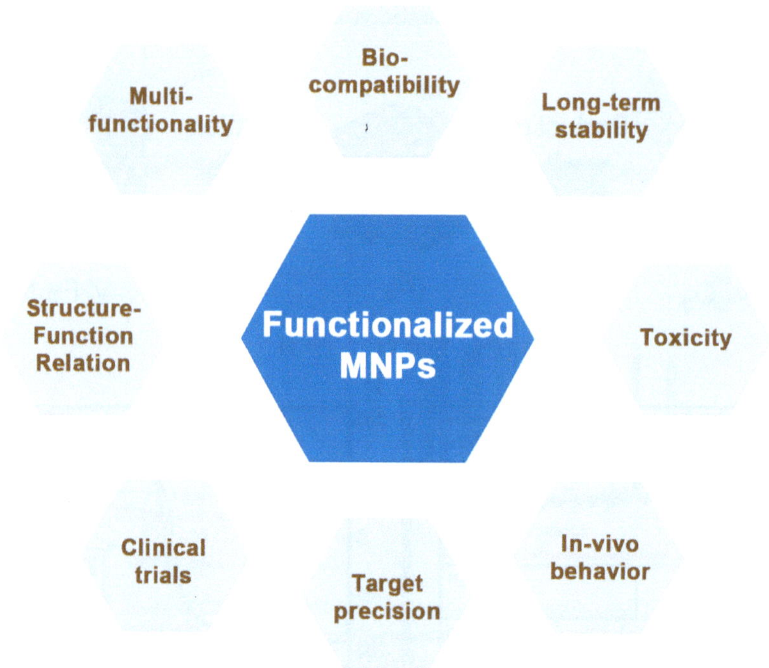

Figure 8. The main parameters and features that need to be addressed by future MNP nanoplatforms for efficient use in biomedical applications and beyond.

As MNPs' prevalence in biomedical applications increases [101], ensuring their biocompatibility and minimizing potential toxicity remains a critical challenge. Therefore, thorough biocompatibility assessments (in vitro and in vivo studies) are imperative. Surface modifications that enhance biocompatibility, such as PEGylation, must be balanced with potential alterations in surface functionality. Maintaining the stability of surface-functionalized MNPs over extended periods is vital for their efficacy. Challenges such as ligand detachment, aggregation, or degradation need to be addressed through robust coating strategies and long-term stability studies. Multi-parametric characterization techniques can shed light on stability issues.

While surface-functionalized MNPs show great promise in preclinical studies, their seamless translation into clinical settings poses significant challenges. Rigorous safety assessments, scalability of production, and regulatory approvals are crucial hurdles that need to be overcome for clinical applications. Achieving precise quantitative control over surface functionalization remains a challenge. Determining the exact number of functional moieties per MNP requires advanced analytical techniques. Techniques like single-particle tracking or advanced spectroscopic methods may offer new insights.

The demand for multifunctional MNPs with multiple surface functionalities adds complexity. Balancing diverse functionalities while maintaining stability and avoiding interference requires innovative design strategies and comprehensive characterization. Unraveling the intricate interplay between surface modifications and functional outcomes is a continual challenge. Advanced computational methods, coupled with detailed experimental studies, are essential to deciphering the complex structure–function relationships. Achieving precise targeting of MNPs to specific cells or tissues remains a challenge, particularly in dynamic biological environments. Incorporating responsive elements into MNPs' coatings that enable active targeting upon specific stimuli could enhance targeting precision [38].

Elucidating the behavior of surface-functionalized MNPs in complex biological environments is crucial. Studying factors like protein corona formation, biodistribution, and clearance pathways will provide insights into their fate after administration inside the human body [10]. In nanotechnology, the protein corona plays a key role in dictating the biological fate and functionality of these nanoscale entities; when nanoparticles, including magnetic nanoparticles (MNPs), come into contact with biological fluids, such as blood or interstitial fluid, they instantaneously interact with a plethora of biomolecules such as proteins present in these environments. Upon exposure to biological fluids, proteins rapidly and spontaneously adsorb onto the surface of the nanoparticles, forming a dynamic and complex layer, often referred to as the protein corona, a complex phenomenon influenced by factors like nanoparticle size, shape, surface charge, and surface chemistry. The protein corona, in turn, fundamentally alters the biological identity of the nanoparticle by mediating interactions with cells, influencing cellular uptake, intracellular trafficking, and biological responses, and determines the fate of the nanoparticle within the body, impacting aspects such as circulation time, distribution in tissues, and potential clearance mechanisms. The concept of protein corona formation represents a dynamic and intricate phenomenon at the interface of nanoparticles and biological systems, with profound influence over the biological behavior and fate of nanoparticles in vivo, underscoring its critical importance in the field of nanomedicine. Researchers are actively working to unravel the complexities of protein corona dynamics to engineer nanoparticles with enhanced biocompatibility and therapeutic efficacy. Iron oxides, for instance, have been shown to remap the immunological tumor environment, especially when interacting with macrophage response through polarization and reprogramming [102].

A closer look at the commercial solutions offered on the market today highlights the presence of quite a few suppliers, such as Dynabeads (Thermo Fisher Scientific, Waltham, MA, USA), Micromod Partikeltechnologie GmbH, NanoMAG-D (Macherey-Nagel, Düren, Germany), Ocean NanoTech MagVigen™ Magnetic Nanoparticles (Ocean NanoTech, San Diego, CA, USA), Bangs Laboratories Magnetic Microspheres (Bangs Laboratories, Inc., Fishers, IN, USA), Ademtech Functionalized Magnetic Particles (Ademtech, Pessac, France), NanoXact (NanoComposix, San Diego, CA, USA), and MagSi Beads (Chemicell, Berlin, Germany). To get a more in-depth estimation of the costs and actual surface coverings offered by one of the mentioned brands, a brief analysis of Micromod's listings shows that 10 mL of MNPs (300 nm; 10 mg/mL) costs just shy of EUR 250, and various functionalizations are possible (dextran, silicate, etc). The cost, however, remains prohibitive, especially when related to NPs' dry weight content. There are quite a few companies (including major brands) offering commercial solutions, many of which can be further tailored to suit demand for specific properties, coatings, concentrations, etc. Each additional request raises the overall price. Therefore, tailoring magnetic formulations for specific applications could be conducted in-house instead, as this approach costs a lot less than commercial solutions and represents one of the key benefits of synthesizing MNPs in-house rather than through procurement services. A brief comparison correlating some of the recent advances in the field of surface-functionalized magnetic nanoparticles (MNPs) with solutions available on the market is given in Table 5.

Table 5. Short comparison correlating commercial solutions available today on the market to the recent advances and the current state-of-the-art.

Application	Recent Advances	Market Solutions
Surface Functionalization Strategies	Advanced strategies like click chemistry, polymer brushes via ATRP, multifunctional coatings.	Established strategies like silane coupling agents and antibody immobilization; advanced strategies emerging in commercial products, like click chemistry-based functionalization kits.
Applications in Drug Delivery	Novel surface coatings and functional groups—enhanced drug loading, controlled release.	MNPs functionalized with specific ligands for targeted drug delivery. Some companies provide customizable options for specific drug payloads.
Magnetic Resonance Imaging (MRI)	Coatings with superior magnetic properties and responsive materials for improved contrast in MRI.	High-quality coatings, ensuring stability and improved imaging performance; some products integrate advanced surface modifications to enhance contrast.
Hyperthermia Treatment	Surface functionalization strategies to optimize heating efficiency for hyperthermia applications.	Commercial options with tailored surface properties to achieve precise heating of targeted tissues within physiologically relevant limits.
Diagnostic Applications	Improved diagnostic capabilities through enhanced binding specificity and signal amplification.	Surface functionalizations optimized for specific diagnostic assays, that may include pre-conjugated antibodies or customizable options for specific biomarker detection.
Multifunctionality	Recent research focuses on integrating multiple functionalities within a single NP platform.	Multifunctional MNPs that combine features like drug delivery, imaging, and targeting within a single particle, designed to streamline complex applications.
Environmental Remediation and Catalysis	Functionalized MNPs are being explored in catalysis and environmental remediation.	An emerging interest in functionalized MNPs for non-biomedical uses, with customizable options for research in these fields.

Integrating multiple functionalities into a single MNP, such as combining imaging and therapy, is a complex challenge aimed at the theranostics realm. Developing streamlined strategies for co-functionalization while preserving each functionality's integrity is an ongoing pursuit. Finally, exploring novel applications of surface-functionalized MNPs beyond biomedicine, such as environmental remediation, catalysis, and energy storage, offers exciting future prospects. Tailoring surface functionalization for non-biomedical contexts requires creative adaptations and a comprehensive understanding of each application's requirements.

Looking forward, the field of surface functionalization of MNPs holds immense potential. Addressing these challenges requires interdisciplinary collaboration, innovative engineering, and meticulous characterization. As the complexities are unraveled, surface-engineered MNPs are poised to continue transforming diverse domains, offering solutions to some of the most pressing challenges in science, technology, and medicine [20,21,103–105].

7. Conclusions

The journey through recent advances in the surface functionalization of magnetic nanoparticles (MNPs) has revealed a captivating landscape where scientific innovation converges with transformative applications. The remarkable synergy between nanotechnology and surface engineering has ushered in a new era, unlocking the full potential of MNPs across diverse domains. Surface functionalization has redefined the capabilities of MNPs in multimodal imaging, drug delivery, and catalysis. The strategic tailoring of MNP surfaces has propelled multimodal imaging, enhancing contrast, enabling targeted imaging, and fostering the integration of different imaging modalities. This convergence has profound implications in disease diagnosis, therapeutic monitoring, and understanding complex biological processes. The horizon of the surface functionalization of MNPs gleams with possibilities. From fine-tuning imaging contrast to precise drug delivery, surface-engineered MNPs stand as catalysts of transformation. Collaboration between disciplines, synergy between theory and experiment, and unwavering commitment to innovation will rewrite the boundaries of what is possible.

Funding: This work was supported by the Romanian Ministry of Research and Innovation through Project No. PN-III-P1-1.1-TE-2021-1657 (TE 84/2022), TE 91/2022 and by the Core Program of the National Institute of Materials Physics, granted by the Romanian Ministry of Research, Innovation and Digitization through the Project PC1-PN23080101.

Institutional Review Board Statement: Not applicable.

Informed Consent Statement: Not applicable.

Data Availability Statement: Not applicable.

Acknowledgments: This work is funded by the Core Program of the National Institute of Materials Physics, granted by the Romanian Ministry of Research, Innovation and Digitalization through the Project PC1-PN23080101 and Projects No. PN-III-P1-1.1-TE-2021-1657 (TE 84/2022) and TE 91/2022.

Conflicts of Interest: The author declares no conflict of interest. The funders had no role in the design of the study; in the collection, analyses, or interpretation of data; in the writing of the manuscript; or in the decision to publish the results.

References

1. Jiang, K.Y.; Zhang, L.L.; Bao, G. Magnetic iron oxide nanoparticles for biomedical applications. *Curr. Opin. Biomed. Eng.* **2021**, *20*, 100330. [CrossRef] [PubMed]
2. Shankar, M.; Kesavan, S.S.; Biswas, K. Exploring the Potentials of Magnetic Nanoscale Material for Different Biomedical Applications: A Review. *Bionanoscience* **2023**. [CrossRef]
3. Matveeva, V.G.; Bronstein, L.M. Magnetic Nanoparticle-Containing Supports as Carriers of Immobilized Enzymes: Key Factors Influencing the Biocatalyst Performance. *Nanomaterials* **2021**, *11*, 2257. [CrossRef] [PubMed]
4. Palade, P.; Comanescu, C.; Radu, C. Synthesis of Nickel and Cobalt Ferrite-Doped Graphene as Efficient Catalysts for Improving the Hydrogen Storage Kinetics of Lithium Borohydride. *Materials* **2023**, *16*, 427. [CrossRef]
5. Mehak; Thummer, R.P.; Pandey, L.M. Surface modified iron-oxide based engineered nanomaterials for hyperthermia therapy of cancer cells. *Biotechnol. Genet.* **2023**. [CrossRef]
6. Pryazhnikov, D.V.; Kubrakova, I.V. Surface-Modified Magnetic Nanoscale Materials: Preparation and Study of Their Structure, Composition, and Properties. *J. Anal. Chem.* **2021**, *76*, 685–706. [CrossRef]
7. Ansari, M.J.; Kadhim, M.M.; Hussein, B.A.; Lafta, H.A.; Kianfar, E. Synthesis and Stability of Magnetic Nanoparticles. *Bionanoscience* **2022**, *12*, 627–638. [CrossRef]
8. Siddique, S.; Chow, J.C.L. Recent Advances in Functionalized Nanoparticles in Cancer Theranostics. *Nanomaterials* **2022**, *12*, 2826. [CrossRef]
9. Comanescu, C. Magnetic Nanoparticles: Current Advances in Nanomedicine, Drug Delivery and MRI. *Chemistry* **2022**, *4*, 872–930. [CrossRef]
10. Senthilkumar, N.; Sharma, P.K.; Sood, N.; Bhalla, N. Designing magnetic nanoparticles for in vivo applications and understanding their fate inside human body. *Coordin. Chem. Rev.* **2021**, *445*, 214082. [CrossRef]
11. Liu, Y.K.; Su, G.M.Y.; Zhang, R.Y.; Dai, R.J.; Li, Z. Nanomaterials-Functionalized Hydrogels for the Treatment of Cutaneous Wounds. *Int. J. Mol. Sci.* **2023**, *24*, 336. [CrossRef] [PubMed]
12. Gambhir, R.P.; Rohiwal, S.S.; Tiwari, A.P. Multifunctional surface functionalized magnetic iron oxide nanoparticles for biomedical applications: A review. *Appl. Surf. Sci. Adv.* **2022**, *11*, 100303. [CrossRef]

13. Gwon, H.; Park, S.; Lu, Q.; Choi, H.J.; Lee, S. Size effect of iron oxide nanorods with controlled aspect ratio on magneto-responsive behavior. *J. Ind. Eng. Chem.* **2023**, *124*, 279–286. [CrossRef]
14. Jimenez-Carretero, M.; Rodriguez-Lopez, J.; Ropero-Moreno, C.; Granada, J.; Delgado-Martin, J.; Martinez-Bueno, M.; Fernandez-Vivas, A.; Jimenez-Lopez, C. Biomimetic magnetic nanoparticles for bacterial magnetic concentration in liquids and qPCR-detection. *Food Control* **2023**, *147*, 109623. [CrossRef]
15. Franzel, L.; Bertino, M.F.; Huba, Z.J.; Carpenter, E.E. Synthesis of magnetic nanoparticles by pulsed laser ablation. *Appl. Surf. Sci.* **2012**, *261*, 332–336. [CrossRef]
16. Liu, J.; Su, D.; Wu, K.; Wang, J.P. High-moment magnetic nanoparticles. *J. Nanopart. Res.* **2020**, *22*, 66. [CrossRef]
17. Carvallo, C.; Fondet, A.; Le Fèvre, R.; Taverna, D.; Guyodo, Y.; Chebbi, I.; Dupuis, V.; Lagroix, F.; Khelfallah, M.; Guigner, J.-M.; et al. Magnetic and structural properties of biogenic magnetic nanoparticles along their production process for use in magnetic hyperthermia. *J. Magn. Magn. Mater.* **2023**, *575*, 170726. [CrossRef]
18. Iacob, N.; Kuncser, A.; Comanescu, C.; Palade, P.; Kuncser, V. Optimization of magnetic fluid hyperthermia with respect to nanoparticle shape-related parameters: Case of magnetite ellipsoidal nanoparticles. *J. Nanopart. Res.* **2020**, *22*, 138. [CrossRef]
19. Pedroso-Santana, S.; Fleitas-Salazar, N. The Use of Capping Agents in the Stabilization and Functionalization of Metallic Nanoparticles for Biomedical Applications. *Part. Part. Syst. Charact.* **2023**, *40*, 2200146. [CrossRef]
20. Lu, A.H.; Salabas, E.E.; Schüth, F. Magnetic Nanoparticles: Synthesis, Protection, Functionalization, and Application. *Angew. Chem. Int. Ed.* **2007**, *46*, 1222–1244. [CrossRef]
21. Majidi, S.; Zeinali, F.; Samad, S.; Farkhani, M.; Soleymani, M.; Akbarzadeh, A. Current methods for synthesis of magnetic nanoparticles. *Artif. Cells Nanomed. Biotechnol.* **2016**, *44*, 722–734. [CrossRef] [PubMed]
22. Gareev, K.; Tagaeva, R.; Bobkov, D.; Yudintceva, N.; Goncharova, D.; Combs, S.E.; Ten, A.; Samochernych, K.; Shevtsov, M. Passing of Nanocarriers across the Histohematic Barriers: Current Approaches for Tumor Theranostics. *Nanomaterials* **2023**, *13*, 1140. [CrossRef] [PubMed]
23. Hao, H.S.; Chen, Y.; Wu, M.Y. Biomimetic nanomedicine toward personalized disease theranostics. *Nano Res.* **2021**, *14*, 2491–2511. [CrossRef]
24. Palade, P.; Comanescu, C.; Kuncser, A.; Berger, D.; Matei, C.; Iacob, N.; Kuncser, V. Mesoporous Cobalt Ferrite Nanosystems Obtained by Surfactant-Assisted Hydrothermal Method: Tuning Morpho-structural and Magnetic Properties via pH-Variation. *Nanomaterials* **2020**, *10*, 476. [CrossRef] [PubMed]
25. Aslam, H.; Shukrullah, S.; Naz, M.Y.; Fatima, H.; Hussain, H.; Ullah, S.; Assiri, M.A. Current and future perspectives of multifunctional magnetic nanoparticles based controlled drug delivery systems. *J. Drug Deliv. Sci. Technol.* **2022**, *67*, 102946. [CrossRef]
26. Hur, J.U.; Shin, J.R.; Han, J.S.; Kim, Y.H.; An, G.S. Self-assembled core-shell Fe_3O_4-Pt nanoparticles via silylation/polymerization-based amino-functionalization. *Colloid Interfac. Sci.* **2022**, *50*, 100655. [CrossRef]
27. Low, L.E.; Lim, H.P.; Ong, Y.S.; Siva, S.P.; Sia, C.S.; Goh, B.H.; Chan, E.S.; Tey, B.T. Stimuli-controllable iron oxide nanoparticle assemblies: Design, manipulation and bio-applications. *J. Control Release* **2022**, *345*, 231–274. [CrossRef]
28. Strassburg, S.; Mayer, K.; Scheibel, T. Functionalization of biopolymer fibers with magnetic nanoparticles. *Phys. Sci. Rev.* **2022**, *7*, 1091–1117. [CrossRef]
29. Darroudi, M.; Gholami, M.; Rezayi, M.; Khazaei, M. An overview and bibliometric analysis on the colorectal cancer therapy by magnetic functionalized nanoparticles for the responsive and targeted drug delivery. *J. Nanobiotechnol.* **2021**, *19*, 399. [CrossRef]
30. Mahajan, R.; Suriyanarayanan, S.; Nicholls, I.A. Improved Solvothermal Synthesis of gamma-Fe_2O_3 Magnetic Nanoparticles for SiO_2 Coating. *Nanomaterials* **2021**, *11*, 1889. [CrossRef]
31. Elahi, N.; Rizwan, M. Progress and prospects of magnetic iron oxide nanoparticles in biomedical applications: A review. *Artif. Organs* **2021**, *45*, 1272–1299. [CrossRef] [PubMed]
32. Ferreira-Faria, I.; Yousefiasl, S.; Macario-Soares, A.; Pereira-Silva, M.; Peixoto, D.; Zafar, H.; Raza, F.; Faneca, H.; Veiga, F.; Hamblin, M.R.; et al. Stem cell membrane-coated abiotic nanomaterials for biomedical applications. *J. Control Release* **2022**, *351*, 174–197. [CrossRef] [PubMed]
33. Das, P.; Jana, N.R. Biomedical Applications of Functional Polyaspartamide-Based Materials. *ACS Appl. Polym. Mater.* **2021**, *3*, 4791–4811. [CrossRef]
34. Klekotka, U.; Zambrzycka-Szelewa, E.; Satula, D.; Kalska-Szostko, B. Stability Studies of Magnetite Nanoparticles in Environmental Solutions. *Materials* **2021**, *14*, 5069. [CrossRef] [PubMed]
35. Zakhireh, S.; Barar, J.; Adibkia, K.; Beygi-Khosrowshahi, Y.; Fathi, M.; Omidain, H.; Omidi, Y. Bioactive Chitosan-Based Organometallic Scaffolds for Tissue Engineering and Regeneration. *Topics Curr. Chem.* **2022**, *380*, 13. [CrossRef] [PubMed]
36. Kwizera, E.A.; Stewart, S.; Mahmud, M.M.; He, X.M. Magnetic Nanoparticle-Mediated Heating for Biomedical Applications. *J. Heat Trans.-T Asme* **2022**, *144*, 030801. [CrossRef] [PubMed]
37. Hu, B.; Rotherham, M.; Farrow, N.; Roach, P.; Dobson, J.; El Haj, A.J. Immobilization of Wnt Fragment Peptides on Magnetic Nanoparticles or Synthetic Surfaces Regulate Wnt Signaling Kinetics. *Int. J. Mol. Sci.* **2022**, *23*, 10164. [CrossRef]
38. Mohapatra, A.; Uthaman, S.; Park, I.K. External and Internal Stimuli-Responsive Metallic Nanotherapeutics for Enhanced Anticancer Therapy. *Front. Mol. Biosci.* **2021**, *7*, 597634. [CrossRef]

39. Kothandaraman, H.; Kaliyamoorthy, A.; Rajaram, A.; Kalaiselvan, C.R.; Sahu, N.K.; Govindasamy, P.; Rajaram, M. Functionalization and Haemolytic analysis of pure superparamagnetic magnetite nanoparticle for hyperthermia application. *J. Biol. Phys.* **2022**, *48*, 383–397. [CrossRef]
40. Duong, H.D.T.; Yoon, S.H.; Nguyen, D.T.; Kim, K.S. Magnetic heating of water dispersible and size-controlled superparamagnetic cobalt iron oxide nanoparticles. *Powder Technol.* **2023**, *427*, 118720. [CrossRef]
41. Ghosal, K.; Chatterjee, S.; Thomas, S.; Roy, P. A Detailed Review on Synthesis, Functionalization, Application, Challenges, and Current Status of Magnetic Nanoparticles in the Field of Drug Delivery and Gene Delivery System. *AAPS PharmSciTech* **2022**, *24*, 25. [CrossRef]
42. Arias-Ramos, N.; Ibarra, L.E.; Serrano-Torres, M.; Yague, B.; Caverzan, M.D.; Chesta, C.A.; Palacios, R.E.; Lopez-Larrubia, P. Iron Oxide Incorporated Conjugated Polymer Nanoparticles for Simultaneous Use in Magnetic Resonance and Fluorescent Imaging of Brain Tumors. *Pharmaceutics* **2021**, *13*, 1258. [CrossRef]
43. Avarand, S.; Morsali, A.; Heravi, M.M.; Beyramabadi, S.A. A quantum chemical study on the magnetic nanocarrier-tirapazamine drug delivery system. *Nanosyst.-Phys. Chem. M* **2021**, *12*, 167–174. [CrossRef]
44. Beagan, A.M.; Alghamdi, A.A.; Lahmadi, S.S.; Halwani, M.A.; Almeataq, M.S.; Alhazaa, A.N.; Alotaibi, K.M.; Alswieleh, A.M. Folic Acid-Terminated Poly(2-Diethyl Amino Ethyl Methacrylate) Brush-Gated Magnetic Mesoporous Nanoparticles as a Smart Drug Delivery System. *Polymers* **2021**, *13*, 59. [CrossRef]
45. Valls-Chivas, A.; Gomez, J.; Garcia-Peiro, J.I.; Hornos, F.; Hueso, J.L. Enzyme-Iron Oxide Nanoassemblies: A Review of Immobilization and Biocatalytic Applications. *Catalysts* **2023**, *13*, 980. [CrossRef]
46. Belleti, E.; Bevilaqua, V.R.; Brito, A.M.M.; Modesto, D.A.; Lanfredi, A.J.C.; Viviani, V.R.; Nantes-Cardoso, I.L. Synthesis of bioluminescent gold nanoparticle-luciferase hybrid systems for technological applications. *Photoch. Photobio Sci.* **2021**, *20*, 1439–1453. [CrossRef] [PubMed]
47. Boosz, P.; Pfister, F.; Stein, R.; Friedrich, B.; Fester, L.; Band, J.; Muhlberger, M.; Schreiber, E.; Lyer, S.; Dudziak, D.; et al. Citrate-Coated Superparamagnetic Iron Oxide Nanoparticles Enable a Stable Non-Spilling Loading of T Cells and Their Magnetic Accumulation. *Cancers* **2021**, *13*, 4143. [CrossRef]
48. Cheah, P.; Qu, J.; Li, Y.; Cao, D.M.; Zhu, X.C.; Zhao, Y.F. The key role of reaction temperature on a polyol synthesis of water-dispersible iron oxide nanoparticles. *J. Magn. Magn. Mater.* **2021**, *540*, 168481. [CrossRef]
49. Correa, T.; Bazylinski, D.A.; Garcia, F.; Abreu, F. A rapid and simple preparation of amphotericin B-loaded bacterial magnetite nanoparticles. *RSC Adv.* **2021**, *11*, 28000–28007. [CrossRef] [PubMed]
50. Dhavale, R.P.; Dhavale, R.P.; Sahoo, S.C.; Kollu, P.; Jadhav, S.U.; Patil, P.S.; Dongale, T.D.; Chougale, A.D.; Patil, P.B. Chitosan coated magnetic nanoparticles as carriers of anticancer drug Telmisartan: pH-responsive controlled drug release and cytotoxicity studies. *J. Phys. Chem. Solids* **2021**, *148*, 109749. [CrossRef]
51. Domagalski, J.T.; Xifre-Perez, E.; Tabrizi, M.A.; Ferre-Borrull, J.; Marsal, L.F. Magnetic nanoparticle decorated anodic alumina nanotubes for fluorescent detection of cathepsin B. *J. Colloid Interf. Sci.* **2021**, *584*, 236–245. [CrossRef] [PubMed]
52. Doswald, S.; Stark, W.J. Preparation of Functionalized Carbon-Coated Cobalt Nanoparticles with Sulfonated Arene Derivatives, a Study on Surface Functionalization and Stability. *Chem.-Eur. J.* **2021**, *27*, 4108–4114. [CrossRef] [PubMed]
53. Ehsanimehr, S.; Moghadam, P.N.; Dehaen, W.; Shafiei-Irannejad, V. PEI grafted Fe_3O_4@SiO_2@SBA-15 labeled FA as a pH-sensitive mesoporous magnetic and biocompatible nanocarrier for targeted delivery of doxorubicin to MCF-7 cell line. *Colloid Surface A* **2021**, *615*, 126302. [CrossRef]
54. Fan, X.M.; Shen, J.J.; Xu, Y.Y.; Gao, J.; Zhang, Y.W. Metabolic integration of azide functionalized glycan on Escherichia coli cell surface for specific covalent immobilization onto magnetic nanoparticles with click chemistry. *Bioresour. Technol.* **2021**, *324*, 124689. [CrossRef]
55. Fernandez-Ponce, C.; Manuel, J.M.; Fernandez-Cisnal, R.; Felix, E.; Beato-Lopez, J.; Munoz-Miranda, J.P.; Beltran, A.M.; Santos, A.J.; Morales, F.M.; Yeste, M.P.; et al. Superficial Characteristics and Functionalization Effectiveness of Non-Toxic Glutathione-Capped Magnetic, Fluorescent, Metallic and Hybrid Nanoparticles for Biomedical Applications. *Metals* **2021**, *11*, 383. [CrossRef]
56. Habra, K.; McArdle, S.E.B.; Morris, R.H.; Cave, G.W.V. Synthesis and Functionalisation of Superparamagnetic Nano-Rods towards the Treatment of Glioblastoma Brain Tumours. *Nanomaterials* **2021**, *11*, 2157. [CrossRef]
57. Jose, R.; Rinita, J.; Jothi, N.S.N. The synthesis and characterisation of curcumin loaded Ag ((1-X)) Ni (X) Fe-2 O-4 for drug delivery. *Mater. Technol.* **2021**, *36*, 339–346. [CrossRef]
58. Kannan, K.; Mukherjee, J.; Mishra, P.; Gupta, M.N. Nickel Ferrite Nanoparticles as an Adsorbent for Immobilized Metal Affinity Chromatography of Proteins. *J. Chromatogr. Sci.* **2021**, *59*, 262–268. [CrossRef]
59. Leitner, N.S.; Schroffenegger, M.; Reimhult, E. Polymer Brush-Grafted Nanoparticles Preferentially Interact with Opsonins and Albumin. *ACS Appl. Bio Mater.* **2021**, *4*, 795–806. [CrossRef]
60. Martin, D.S.; Oropesa-Nunez, R.; de la Torre, T.Z.G. Evaluating the Performance of a Magnetic Nanoparticle-Based Detection Method Using Circle-to-Circle Amplification. *Biosensors* **2021**, *11*, 173. [CrossRef]
61. Moskvin, M.; Huntosova, V.; Herynek, V.; Matous, P.; Michalcova, A.; Lobaz, V.; Zasonska, B.; Slouf, M.; Seliga, R.; Horak, D. In vitro cellular activity of maghemite/cerium oxide magnetic nanoparticles with antioxidant properties. *Colloid Surface B* **2021**, *204*, 111824. [CrossRef] [PubMed]

62. Nayeem, J.; Al-Bari, M.A.A.; Mahiuddin, M.; Rahman, M.A.; Mefford, O.T.; Ahmad, H.; Rahman, M.M. Silica coating of iron oxide magnetic nanoparticles by reverse microemulsion method and their functionalization with cationic polymer P (NIPAm-co-AMPTMA) for antibacterial vancomycin immobilization. *Colloid Surface A* **2021**, *611*, 125857. [CrossRef]
63. Reyes-Ortega, F.; Delgado, A.V.; Iglesias, G.R. Modulation of the Magnetic Hyperthermia Response Using Different Superparamagnetic Iron Oxide Nanoparticle Morphologies. *Nanomaterials* **2021**, *11*, 627. [CrossRef]
64. Rezaei, A.; Morsali, A.; Bozorgmehr, M.R.; Nasrabadi, M. Quantum chemical analysis of 5-aminolevulinic acid anticancer drug delivery systems: Carbon nanotube, -COOH functionalized carbon nanotube and iron oxide nanoparticle. *J. Mol. Liq.* **2021**, *340*, 117182. [CrossRef]
65. Veloso, S.R.S.; Silva, J.F.G.; Hilliou, L.; Moura, C.; Coutinho, P.J.G.; Martins, J.A.; Testa-Anta, M.; Salgueirino, V.; Correa-Duarte, M.A.; Ferreira, P.M.T.; et al. Impact of Citrate and Lipid-Functionalized Magnetic Nanoparticles in Dehydropeptide Supramolecular Magnetogels: Properties, Design and Drug Release. *Nanomaterials* **2021**, *11*, 16. [CrossRef]
66. Wu, K.; Liu, J.M.; Saha, R.; Ma, B.; Su, D.Q.; Chugh, V.K.; Wang, J.P. Stable and Monodisperse Iron Nitride Nanoparticle Suspension for Magnetic Diagnosis and Treatment: Development of Synthesis and Surface Functionalization Strategies. *ACS Appl. Nano Mater.* **2021**, *4*, 4409–4418. [CrossRef]
67. Zare, M.; Sarkati, M.N. Chitosan-functionalized Fe_3O_4 nanoparticles as an excellent biocompatible nanocarrier for silymarin delivery. *Polym. Adv. Technol.* **2021**, *32*, 4094–4100. [CrossRef]
68. Zarinwall, A.; Asadian-Birjand, M.; Seleci, D.A.; Maurer, V.; Trautner, A.; Garnweitner, G.; Fuchs, H. Magnetic Nanoparticle-Based Dianthin Targeting for Controlled Drug Release Using the Endosomal Escape Enhancer SO1861. *Nanomaterials* **2021**, *11*, 1057. [CrossRef]
69. Zhalechin, M.; Dehaghi, S.M.; Najafi, M.; Moghimi, A. Magnetic polymeric core-shell as a carrier for gradual release in-vitro test drug delivery. *Heliyon* **2021**, *7*, e06652. [CrossRef]
70. Zuk, M.; Gaweda, W.; Majkowska-Pilip, A.; Osial, M.; Wolski, M.; Bilewicz, A.; Krysinski, P. Hybrid Radiobioconjugated Superparamagnetic Iron Oxide-Based Nanoparticles for Multimodal Cancer Therapy. *Pharmaceutics* **2021**, *13*, 1843. [CrossRef]
71. Ali, T.H.; Mandal, A.M.; Heidelberg, T.; Hussen, R.S.D. Sugar based cationic magnetic core-shell silica nanoparticles for nucleic acid extraction. *RSC Adv.* **2022**, *12*, 13566–13579. [CrossRef]
72. Behr, J.; Carnell, L.R.; Stein, R.; Pfister, F.; Friedrich, B.; Huber, C.; Lyer, S.; Band, J.; Schreiber, E.; Alexiou, C.; et al. In Vitro Setup for Determination of Nanoparticle-Mediated Magnetic Cell and Drug Accumulation in Tumor Spheroids under Flow Conditions. *Cancers* **2022**, *14*, 5978. [CrossRef]
73. Bin Jang, S.; Jin, S.M.; Kim, H.S.; Jeong, Y.Y.; Lee, S.J.; Hahn, S.; Lee, H.; Lee, H.S.; Kim, J.H.; Lee, D.Y. DAMP-modulating nanoparticle for successful pancreatic islet and stem cell transplantation. *Biomaterials* **2022**, *287*, 121679. [CrossRef]
74. Gupta, R.; Gupta, P.; Footer, C.; Stenning, G.B.G.; Darr, J.A.; Pancholi, K. Tuneable magnetic nanocomposites for remote self-healing. *Sci. Rep.* **2022**, *12*, 10180. [CrossRef]
75. Ikram, H.; Al Rashid, A.; Koc, M. Synthesis and characterization of hematite (alpha-Fe_2O_3) reinforced polylactic acid (PLA) nanocomposites for biomedical applications. *Compos. Part C-Open* **2022**, *9*, 100331. [CrossRef]
76. Liu, C.H.; Lin, C.H.; Chen, Y.J.; Wu, W.C.; Wang, C.C. Multifunctional magnetic nanocarriers for delivery of siRNA and shRNA plasmid to mammalian cells: Characterization, adsorption and release behaviors. *Colloid Surface B* **2022**, *219*, 112861. [CrossRef]
77. Miola, M.; Verne, E. In situ reduction of Ag on magnetic nanoparticles with gallic acid: Effect of the synthesis parameters on morphology. *Nanomedicine* **2022**, *17*, 499–511. [CrossRef]
78. Mun, H.; Chaban, Y.; Tabish, T.A.; Thorat, N.; Cowieson, N.; Owen, C.D.; Townley, H.E. CD44 and CD221 directed magnetic cubosomes for the targeted delivery of helenalin to rhabdomyosarcoma cells. *Nano Res.* **2023**, *16*, 2915–2926. [CrossRef]
79. Mushtaq, S.; Shahzad, K.; Rizwan, M.; Ul-Hamid, A.; Abbasi, B.H.; Khalid, W.; Atif, M.; Ahmad, N.; Ali, Z.; Abbasi, R. Magnetoelectric core-shell $CoFe_2O_4$@$BaTiO_3$ nanorods: Their role in drug delivery and effect on multidrug resistance pump activity in vitro. *RSC Adv.* **2022**, *12*, 24958–24979. [CrossRef]
80. Nayak, J.; Prajapati, K.S.; Kumar, S.; Vashistha, V.K.; Sahoo, S.K.; Kumar, R. Thiolated β-cyclodextrin modified iron oxide nanoparticles for effective targeted cancer therapy. *Mater. Today Commun.* **2022**, *33*, 104644. [CrossRef]
81. Rocha, J.M.V.; de Souza, V.B.; Panunto, P.C.; Nicolosi, J.S.; da Silva, E.D.; Cadore, S.; Londono, O.M.; Muraca, D.; Tancredi, P.; de Brot, M.; et al. In vitro and in vivo acute toxicity of a novel citrate-coated magnetite nanoparticle. *PLoS ONE* **2022**, *17*, e0277396. [CrossRef]
82. Swain, S.K.; Phaomei, G.; Tripathy, S.K.; Yaiphaba, N.; Devi, R.B.; Nayak, S.; Parida, B.B. Effect of beta-cyclodextrin decoration on structural, optical and magnetic properties of luminescent magnetic nanoparticles and its application as a drug carrier. *J. Mol. Struct.* **2022**, *1247*, 131330. [CrossRef]
83. Szymczyk, A.; Drozd, M.; Kaminska, A.; Matczuk, M.; Trzaskowski, M.; Mazurkiewicz-Pawlicka, M.; Ziolkowski, R.; Malinowska, E. Comparative Evaluation of Different Surface Coatings of Fe_3O_4-Based Magnetic Nano Sorbent for Applications in the Nucleic Acids Extraction. *Int. J. Mol. Sci.* **2022**, *23*, 8860. [CrossRef]
84. Tang, X.Z.; Manamanchaiyaporn, L.; Zhou, Q.; Huang, C.Y.; Li, L.H.; Li, Z.Q.; Wang, L.C.; Wang, J.N.; Ren, L.; Xu, T.T.; et al. Synergistic Integration and Pharmacomechanical Function of Enzyme-Magnetite Nanoparticle Swarms for Low-Dose Fast Thrombolysis. *Small* **2022**, *18*, 2202848. [CrossRef]

85. Toyos-Rodriguez, C.; Llamedo-Gonzalez, A.; Pando, D.; Garcia, S.; Garcia, J.A.; Garcia-Alonso, F.J.; De la Escosura-Muniz, A. Novel magnetic beads with improved performance for Alzheimer's disease biomarker detection. *Microchem. J.* **2022**, *175*, 107211. [CrossRef]
86. Treder, N.; Roszkowaka, A.; Oledzka, I.; Baczek, T.; Plenis, A. Effects of Fe_3O_4 Magnetic Nanoparticle Functionalization with Ionic Liquids and a Double-Chained Surfactant on the Pretreatment of Plasma Samples during Drug Extraction. *Anal. Chem.* **2022**, *94*, 16587–16595. [CrossRef]
87. Urquizo, I.A.F.; Garcia, T.C.H.; Loredo, S.L.; Galindo, J.T.E.; Casillas, P.E.G.; Barron, J.C.S.; Gonzalez, C.C. Effect of Aminosilane Nanoparticle Coating on Structural and Magnetic Properties and Cell Viability in Human Cancer Cell Lines. *Part. Part. Syst. Charact.* **2022**, *39*, 2200106. [CrossRef]
88. Valdivia, V.; Gimeno-Ferrero, R.; Leal, M.P.; Paggiaro, C.; Fernandez-Romero, A.M.; Gonzalez-Rodriguez, M.L.; Fernandez, I. Biologically Relevant Micellar Nanocarrier Systems for Drug Encapsulation and Functionalization of Metallic Nanoparticles. *Nanomaterials* **2022**, *12*, 1753. [CrossRef]
89. Zohreh, N.; Karimi, N.; Hosseini, S.H.; Istrate, C.; Busuioc, C. Fabrication of a magnetic nanocarrier for doxorubicin delivery based on hyperbranched polyglycerol and carboxymethyl cellulose: An investigation on the effect of borax cross-linker on pH-sensitivity. *Int. J. Biol. Macromol.* **2022**, *203*, 80–92. [CrossRef]
90. Beltrame, J.M.; Ribeiro, B.B.P.; Guindani, C.; Candiotto, G.; Felipe, K.B.; Lucas, R.; Zottis, A.D.; Isoppo, E.; Sayer, C.; de Araujo, P.H.H. Coating of SPIONs with a Cysteine-Decorated Copolyester: A Possible Novel Nanoplatform for Enzymatic Release. *Pharmaceutics* **2023**, *15*, 1000. [CrossRef]
91. Chauhan, M.; Basu, S.M.; Qasim, M.; Giri, J. Polypropylene sulphide coating on magnetic nanoparticles as a novel platform for excellent biocompatible, stimuli-responsive smart magnetic nanocarriers for cancer therapeutics. *Nanoscale* **2023**, *15*, 7384–7402. [CrossRef] [PubMed]
92. Jalali, S.; Moghadam, P.N.; Shafiei-Irannejad, V. Synthesis of Magnetic Nanocarrier Conjugated by Folate Based on Tragacanth and In Vitro Investigation of their Efficiency on Breast Cancer Cells. *Starch-Starke* **2023**, *75*, 2200092. [CrossRef]
93. Liu, L.; Yang, Z.J.; Liu, C.B.; Wang, M.Y.; Chen, X. Preparation of PEI-modified nanoparticles by dopamine self-polymerization for efficient DNA delivery. *Biotechnol. Appl. Biochem.* **2023**, *70*, 824–834. [CrossRef] [PubMed]
94. Nguyen, T.N.; Tran, Q.H.; Terki, F.; Charnay, C.; Dumail, X.; Reibel, C.; Cazals, G.; Valette, G.; Jay-Allemand, C.; Bidel, L.P.R. Aggregation of magnetic nanoparticles functionalized with trans-resveratrol in aqueous solution. *Discov. Nano* **2023**, *18*, 64. [CrossRef] [PubMed]
95. Kovrigina, E.; Poletaeva, Y.; Zheng, Y.; Chubarov, A.; Dmitrienko, E. Nylon-6-Coated Doxorubicin-Loaded Magnetic Nanoparticles and Nanocapsules for Cancer Treatment. *Magnetochemistry* **2023**, *9*, 106. [CrossRef]
96. Baldea, I.; Petran, A.; Florea, A.; Sevastre-Berghian, A.; Nenu, I.; Filip, G.A.; Cenariu, M.; Radu, M.T.; Iacovita, C. Magnetic Nanoclusters Stabilized with Poly[3,4-Dihydroxybenzhydrazide] as Efficient Therapeutic Agents for Cancer Cells Destruction. *Nanomaterials* **2023**, *13*, 933. [CrossRef]
97. Urquizo, I.A.F.; Tozcano, D.I.M.; Gómez, L.E.V.; Pérez, J.A.R.; González, C.C. Enhancing the Cytocompatibility of Cobalt-Iron Ferrite Nanoparticles Through Chemical Substitution and Surface Modification. *Adv. Mater. Interfaces* **2023**, *10*, 2300206. [CrossRef]
98. Toderascu, L.I.; Sima, L.E.; Orobeti, S.; Florian, P.E.; Icriverzi, M.; Maraloiu, V.-A.; Comanescu, C.; Iacob, N.; Kuncser, V.; Antohe, I.; et al. Synthesis and Anti-Melanoma Activity of L-Cysteine-Coated Iron Oxide Nanoparticles Loaded with Doxorubicin. *Nanomaterials* **2023**, *13*, 621. [CrossRef]
99. Xu, X.; Xiang, H.J.; Wang, Z.J.; Wu, C.J.; Lu, C.C. Doping engineering and functionalization of iron oxide nanoclusters for biomedical applications. *J. Alloys Compd.* **2022**, *923*, 166459. [CrossRef]
100. Upadhyay, K.; Tamrakar, R.K.; Thomas, S.; Kumar, M. Surface functionalized nanoparticle A boon to biomedical science. *Chem.-Biol. Interact.* **2023**, *380*, 110537. [CrossRef]
101. Popova, V.; Dmitrienko, E.; Chubarov, A. Magnetic Nanocomposites and Imprinted Polymers for Biomedical Applications of Nucleic Acids. *Magnetochemistry* **2023**, *9*, 12. [CrossRef]
102. Mulens-Arias, V.; Rojas, J.M.; Barber, D.F. The Use of Iron Oxide Nanoparticles to Reprogram Macrophage Responses and the Immunological Tumor Microenvironment. *Front. Immunol.* **2021**, *12*, 693709. [CrossRef] [PubMed]
103. Bilal, M.; Iqbal, H.M.N.; Adil, S.F.; Shaik, M.R.; Abdelgawad, A.; Hatshan, M.R.; Khan, M. Surface-coated magnetic nanostructured materials for robust bio-catalysis and biomedical applications-A review. *J. Adv. Res.* **2022**, *38*, 157–177. [CrossRef] [PubMed]
104. Bohara, R.A.; Thorat, N.D.; Pawar, S.H. Role of functionalization: Strategies to explore potential nano-bio applications of magnetic nanoparticles. *RSC Adv.* **2016**, *6*, 43989–44012. [CrossRef]
105. Mittal, N.; Kundu, A.; Pathania, A.R. A review of the chemical synthesis of magnetic nano-particles and biomedical applications. *Mater. Today Proc.* **2023**. [CrossRef]

coatings

MDPI

Review

Pitfalls and Challenges in Specific Absorption Rate Evaluation for Functionalized and Coated Magnetic Nanoparticles Used in Magnetic Fluid Hyperthermia

Nicusor Iacob

National Institute of Materials Physics, 077125 Magurele, Ilfov, Romania; nicusor.iacob@infim.ro

Abstract: In recent decades, magnetic hyperthermia (MH) has gained considerable scientific interest in cancer treatment due to its ability to heat tumor tissues deeply localized inside the body. Functionalizing magnetic nanoparticles (MNPs) with vector molecules via specific organic molecules that coat the particle surface has enabled targeting particular tissues, thereby increasing the specificity of MH. MH relies on applying radiofrequency (RF) magnetic fields to a magnetic nanoparticle distribution injected in a tumor tissue. The RF field energy is converted into thermal energy through specific relaxation mechanisms and magnetic hysteresis-driven processes. This increases the tumor tissue temperature over the physiological threshold, triggering a series of cellular apoptosis processes. Additionally, the mechanical effects of low-frequency AC fields on anisotropic MNPs have been shown to be highly effective in disrupting the functional cellular components. From the macroscopic perspective, a crucial parameter measuring the efficiency of magnetic nanoparticle systems in MH is the specific absorption rate (SAR). This parameter is experimentally evaluated by different calorimetric and magnetic techniques and methodologies, which have specific drawbacks and may induce significant errors. From a microscopic perspective, MH relies on localized thermal and kinetic effects in the nanoparticle proximity environment. Studying MH at the cellular level has become a focused research topic in the last decade. In the context of these two perspectives, inevitable questions arise: could the thermal and kinetic effects exhibited at the cellular scale be linked by the macroscopic SAR parameter, or should we find new formulas for quantifying them? The present work offers a general perspective of MH, highlighting the experimental pitfalls encountered in SAR evaluation and motivating the necessity of standardizing the devices and protocols involved. It also discusses the challenges that arise in MH performance evaluation at the cellular level.

Keywords: SAR evaluation pitfalls; cellular magnetic hyperthermia; SAR methodologies; magnetic nanoparticles; nanoparticle mechanical effects

1. Magnetic Hyperthermia as a Hope in Cancer Therapy

1.1. Introduction

Magnetic hyperthermia (MH) has emerged as a new approach to cancer therapy, raising hopes of finding an effective solution to this disease, which has spread rapidly throughout the world in the last century. On a microscopic level, cancer is induced by changes at the DNA level [1,2]. As a result, the regulatory processes that control cell growth and proliferation in different tissues are altered. DNA is a very complex and dynamic information system that maintains the proper functioning of living beings in the finest detail. It has self-error cleaning capabilities through specific repair mechanisms [3].

Damage to DNA information sequences caused by external factors can overcome these informational repair mechanisms, and a wave of errors can propagate through the system. This contributes to the birth of malignant cells in specific tissues [2,4]. One of the immune system's roles is to recognize and neutralize these abnormal cells, but its ability can also be overwhelmed by the amplitude of the phenomenon's occurrence. For this reason, cancer treatment is focused on two aspects: the genetic mechanisms that trigger the repair of DNA damage [5,6] and the destruction of cancer cells by cytostatic drugs or external factors such as heat. It is known that cancer cells are less resistant than healthy cells to a temperature increase above the physiological threshold [7]. Heat can be delivered to tumor tissue in various therapeutic approaches: ultrasonic ablation [8,9], laser ablation [10,11], or magnetic hyperthermia [12–16]. For the successful killing of all cancer cells in a tumor, homogeneous heat distribution and the optimal timing of the heating steps are essential [17]. If some parts of the diseased tissue remain alive, the process of cell proliferation will continue, and the tumor will grow again.

1.2. MH Application

MH is based on injecting a precise, dosed volume of physiological liquid containing well-dispersed MNPs. This may be performed directly into the diseased tissue or in the bloodstream if the nanoparticles are vectorized with specific molecules for tumor targeting. Concomitantly, an alternative magnetic field in the radiofrequency range is applied, and through intrinsic specific relaxation mechanisms or hysteresis loss, nanoparticles will dissipate heat in the surrounding biological environment. MH can be applied to various types of deep-seated tumors, thus gaining significant scientific interest. Magnetic hyperthermia can also be an adjuvant therapy along with other cancer treatment approaches, such as chemotherapy and radiotherapy [18,19]. Furthermore, MNPs may be used as nano-delivery platforms for specific drug molecules (cytostatic), allowing for the integration of the cytotoxic effects with those of hyperthermia [20–22]. Drug molecules used in cancer therapy cannot be directly chemically attached to the nanoparticle, and an organic layer with high affinity for specific ligands and drug molecules should cover the nanoparticle surface [23].

A key requirement for materials used in cancer treatment is biocompatibility. All final material products (nanoparticles, coating layer, and functional molecules) must be compatible with the human body. A wide class of biocompatible organic molecules (chitosan, dextran, different lipids and fatty acids, polyacrylic acids, polydopamine, starch, etc.) has been developed [24,25] providing chemical support for drug or marker molecules. In addition to the biocompatibility condition, magnetic hyperthermia must be applied with some limitations given by the negative effects of alternating magnetic fields in the human body. Radiofrequency magnetic fields induce electric currents in biological tissues, which may increase the local temperature. Brezovich, in 1998, established a criterion for the permissible limits of the direct application of AC magnetic fields to the human body: the product between frequency and intensity of the applied field (f·H) should not exceed the value of $4.85 \times 10^8\ Am^{-1}s^{-1}$ [26]. Furthermore, more permissive limits have been elaborated: $f{\cdot}H = 5 \times 10^9$ [27] and $f{\cdot}H = 9 \times 10^9\ Am^{-1}s^{-1}$ [28]. A recent work [29] recommended changing the Brezovich criterion with the $SAR_{max}\ [Wg^{-1}]$ parameter, as the maximum power absorbed per unit mass of a specific tissue. Theoretical justification is given by the calculation of the eddy-current-induced power in a muscle tissue in cylindrically symmetric and homogenous magnetic field approximation: $P_v(r) = \frac{1}{2}\sigma\pi^2{\mu_0}^2{H_0}^2f^2r^2$, where σ—electrical conductivity of the tissue; μ_0—vacuum permeability; and r—cylinder radius. The calculus relied on the magnetic field parameters used in clinical experiments described in [29], where tolerable power values for different body parts where higher than those predicted by the Brezovich criterion. Accordingly, the $P_v(r)$ and $SAR_{max} = P_v(r)/\rho_{muscle}$

(ρ_{muscle}—muscle density) calculated values for torso, neck, and thigh were not as divergent as the product $f \cdot H_0$, which was three to five times higher for the thigh than for the torso. In this calculation, the radius of the torso is considered 14 cm and that of neck and thigh 7 cm.

1.3. Cellular MH

At the cellular level, MH triggers apoptotic effects by localized thermal transfer mechanisms between MNPs and cell components (membrane, extracellular matrix, cytoskeleton, cytoplasm, nucleus, etc.). It is known that some parts of the cell are more sensitive to heat than others (nucleus and mitochondria) [30]. Hence, MH effects may be very peculiar at this microscopic level. Heat transfer from a particle to the nucleus may induce cell death faster than if all cellular volume is heated over the physiological limit. Cellular MH is strongly influenced by MNP localization inside the cell or in the extracellular matrix.

1.3.1. Nanoparticle Uptake and Localization

Mammalian cells possess complex uptake mechanisms adequate for different particle sizes and morphology [31,32]. Relatively small particles (<150 nm) are internalized through a specific mechanism of clathrin-mediated endocytosis, while those 250 nm to 3 μm in size are internalized through micropinocytosis and phagocytosis [33]. Different particle shapes make distinct angles between the cellular membrane and particles at cell adhesion sites, influencing their internalization [33]. It seems that particles with shape anisotropy are more easily internalized by cells than spherical ones [34].

Magnetic hyperthermia experiments performed in vitro on DX3 human melanoma cells incubated with citric acid-coated iron oxide nanoparticles and exposed to a variable magnetic field of up to 16.1 kA/m in strength at 950 kHz for 2 h, showed high internal uptake of nanoparticles, as demonstrated by transmission electron microscopy (TEM) measurements [35]. Another work [36] studied magnetic hyperthermia in vitro in a glial microtumor phantom incubated with polyacrylic acid-coated and lauric acid-coated Fe_3O_4 nanoparticles. TEM studies showed the partial internalization of the nanoparticles into vesicles distributed in the cytoplasm and the formation of MNP clusters attached to the cell membrane. The effects of magnetic hyperthermia treatment, performed at a frequency of 560 kHz and a field strength of 24 kAm^{-1}, were compared with those induced in cells by conventional heating in a water bath. Both hyperthermia and the water bath-based classical method induced the apoptosis process as measured by viability tests, but in the case of magnetic hyperthermia, the local damage at the cellular level was more pronounced than in the classical case, possibly due to the mechanical vibration of the nanoparticles under AC magnetic field excitation.

1.3.2. Extracellular/Intracellular Thermal and Mechanical MH Effects

Nanoparticles can be driven to specific locations within the cell, but they can also remain trapped in the extracellular matrix [37,38]. This led to the idea of investigating the local efficiency of MH in the intra- or extracellular space (Figure 1).

Comparative intracellular and extracellular MH experiments were performed on SK-Hep1 hepatocellular carcinoma cells incubated with polystyrene sulfonic acid-coated MNPs immediately after incubation and after 24 h. The MH results showed that nanoparticle localization in the extracellular matrix was more efficient than internalization in the cytoplasm [39]. In addition, other in vitro experiments have shown that the MH process is more effective at the extracellular level than at the intracellular level [40]. In contrast, in vivo experiments on tumors induced in mice showed that MH was more effective in the intracellular space than in the extracellular matrix, even when the temperature reached in the intracellular space was lower than outside the cell [41]. This was explained by the high temperature reached in the vicinity of the nanoparticles (several tens of degrees), which has a strong impact on the integrity of the organelle membranes by inducing local damage. Ratiometric luminescence thermometers based on Sm^{3+}/Eu^{3+} were developed to detect local temperature increases on the surface of nanoheaters or in specific parts of the cell. Significant temperature differences were found between the nanoparticles and the cellular environment in their immediate vicinity. This suggests that some functional parts of the cell are more sensitive than others, requiring small amounts of heat to trigger the chain of apoptotic processes [30]. Other strategies have used MNPs as immobilizer nanoheaters attached to the cell membrane to induce physical damage (pores) capable of allowing for the passage of drug molecules into the cytoplasmic space [42].

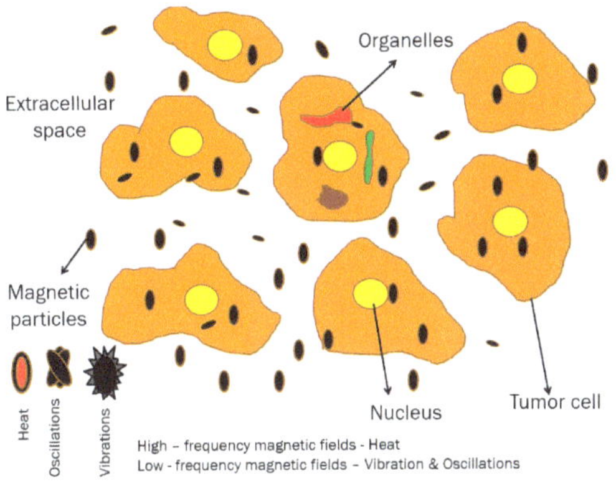

Figure 1. Localization of magnetic nanoparticles in both intra- and extracellular space. Under high- and low-frequency magnetic fields, particles generate heat, mechanical vibrations, and oscillations.

Specific assembly behavior was observed in a cellular MH experiment [43] where nanoparticles were aligned in chains. Furthermore, mechanical effects induced by vibration or oscillation under an AC magnetic field were highlighted by microscopic examination of nuclear debris after MH. Similar works [44–46] showed that low-frequency AC and dynamic magnetic fields induce mechanical forces mediated by MNPs of specific shapes (disks of 60 nm in thickness and 1 μm in diameter) inside the cytoplasm, which act on the cell membrane, nuclei, or various organelles, leading to cell disruption. The vibration and oscillation effects of nanoparticles are usually exploited in MH under low-frequency fields, where highly anisotropic (rod-shaped) particles of appreciable size (200 nm in length) have been designed for operation in 35 kHz magnetic fields [47,48]. Nanoplates of 1 μm in diameter have also attracted attention for use in very low and weak fields (20 Hz and 30 mT), showing lethal effects on cancer cells without significant heat release [49]. Dieny et al. [50] explained why superparamagnetic nanoparticles (SPM-NPs) cannot be used in this approach. SPM-NPs develop, under low AC fields, mechanical forces of the order of femtonewtons (fN), but those involved in biological processes vary between piconewtons (pN) and hundreds of piconewtons (pN). In in vitro experiments, U87 glioblastoma cells (brain cancer cells) were incubated with gold-coated $Ni_{80}Fe_{20}$ microdisks of 1.3 μm in diameter and 60 nm in thickness. After the U87 cells absorbed the nanodisks, a magnetic field of 400 mT with frequency ranges between 12 and 20 Hz was applied for 30–45 min. About 80% of the U87 cells were destroyed, and those who survived became nearly spherical, indicating that their cytoskeleton was damaged. Based on the same microdisks and cell type, in vivo experiments were performed without success, with the main cause being low diffusivity of the MNPs into the tumor. As was noted in [50], other in vivo experiments (on mice) using similar discoidal particles and types of cancer cells proved positive results.

1.4. MNP Synthesis

Few chemical methods allowing for the preparation of high-performance MNPs for MH exist. One is the thermal decomposition method, a particularly effective approach to synthesizing MNPs with high crystallinity, a narrow size distribution, and regular shapes. This method involves the thermal decomposition of organometallic precursors in organic solvents in an inert atmosphere [51]. It was proved that adding molecular oxygen enhances the magnetic properties of nanoparticles and improves the method's reproducibility [51]. The influence of synthesis parameters on nanoparticle size can be found in [52]. There, it has been demonstrated that the heating rate plays a pivotal role in the synthesis of monodisperse spherical iron oxide nanoparticles. A lower heating rate was found to promote polydispersity and size increase, while higher rates resulted in a broad nanoparticle distribution. The annealing time, dependent on the temperature plateau, was found to be crucial to obtaining a monodisperse particle distribution. Co-doped $Fe_{30}O_4/\gamma$-Fe_2O_3 core-shell nanoparticles with high SAR values (up to ~9300 Wg^{-1} for 14.6 nm particle size) were synthesized by thermal decomposition, proving the impact of inert flow gas on nanoparticle size and morphology [53]. However, the primary disadvantage of the thermal decomposition method is related to the milligram scale of nanoparticle production. In contrast, solvothermal synthesis represents a compelling alternative.

The solvothermal method involves heating a metallic salt solution in an appropriate solvent within a stainless-steel autoclave, along with nucleating agents and surfactants. Gavilan et al. [54] developed a scalable protocol for the preparation of high-crystallinity and monodisperse MNPs of various shapes (cubic, faceted, and spherical) at the gram scale by the optimization of the solvothermal method. The protocol utilizes alcohol as the solvent and employs shape-directing agents. The SAR values obtained, in their work, for 19 nm cubic MNPs exceeded 500 Wg^{-1}, up to 10-fold higher than that for similar-size MNPs synthesized by coprecipitation. In the alternative, if water is used as the solvent, the method is designated as hydrothermal.

Another highly used route is the microemulsion method, which relies on the coexistence of two immiscible liquid phases, water and oil. In this system, oil micelles form in water and act as chemical reactors. When a second emulsion or base is added, the precipitate forms, causing the micelles to collide, break apart, and consolidate, facilitating the growth of nanocrystals inside them. This method allows for the production of MNPs in the size range from 1 to 100 nm. By changing the oil phase, the reaction conditions and the type and amount of surfactant, different forms of MNPs can be obtained. An eco-friendly, oil-in-water microemulsion-based method was utilized to synthesize $Zn_{0.4}Fe_{2.6}O_4@SiO_2$ biocompatible clusters, which showed high SAR values (up to 2600 Wg^{-1}) in vitro magnetic hyperthermia experiments [55].

1.5. Magnetic Nanoparticle and Necrotic Tissue Clearance

Eliminating MNPs from the body is a complex process that depends on the particle size. The elimination of particles larger than 200 nm is primarily performed through the reticuloendothelial system (RES), with the liver, spleen, and lungs being the primary organs involved [56]. Particles smaller than 10 nm are eliminated through renal clearance [57]. In the range of 10–40 nm, particles are transported by the blood circulation, cross the capillary walls, and are phagocytized by macrophages [58]. For sizes over 50 nm, particles may evade the reticuloendothelial system, resulting in a prolonged circulation period [58]. An in vivo study [59] involving the injection of citrate-coated manganese ferrite into mice proved that the presence of nanoparticles in organs over time lasted 60 days in the liver, 30 days in the spleen, and about 12 h in the heart, lungs, kidneys, and blood. With regard to the elimination of necrotic cells resulting in MH, the predominant mechanism is phagocytosis [60], responsible for engulfing particles measuring over 0.5 μm (necrotic debris).

2. Physical Mechanisms Involved in Magnetic Hyperthermia

The principle behind magnetic hyperthermia is the generation of heat in nanoparticle systems, typically dispersed in fluid phases, under the influence of alternating magnetic excitation. This process depends on the magnetic mechanisms involved (hysteresis loss and superparamagnetic relaxation: Neel and Brownian relaxation) and the possible Joule effect in metallic nanoparticles due to the electrical currents induced by magnetic field oscillation.

Two categories of magnetic nanoparticles are distinguished: single-domain and multidomain nanoparticles. Bulk magnetic materials are divided into magnetic domains where magnetic spins are all oriented in a specific direction. The magnetic domains are separated by walls in which the spins gradually orient from one direction to another, corresponding to the two adjacent domains. At a specific size, a particle may experience a single magnetic domain where spins are all aligned in a particular direction, named the magnetic easy axis, defined by the magnetic anisotropy energy. Over a specific temperature (blocking temperature—T_B), spins may fluctuate coherently between the two directions of the easy axis—a phenomenon known as superparamagnetic behavior [61–63]. For magnetic single-domain nanoparticles subjected to AC fields, the mechanisms responsible for heat generation are hysteresis loss, described by the Stoner–Wohlfarth model, when the particles are in the magnetic frozen regime ($T < T_B$), and superparamagnetic relaxation, described by the Rosensweig model ($T > T_B$).

The Stoner–Wohlfarth model is a theoretical framework that elucidates the phenomenon of magnetization in single-domain nanoparticles. In this model, magnetic moments spontaneously align with preferential directions, characterized by effective anisotropy energy. The magnetic anisotropy energy can be defined as the energy required to rotate the magnetic moments from the easy axis (EA) direction to a direction that at 90° with respect to the EA, also named the hard magnetization direction.

The hysteresis loss mechanism in magnetic single-domain nanoparticles in the AC magnetic regime (Figure 2) occurs when the anisotropy energy, KV (K—anisotropy constant; V—particle's volume), is greater than the thermal energy, k_BT (k_B—Boltzmann constant; T—temperature). This happens when the particles are in the frozen magnetic state with all spins aligned with the easy axis and do not rotate as a whole rigid body. This is the case of high-viscosity fluids or when the particle's rotation is blocked by strong magnetic or physical interactions (e.g., dipolar interactions or organic molecules chain matrices). In the particular case of applied magnetic field aligned with the EA ($\alpha = 0$), the hysteresis loop is perfectly a square, and the coercive field is identical to the anisotropy field: $H_c = 2K_{eff}/\mu_0 M_s$ (Figure 2).

In the case of multidomain nanoparticles, the mechanism of heat generation is given by the hysteresis loss, but in contrast with single-domain particles, the coercivity is lower; hence, heat production is not as efficient. Nevertheless, experimental research studies on multidomain nanoparticle performance in MH [64] have been reported.

The power dissipation in non-interacting SPM-NPs dispersed in a liquid phase and magnetically excited by RF magnetic fields is described by the Rosensweig model [65]:

$$P = \mu_0 \chi_0 \pi f H_0^2 \cdot \frac{\omega\tau}{1 + (\omega\tau)^2} \quad (1)$$

where, $\mu_0 = 1.25 \times 10^{-6}\ \mathrm{N \cdot A^{-2}}$—vacuum magnetic permeability; H_0—field amplitude, χ_0—equilibrium susceptibility; f and H_0—magnetic field frequency and strength; and

τ—the effective relaxation time integrating contributions from Neel and Brownian relaxation processes (Figure 2):

- Neel relaxation when $T > T_B$ and magnetic moments fluctuate statistically and coherently around the nanoparticle's easy axis between the two energy minima. Neel relaxation is defined by relaxation time: $\tau = \tau_0 \exp(KV/k_B T)$, where τ_0 is a time constant, with $\tau_0 = 10^{-10}$ s. Neel relaxation phenomena take place in both dry and fluid-suspended SPM-NPs.
- Brownian relation when the magnetic moments are strongly bound to the nanoparticle (the case of high values of KV) and cannot be driven by the AC field. In this case, the particle rotates as a whole against the fluid viscosity resistance, being characterized by a relaxation time defined as $\tau_B = 3\eta V_H/k_B T$, related to the fluid viscosity (η) and particle's hydrodynamic volume.

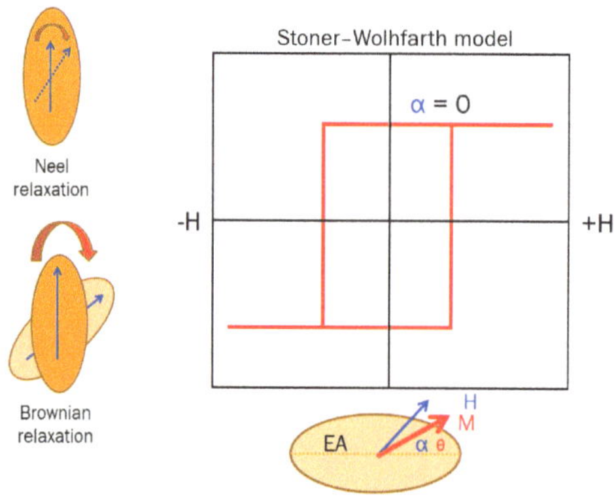

Figure 2. Neel and Brownian relaxation mechanisms in SPM-NPs. Hysteresis loop in Stoner–Wohlfarth model for applying magnetic field parallel with EA ($\alpha = 0$).

Hence, an effective relaxation time can be defined as $\tau = \tau_B \cdot \tau_N / (\tau_B + \tau_N)$.

An additional mechanism that theoretically may contribute to heat generation in MH is given by the magnetic field oscillations within metallic nanoparticles that induce electrical currents. Therefore, the electrical resistive loss may increase the particle temperature. This phenomenon depends on the magnetic and electrical properties of the material and the applied magnetic field parameters. The maximum power dissipation of eddy currents in fine metallic particles happens when the particle size is of the order of skin depth inside the particle [66]. Skin depth is defined by the depth in the material where the field is attenuated by 1/e and is expressed by the relation [67] $\delta = 1/\sqrt{\pi f \mu \sigma}$, where f—the field frequency; μ—magnetic permeability; and σ—electrical conductivity. For instance, the skin depth at MH frequencies in magnetic materials (low electrical conductivity) are of the order of microns, which is far from the usual particle size used in MH [68]. This is why this heat mechanism is not considered in magnetic hyperthermia.

3. SAR Evaluation Methods in Magnetic Hyperthermia

3.1. Specific Absorption Rate (SAR) Bioheat Equation

Almost all the attention in the MH research field used to be focused on single-domain MNPs, mainly due to the heat efficiency of superparamagnetic relaxation phenomena. Even if MH may have highly specific localized effects, as experiments have proved, its standard approach works in the approximation of homogeneous heat distribution in the tumor tissue. In this way, a physical quantity, called specific absorption rate (SAR), was introduced to quantify the amount of power released by MH mechanisms in the tissue mass unit:

$$SAR = P_{abs}/m = \left[W \cdot g^{-1}\right] \quad (2)$$

The SAR must have enough high values to compensate for the heat loss driven by the physiological thermoregulation processes, which try to reestablish the temperature at normal limits. MH may be numerically modeled in vivo by the so-called bioheat transfer equation (BHTE) [69]:

$$\nabla \cdot k \nabla T + q_p + q_m - W c_b (T - T_a) = \rho c_p (\partial T/\partial t) \quad (3)$$

where $T(^{\circ}C)$ and $T_a(^{\circ}C)$—the local temperature recorded inside the diseased tissue and the arterial temperature, respectively; $c_b(J/kg/^{\circ}C)$ and $c_p(J/kg/^{\circ}C)$—specific heat of the blood and of the tumor, respectively; $k(w/m/^{\circ}C)$—thermal conductivity of the tumor tissue; and $W(kg/m^3/s)$—the blood flow rate. The terms $q_p(w/m^3)$ and $q_m(w/m^3)$ quantify the power generated by the MH process and by the metabolism processes, respectively. MH can be optimized in relation with the q_p parameter for biological accepted limits. The term q_p is directly related to the SAR parameter by the relation

$$SAR = P_{abs}/m = c \times \Delta T/\Delta t = q_p/\rho \quad (4)$$

where ρ—the tumor tissue's density; c—the specific heat of its density; and $\Delta T/\Delta t$—the temperature increase rate. Another formulation of the SAR can be related to the mass of magnetic material spread into the tumor volume and is known under the name of specific loss power (SLP) [70]:

$$SLP = P_{abs}/m_{NP} = c \cdot m/m_{NP} \times \Delta T/\Delta t \quad (5)$$

where m is the tissue mass and m_{NP} is the mass of the MNPs contained by the investigated tissue.

The SAR factor can be evaluated in in vivo experiments (by monitoring the temperature increments by specific methods such as ultrasound echo measurement [71] or thermosensitive light emission effect [72]), in in vitro experiments, and directly in ferrofluid samples, where the temperature may be recorded with a simple optical fiber thermometer. Most SAR evaluation measurements are performed by calorimetry techniques in ferrofluid samples containing "fresh" synthesized nanoheater systems dispersed in liquids (water, physiological serum, or different oily phases). They are subjected to oscillating magnetic fields (usually in the radiofrequency range (RF) 50–1000 kHz). As a standard method, the temperature increase in ferrofluid samples is measured with optical fiber thermometers (metallic sensors are not allowed in RF fields), but IR imaging is also used [73]. The ferrofluid samples are placed inside circular coils connected to RF generators, which may be commercial (most of them) or home-made. Depending on the coil geometry and setting of the inductor capacitors, the working frequency can be adjusted. The SAR evaluation methods relied on calorimetric measurements involving recording time–temperature heating curves during a temperature range that includes the physiological point. Considering that the heat dissipated in the MH process is strongly dependent on the field parameters (f, H_0), a more specific loss power term, called intrinsic loss power (ILP), can be expressed independently of these parameters [74] as

$$ILP = SLP/f \cdot H^2 \left[H \cdot m^2/kg\right] \quad (6)$$

This is very useful in the evaluation of heat performance in the case of SMNPs where dissipated power depends on the square of field intensity according to the Rosensweig model. In the case of ferrofluids, SLP may be expressed as [75]

$$SLP = (\rho_{FF}/\eta \times \rho_{NP}) \times SAR \tag{7}$$

where η represents the ferrofluid volume fraction, and ρ_{FF} and ρ_{NP} are ferrofluid and nanoparticles densities.

Along with calorimetric methods, magnetic methods may also provide information about the heat efficiency of nanoparticles in MH application. The SAR may be seen as the product between the frequency of the applied magnetic field and the energy released in dispersion media during a field oscillation cycle. This energy can be evaluated through dynamical hysteresis measurements integrating the area of the magnetization loop over a complete field oscillation. Hence, the SAR may be written as [76]

$$SAR = f \cdot A = f \cdot \int_{-H_0}^{+H_0} \mu_0 M(H) dH \tag{8}$$

Another SAR evaluation magnetic method uses susceptibility measurements [77]. The volume of experimental SAR evaluation data has increased tremendously in the last decades, and a huge scientific effort has been made to evaluate the nanoparticle performance in MH. In the following, calorimetric and magnetic techniques and methodologies for evaluating the SAR parameter in magnetic fluid samples will be mentioned with a few concrete examples, highlighting their main advantages and disadvantages.

3.2. Calorimetric Methods in SAR Evaluation

3.2.1. General Aspects

As mentioned above, RF induction devices are commercially available, usually equipped with coils of different geometry parameters (diameter, length, and pitch). Depending on the coil geometry and capacitor configuration, these RF devices allow one to work with multiple frequencies and sample volumes and shapes. These coils are cooled with water or other special cooling liquids. Typical SAR measurements involve small volumes of ferrofluid enclosed in vials of a maximum of a few milliliters placed in the inner space of the RF coil. The oscillating magnetic field generated by the coil activates relaxation or hysteresis mechanisms in nanoparticles, leading to a temperature increase in the sample volume that is time-measured with an optical fiber thermometer. The shape of the heating curve acquired during an MH experiment is dictated by the competition between heating rates and loss rates (induced by thermal conduction, natural convection, and radiative processes).

Most of the experimental MH setups do not provide adiabatic conditions during the measurements, and this may induce imprecisions in SAR evaluation. Adiabatic environments around the ferrofluid samples are not trivial to build, particularly when the sample volume is very small (0.5–2 mL). Even sample holder walls may store consistent amounts of heat generated during the MH experiment under non-adiabatic conditions. Another important issue in MH experiments is related to large temperature gradients, especially generated when the induction coils are cooled by water at low temperatures. In [78], numerical simulations performed with commercial software Comsol Multiphysics 5.2 demonstrated the influence of cooling water temperature and water flow on the temperature environment inside the coil. If cooling water is used under 15 °C (usually from a standard tap), the water vapors from the surrounding air may reach the condensation temperature point (dew point) around the coil, and water drops on the coil surface may appear. In this case, the surrounding thermal transfer conditions are drastically changed. This can happen particularly in summer, when air humidity is high. For example, if the air temperature is 25 °C and indoor humidity is about 45%, the dew point will be 13.8 °C [79]. Technical solutions for improving the environmental thermodynamic parameters require enclosing the MH experimental setup in specially sealed walls where the inside air may be removed or dried before the experiments. However, even with these special experimental arrangements, in the case of using cooled water at low temperatures, high thermal gradients remain a major issue in evaluating the SAR, especially in the case of slow MH heat rates. Cooling the RF coils with water chillers at a precise temperature may solve the issue, but additional costs are involved. However, the cooling water temperature should be optimal, depending on the values of RF currents through the coil. Therefore, to avoid the heat thermal gradients around the sample, the cooling process should not allow the temperature to increase inside the coil metal. Numerical simulations found the optimal temperature for water cooling at 20 °C for the solenoid coil and the field parameters involved [80]. The used optical fiber thermometers acquire data from a single point (most likely from the middle of the sample's volume), even if the temperature at the sample extremities could be a few degrees lower. On the other hand, the coil geometry may induce inhomogeneities in the field distribution inside the sample, which, in turn, may induce heat nonuniformities. In this respect, RF coils with diameters fitting the sample's geometry, optimal length, and small pitches between turns should be used to ensure relatively uniform field distributions within the sample. Numerical simulations regarding the influence of coil geometry on heating efficiency in MH in vivo experiments were reported in [81]. In the work, various types of coils (helical, Helmholtz, and pancake coils) and field parameters were simulated, showing variation in field homogeneity inside the simulated model. A solution for this issue may involve using a single RF coil for different ferrofluid samples placed in the same type of vials, which should be identically positioned related to the coil geometry. The optical fiber sensor should also be placed carefully in the same position inside the ferrofluid volume for all samples. Some of these technical considerations were investigated by Makridis et al. in an experimental protocol [82]. They performed a comprehensive investigation of the influence of various experimental factors (MNP type, ferrofluid concentration, magnetic field inhomogeneities, coil geometry, etc.) and evaluation methods on SAR values. In this work, 20 nm Fe_3O_4 nanoparticles (SPM-NPs) and 40 nm Fe_3O_4 nanoparticles (FM-NPs) were subjected to MH experiments using three different commercial RF devices in a 10 min heating and 10 min cooling protocol. For FM-MNPs, the influence of the sample volume on SLP was investigated by keeping the same concentration (1 mg mL^{-1}) but changing the volume from 0.5 mL to 2 mL. For 1 mL, SLP increased with the sample volume, and over this value, SLP stabilized at 400 Wg^{-1}. Magnetic field homogeneity effect on SLP was investigated in the case of SPM-NPs using four coils of different geometries (No. of turns/diameter: 9 turns/50 mm; 9 turns/60 mm; 17 turns/50 mm; 17 turns/60 mm) for the same field parameters (10.4 kA^{-1} and 300 kHz). SLP values ranged from 35 Wg^{-1} to 41 Wg^{-1}, indicating relative homogeneity in the coil central region (1 cm^2). The optical fiber position in the ferrofluid sample also influences the SLP values, as it was proved in the above work. Accordingly, the recorded temperature varied from the highest value recorded in the middle of the sample to the bottom of the sample by 5%, to the sample surface by 7%, to the left side of the sample by 5% and to the right side of the sample by 2%. The effects of the size and shape of the sample vessel on SLP were also investigated in four cases: three vessels made of glass (with the same specific capacity of 800 J kg^{-1} K^{-1} and one vessel made from plastic with specific heat capacity of 1900 J kg^{-1} K^{-1}). All vessels had different shapes. The field parameters were 765 kHz and 20 kAm^{-1}. Each vessel had the sample (1 mL at 1 mg mL^{-1}) partially or fully contained in the coil inner space. The glass vessels had their sample's volume of 63%, 100%, and 100% contained inside the coil, and the plastic vessel had only 21% the sample volume inside the coil. This was reflected in the low SLP value of around 40 Wg^{-1} obtained for the sample contained by the plastic vessel compared with the glass vessels, where the SLP was found to be between 350 and around 500 Wg^{-1}. In the case of those two ferrofluid samples contained by glass vessels fully inside the coil, the SLP values varied between 350 and 400 Wg^{-1}. The maximum SLP of 500 Wg^{-1} was recorded for the ferrofluid samples contained in a proportion of 63% inside the coil. In this case, the vessel shape could be the cause.

These peculiar aspects commented here are not discussed in the MH literature enough, but they are experimental pitfalls, influencing the reported SAR results. Therefore, SAR evaluation by calorimetric techniques represents a challenge in these technical circumstances. Special experimental setups and innovative methodologies capable of mitigating their effects are required to allow comparable SAR results to be reported. Below are described different experimental approaches implemented in SAR evaluation to avoid or to compensate for heat loss.

3.2.2. Initial Slope (IS) Method (Sketched in Figure 3a)

The most used technique for trying to avoid the effects of heat loss is to record the temperature increase just for a small period of time at the beginning of the MH process with the assumption that the sample "does not have enough time" to lose heat in the surrounding environment. This technique analyzes only the initial part of the heating curve, which is properly fitted with the linear approach to extract the maximum heating slope. Hence, Equation (4) becomes

$$SAR = c \times |\Delta T/\Delta t|_{t \to 0} = c \times p_{initial} \quad (9)$$

The crucial condition involved here is related to the thermodynamic equilibrium between the sample and the external environment during the MH measurement. Multicore flower-like iron oxide MNPs of 23 nm and 56 nm coated with oleic acid subjected to 23.8 kAm^{-1} and 202 kHz MH experiments showed SAR values of 305 and 1540 Wg^{-1}, respectively [83].

3.2.3. Box–Lucas (BL) Method (Sketched in Figure 3b)

Another very popular method used for heat slope evaluation ($\Delta T/\Delta t$) is based on the fitting of the heating curve by the Box–Lucas model: $T = T_0 + T_{max}(1 - e^{-Bt})$, where T_0—the equilibrium temperature; T_{max}—the maximum heating temperature; t—the heating time; and B—a fitting constant. The product ($T_{max} \times B$) gives the maximum heating slope. Tithito et al. [84] reported SAR values of 9.44 Wg^{-1} calculated with the BL method for the field parameters of 6.03 kAm^{-1} and 130 kHz and (Mn, Zn) Fe_2O_4 SPM-NPs integrated in an apatite-based composite material. In [85], calorimetric MH experiments on monodisperse, 15 nm sized oleic acid-coated $CoFe_2O_4$ cubic nanoparticles were performed by using a commercial device (DM100—nBnanoscale Biomagnetics, Quantum design Europe, Les Ulis, France) and the BL protocol. The heating curves were recorded for a strength field of 15.89 kAm^{-1} and different frequencies (252 kHz, 323 kHz, and 397 kHz). Using the BL and IS methods SAR calculated values showed considerable difference between these two methods. For instance, at 323 kHz, SAR_{BL} was 104.18 Wg^{-1}, and SAR_{IS} was 65.86 Wg^{-1}.

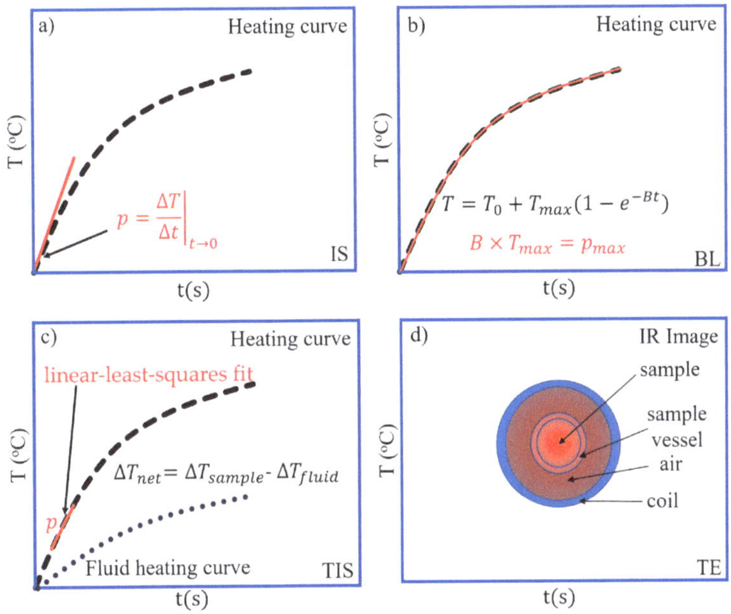

Figure 3. Sketches of the heating slope methods: (**a**) initial slope (IS) method; (**b**) Box–Lucas (BL) method; (**c**) temperature increase slope (TIS) method; (**d**) thermographic (TE) method.

3.2.4. Temperature Increase Slope (TIS) Method (Sketched in Figure 3c)

Another heating slope method was developed in [86]. In the work, the authors demonstrated consistent errors in SLP measurements even though the field parameters and sample's volume and shape were identical for each experiment. Coupled effects between the magnetic field inhomogeneities generated by the coil geometry and particle distributions in the sample volume were identified as main sources of errors. Therefore, they implemented a method that evaluates SLP from the maximum slope of the MH heating curve in few steps: (a) subtract the initial sample temperature; (b) subtract the temperature of the blank sample (fluid without nanoparticles) recorded under identical MH conditions; (c) make a linear–least-squares fit of an arbitrary chosen range from the temperature increase curve; (d) validate the quasi-adiabatic conditions met by the chosen time–temperature range through the inspection of the first derivative of the net temperature increase: $[\Delta T_{net}(t_{n+1}) - \Delta T_{net}(t_n)]/(t_{n+1} - t_n)$, where $t_{n+1} - t_n = 0.4$ s—the time step resolution. Along with the mean value of the rate of temperature rise, these values should not exceed 5% of the net temperature increase slope. The method was validated by MH experiments performed on different dextran- and citrate-coated magnetic water-based colloidal systems in a range of RF frequencies (150–375 kHz) and field intensities (4–44 kAm^{-1}). The average SLP reported values ranged between 108 and 383 Wg^{-1}.

3.2.5. Thermographic Evaluation (TE) Method (Sketched in Figure 3d)

The thermographic approach of MH was investigated in experiments were glucose-coated iron oxide 25 nm nanoparticles dispersed into different polyacrylamide gels, for emulating intracellular viscosity, were subjected to RF magnetic fields of 32 kAm^{-1} and 350 kHz [73]. The temperature behavior (heating and cooling curves) of the samples was monitored through a commercial thermographic camera before, during, and after the MH experiment. A sample of discoidal shape of 13 mm in diameter was inserted inside a holder adjusted in the inner space of an RF coil. The method evidenced a difference in the maximum temperature increase between the center and periphery of the sample of around 5 °C. The radial temperature distribution offers information about lateral thermal gradients.

3.2.6. Adiabatic Heating Curve Methods (Sketched in Figure 4a,b)

Iacob et al. [69,75] proposed two simple methodologies involving the recording of time–temperature behavior in both the MH heating regime and the cooling regime, where the sample is subjected to natural convection after the RF magnetic field is turned off. This can be performed continuously, where the temperature is recorded in the entire ranges of the heating and cooling processes (heating and cooling curves—HCCs), or in steps, where the temperature is recorded in successive short intervals in both heating and cooling regimes in correspondence to the intervals of the application of the magnetic fields (heating and cooling steps—HCSs).

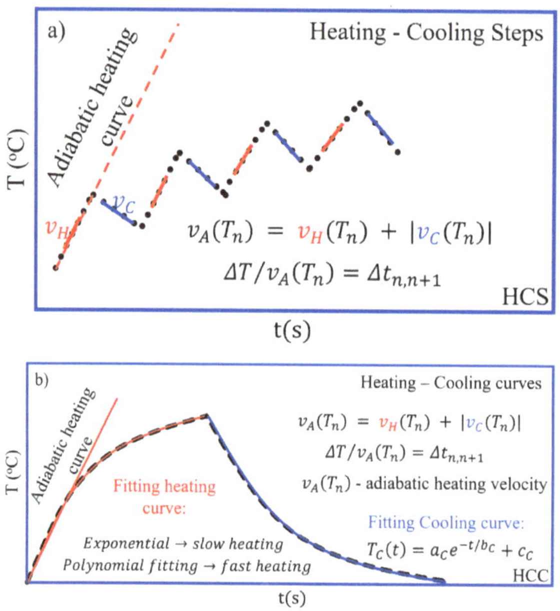

Figure 4. (**a**) Sketch of heating cooling step method and (**b**) sketch of heating cooling curve method.

In the case of the HCS method (Figure 4a), the experimental points are fitted with linear functions in order to extract the heating and cooling velocities ($v_H(T)$ and $v_C(T)$) in correspondence to the temperature points [75]. The continuous dependence of $v_H(T)$ and $v_C(T)$ is further obtained by the proper fitting of the experimental points. The next step is to numerically compute the adiabatic heating velocity as $v_A(T_n) = v_H(T_n) + |v_C(T_n)|$ at all temperature points equally separated in arbitrary mode ($T_{n+1} - T_n = \Delta T =$ constant) in the experimental range ($[T_0, T_1, \ldots, T_n]$). The $\Delta T/v_A(T_n) = \Delta t_{n,n+1}$ ratio gives the time required for the temperature to increase between two consecutive points. The summation of all these time intervals $\sum_{n=0}^{n} \Delta t_{n,n+1}$ provides the time during which the temperature increases from T_0 to T_n. The generation of the adiabatic heating curve $T_n(t_n)$ is, therefore, possible by the inverse representation of the $t_n(T_n)$ values. Based on the experimental $v_A(T_n)$ values, the volume fraction dependence of the input power (q_P) from the bioheat equation was calculated in the MH temperature range. Accordingly, q_P ranged between 0.05 and 0.5 Wcm^{-3}.

In the case of the HCC method (Figure 4b), a more versatile way for calculating the adiabatic heating behavior based on experimental data was developed [69]. The heating and cooling temperature profiles are acquired continuously in the MH process and during natural convection after the magnetic RF field is turned off. The cooling temperature curve that contains information about heat loss due to thermal gradients around the ferrofluid sample can always be fitted by an exponential function: $T_C(t) = a_C e^{-t/b_C} + c_C$, with a_C, b_C, and c_C the fitting parameters. The heating curve may be fitted in the general case by polynomial functions, i.e., $T_H(t) = a_m t^m + a_{m-1} t^{m-1} + \cdots a_0$, with a_m the coefficients, but for low heating rates, it can be fitted by exponential functions: $T_H(t) = a_H e^{-t/b_H} + c_H$, with a_H, b_H, and c_H the fitting parameters ($b_H < 0$). In the case of low heating rates (under 0.1 °C/s), the derivative of the exponential heating and cooling profiles gives the temporal heating and cooling rates ($v_H(t)$ and $v_C(t)$). Next, by using a simple mathematical trick of eliminating the time variable that is either numerical or analytical between $v_H(t)$ and $T_H(t)$, and $v_C(t)$ and $T_C(t)$, the correspondence between heating and cooling rates (($v_H(T)$ and $v_C(T)$, respectively) and temperature is thus obtained: $v_H(T) = (c_H - T)/b_H$ and $v_C(T) = (c_C - T)/b_C$. By following the same procedure as in the temperature step approach, the experimental adiabatic heating curve $T_H{}^*(t)$ is constructed. In the case of high heating rates, where the heating curves are fitted by polynomial functions, the correspondence $t_n(T_n)$ can be simply found by solving the equation $a_4 t_n{}^4 + a_3 t_n{}^3 + \cdots a_0 = T_n$, where $(a_0, \ldots, a_4)$ are polynomial coefficients. The heating velocity $v_H(t_n)$ is obtained by numerically evaluating the derivative of the polynomial function at each t_n. But t_n corresponds to T_n, and the $v_H(T_n)$ dependence is found. Next, the adiabatic heating rate $v_A(T_n)$ and $T_H{}^*(t)$ are easily computed. In this case, the less linear profile of the adiabatic heating curve is evidenced. If the Rosensweig model (Equation (1)) is computed with the physical parameters of the real nanoparticle system as input values, the dissipated power P_n can be calculated at each consecutive point T_n of the experimental temperature range. Therefore, the heat dissipated becomes $Q = P_n \cdot \Delta t_{n,n+1} = m \cdot c \cdot \Delta T_{n,n+1}$, and $t_n(T_n)$, the time needed for the temperature to increase from T_0 to T_n, is obtained by summation: $t_n(T_n) = \sum_0^n \Delta(t_n, t_{n+1})$. In this way, a theoretical heating curve can be generated by the inverse representation of $t_n(T_n)$. These two methodologies, relied on the continuous and step MH approaches, are strongly validated by the overlapping of both heating profiles: the experimental adiabatic and the theoretical one in the case of SPM nanoparticle systems. Nevertheless, poorer overlapping is observed in the case of high heating rates (the samples with high nanoparticle concentration), possibly induced by changes in SPM nanoparticles' behavior resulting from magnetic dipolar interactions or due to thermal inertia that induces deviation from linear heat transfer behavior. Both experimental approaches used oleic acid-coated Fe_3O_4 SP-MNPs dispersed in a polar fluid (transformer oil [87]) in low- and high-volume fractions (0.004 and 0.15). The sample vials were positioned in a PVC tube with vacuum walls centered in a commercial 235 kHz RF coil setup. The main advantage of these methods is given by the simple construction of the experimental setup. The mathematical approaches are also not complicated.

3.2.7. Modified Law of Cooling (MLC) Method (Sketched in Figure 5)

A similar method that accounts for heat loss is described in [88]. Accordingly, from the theoretical perspective, the method is based on the modified law of cooling:

$$dQ/dt = mC_p dT/dt + UA(T - T_{env}) \qquad (10)$$

where Q—the thermal energy; m—the ferrofluid mass; C_p—the specific heat of the ferrofluid; U—the heat transfer coefficient; A—the effective surface area of the sample; T—the measured temperature; and T_{env}—the environment temperature. The first term on the right side of Equation (10) represents the experimental heat recorded during MH, and the second term is the heat lost to the environment. Divided by mC_p, Equation (10) may be written as

$$dT'/dt = dT/dt + k(T - T_{env}) \quad (11)$$

where T′—the adiabatic temperature (if the heat loss did not exist); $k = UA/mC_p$—a time constant. The k value is determined by the condition $dT/dt = 0$, the case when the RF field is turned off. By fitting the experimental heating curve with Equation (11), the adiabatic heating curve is subtracted. For more accurate SAR results, the reference heating curve from the blank ferrofluid sample was subtracted. By using the MLC model, Chalkidou et al. reported SAR values up to 400 Wg^{-1} in MH in vitro experiments.

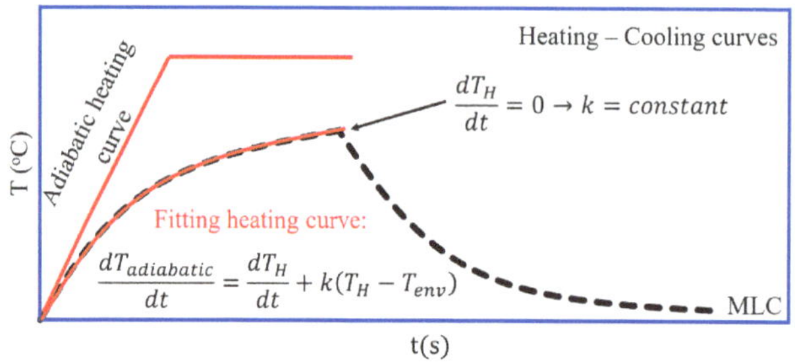

Figure 5. Sketch of MLC method.

3.2.8. Effective Thermal Conductance (ETC) Method (Sketched in Figure 6)

A general equation relies on a developed thermodynamic approach for determining the SLP factor containing terms corresponding to different thermodynamic regimes: adiabatic approximation, and non-adiabatic and non-radiating conditions; the isothermal case may be found in [89]

$$SLP = \frac{1}{m_{np}} C_{susp} \frac{dT}{dt} + \frac{1}{m_{np}} \in (T - T_{air}) + \frac{1}{m_{np}} \sigma\eta A_t \left(T^4 - T_{air}^4\right) \quad (12)$$

where m_{np}—the nanoparticle mass; C_{susp}—the heat capacity of the nanoparticle suspension; $\in$—the effective thermal conductance of the sample's surrounding environment; σ—the Stefan–Boltzmann constant; η—the emissivity; and A_t—the total sample surface. HM experiments performed on Fe_3O_4 and $MgFe_2O_4$ nanoparticles of different sizes dispersed in water under non-adiabatic and radiating conditions validated the equation with high accuracy. Curves for heating under the RF field conditions of 70.5 kHz and 70 Oe and cooling curves were recorded, and by their proper fitting, the $\in$ parameter and finally the SLP were obtained. The heating curves were fitted by Equation (12), and the cooling curves were fitted by the equation

$$T(t) = T_{air} + \Delta T_{max} e^{-\frac{\in}{C_{susp}} t} \quad (13)$$

which is written for the cooling process when the magnetic field is turned off. In this case, SLP = 0, and $\Delta T_{max} = T_{max} - T_{air}$ is the temperature difference between the sample and the environment at the moment when the field is suppressed. In respect of the size, the reported SLP values were greater for Fe_3O_4 nanoparticles (e.g., 0.86 Wg^{-1} for 13.4 nm particle size) than for $MgFe_2O_4$ nanoparticles (1.6 Wg^{-1} for 12.7 nm particle size).

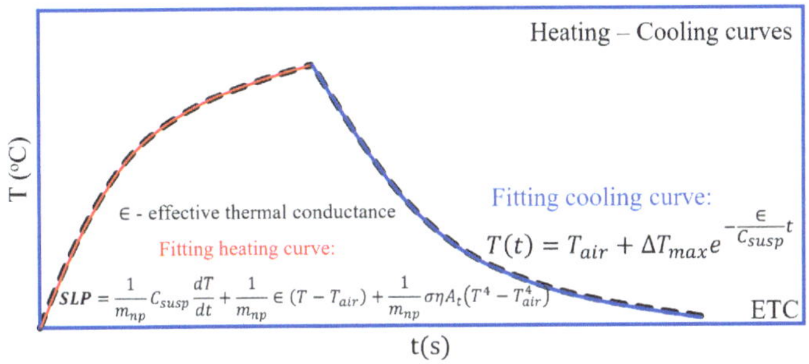

Figure 6. Sketch of ETC method.

3.2.9. Evaluation of Nanoheater Temperature (ENT) Method (Sketched in Figure 7)

Another SAR evaluation method under non-adiabatic conditions that computes heat loss based on recording both heating and cooling curves was developed in [90]. In the study, the MH setup is completed with a water shell surrounding the magnetic fluid sample based on Fe_3O_4 nanoparticles in order to protect the sample against possible short temperature variations from the environment. The entire holder is placed in a home-made RF device working at 100 kHz. The SAR is evaluated through a developed methodology relying on solving a set of coupled differential equations describing the heat exchange between MH setup components:

$$P - \frac{c_s m_s}{\tau_s}[T_s(t) - T_w(t)] = c_s m_s \frac{dT_s(t)}{dt} \tag{14}$$

$$\frac{c_s m_s}{\tau_s}[T_s(t) - T_w(t)] - \frac{(\tau_{w1} + \tau_{w2}) c_w m_w}{\tau_{w1}\tau_{w2}}[T_w(t) - T_a] = c_w m_w \frac{dT_w(t)}{dt} \tag{15}$$

where P—the power term; c_s and m_s—specific heat and mass of the heat source (nanoparticles); c_w and m_w—specific heat and mass of water; τ_s—the time constant of the heat exchange between the nanoparticles and water; T_s—the nanoparticles' temperature; T_w—the temperature of water; τ_{w1} and τ_{w2}—the time constants of exponential functions that fit the cooling curve of the magnetic fluid sample under natural convection conditions; and $T_a = 23.5$ °C—the equilibrium temperature of the environment. The methodology exploits the thermal equilibrium condition attained during the MH process when the heat released by the nanoparticles is completely lost in the environment. In this situation, the system of equations is reduced to

$$P = \frac{(\tau_{w1} + \tau_{w2}) c_w m_w}{\tau_{w1}\tau_{w2}}[T_M - T_a] \tag{16}$$

where T_M—maximum temperature reached in the heating process when the sample enters the thermal equilibrium. It brings some complexity regarding the experimental setup and mathematical methodology, also involving a calibration procedure, but has, as a main advantage, the possibility of estimating the temperature of the magnetic nanoheaters dispersed in the fluid sample. Based on this model and by using two Fe_3O_4 nanoparticle samples, one of 50–100 nm particle size and other of 10–20 nm particle size, the calculated SAR has an approximatively linear dependence in respect of the field amplitude for both samples, reaching higher values for the larger particles (about 170 Wg^{-1} for 90 mT).

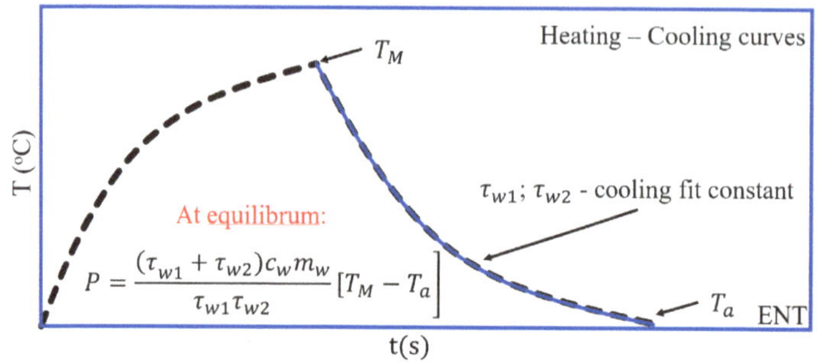

Figure 7. Sketch of ENT method.

3.2.10. Peak Analysis (PA) Method (Sketched in Figure 8)

Based on a heat diffusion equation, the SLP was calculated by a new method called the Peak Analysis Method (PAM) using a 1D temperature diffusion model and a zig-zag protocol of intermittent heating and cooling steps [91] (similar to [75]). Accordingly, the temperature variation in the 1D of the MH system can be written as $\partial \Delta T/\partial t = \propto \left(\partial^2 \Delta T/\partial x^2\right) + S$, where $\propto = k/\rho c$—the thermal diffusivity (k—the thermal conductivity; ρ—the sample density; and c—the heat capacity); $S = SLP/\rho c$—the heating source; and $\Delta T = T - T_{env}$. The method calculates SLP at the point (peak) of the transition between the heating and cooling curves. Around this peak, the effects of loss are minimized, and the temperature profile is practically unchanged. Therefore, the source heat may be written as $S = |\partial \Delta T/\partial t|_{heating} - |\partial \Delta T/\partial t|_{cooling}$. The experimental validation of this approach was performed by using three devices with the same field parameters (~165 kHz and 35 mT) with which a 1 mg/mL magnetite nanoparticle suspension was heated. The temperature was acquired with an optical fiber thermometer. The SLP values were compared with those obtained from a single heating–cooling cycle experiment recorded on the same devices. The SLP values (around 300 Wg^{-1}) were consistent for all three devices in the zig-zag protocol, drastically reducing the errors between the devices and the measuring time of the SLP evaluation.

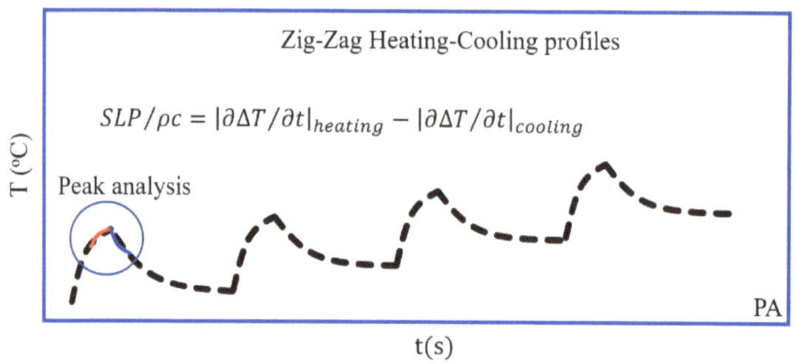

Figure 8. Sketch of PA method.

3.2.11. Pulse Heating (PH) Method (Sketched in Figure 9)

In [92], an alternative method for SAR evaluation in MH experiments claims high precision due to non-transient measurements. The experimental setup is complicated, providing the almost adiabatic conditions accomplished by maintaining a controlled thermal equilibrium between the environment and the fluid sample. The method is based on applying consecutive AC magnetic pulses that generate heating ramps in a ferrofluid sample. It is assumed that the entire the heat generated during a heating pulse remains in the sample, allowing for the measurement of the adiabatic temperature increment (ΔT). The method has, along with precision, the advantages of measuring low SAR values. This was experimentally proved by heating, under 108 kHz and 1.1 kAm^{-1} field conditions, a ferrofluid sample of Fe_3O_4 nanoparticles of 50 nm in hydrodynamic diameter and 100 mg mL^{-1} concentration. Each heating pulse was delivered in 600 s and generated a temperature slope of 2×10^{-4} °C/s, corresponding to a low SAR value of 0.217 Wg^{-1}.

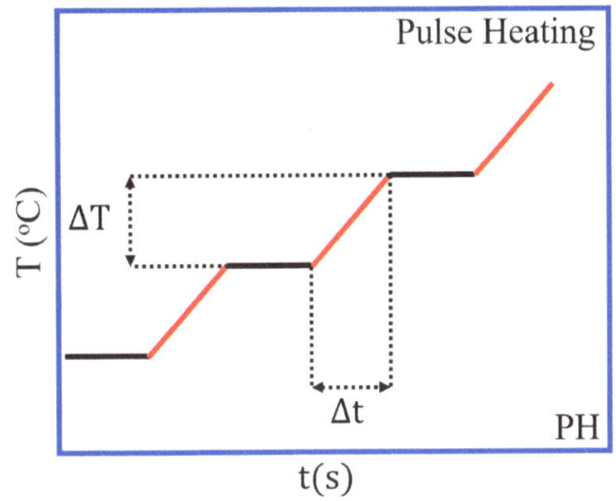

Figure 9. Sketch of PH method.

3.2.12. Discussion

The heat slope (IS, BL, and TIS) methods do not imply a complex experimental setup and do not account for heat loss, which is very important in MH experiments that involve very low heating rates. IS and BL are quite simple-to-compute methods, but the TIS method requires some computational effort. However, all these methods work very well in the case of significant heating rates. Both the BL and TIS methods offer higher accuracy compared with the IS method, but the BL method may be preferable due to simplicity. Good thermal insulation of the sample is recommended for all these methods.

The HCS and HCC methods do not involve complicated experimental conditions and computation. They allow for the reconstruction of the adiabatic heating curve and work very well for low heating rates (under 0.1 °C/s). At high heating rates, the methodology cannot accurately predict the entire MH adiabatic heating behavior. However, the HCC method is easier to implement. The MLC method is a similar method that allows for the reconstruction of the experimental adiabatic MH profile and is not hard to implement experimentally. The method proved to have high accuracy. The ETC method has been experimentally proved to have high accuracy in SLP evaluation, but its disadvantage is the inability to reconstruct the experimental adiabatic curve. However, this aspect may be useful only when nonlinear heating behaviors exist. The ENT methodology developed can calculate the temperature of nanoparticles dispersed in ferrofluid. The complexity of the experimental setup and evaluation methodology gives the method a disadvantage. Based on applying consecutive AC magnetic pulses, the PH method claims high precision even in the case of low SAR values (as experimentally proved). The disadvantage of the PH method is given by the difficulty of its experimental implementation. The PA method, along with the zig-zag protocol, demonstrated high accuracy in SLP evaluation, compared with other methods, but not easy experimental implementation. The TE method is a simple and interesting method with the advantage of the recording radial thermal gradients. For high-accuracy results, the method requires quasi-adiabatic protection of the sample and a high-resolution camera. However, there is a problem with the implementation of thermal protection at the sample side towards the IR camera. A low-thickness sample would have the advantage of eliminating the volume thermal gradients.

Regarding the calorimetric evaluation methods, there is an important issue related to the accuracy of the reported SAR results. Different evaluation methods involving different experimental setups and mathematical approaches provide SAR values that are more or less accurate. The errors induced by the method principles cumulate with errors induced by the thermal transfer between setup components and the magnetic fluid properties [82,91,93]. A possible solution may be the standardization of these calorimetric methods, but this cannot be successful without the standardization of the physical equipment involved in MH experiments. In the case of well-established experimental techniques, such as X-ray diffraction, electron microscopy, and magnetometry, standardized commercial devices are utilized. The question, therefore, arises as to why standardized devices for SAR measurements are not employed in the context of magnetic hyperthermia. While there are indeed such devices on the market, they are limited in terms of the range of models and available options. It is very important that for the same field parameters set in every MH experiment performed around the globe, the RF coil has the same geometry and surrounding environment. This can be accomplished only with standardized MH equipment. This equipment should come with integrated working protocols (regarding sample concentration, ferrofluid stability, and type of coating), calculation methodologies (proper methods for specific heating rates), and possible calibration samples.

3.3. Magnetic Methods in SAR Evaluation

Dynamic magnetic hysteresis measurements performed in the MH frequency regime bring high accuracy in SAR evaluation [76,94–99], but commercial devices that deliver proper field intensities are not available. Instead, diverse locally implemented solutions have been built with good results [94,95,98]. Comparative analyses with calorimetric methods have been performed, indicating good agreement [94,95]. The advantages of this technique are mainly given by the speed and accuracy of measurements. Drawbacks are given by the impossibility of measuring hysteresis loops in the case of SMNPs. For that, AC susceptometers working at MH frequencies have been built in different laboratories because, commercially, they are not available [77,100,101]. Most commercial devices (e.g., PPMS by Quantum Design) for magnetic measurements allow for χ'' measurements as a function of temperature and field parameters ($f \leq 10$ kHz and $H_0 \leq 1.2$), which are not useful in the range of MH. Each magnetometer technique has advantages and disadvantages regarding the type of nanoparticles that can be investigated. These techniques also are not standardized, allowing for errors between different experimental setups that may arise from the quality of electronic components, calibration procedures, and working protocols. Hence, building standardized AC magnetometers/susceptometers for MH application on a large scale is clearly a necessity.

4. Effects of Coating Layer on SAR

The hydrodynamic volume of a particle is strongly influenced by the thickness of the organic coating layer, thereby reducing the magnetic dipolar interactions and avoiding particle agglomeration. This is very important in preserving the superparamagnetic behavior in fine MNP assemblies. On the other hand, the organic layer may suppress Brownian relaxation. The organic layer may also chemically interact with the magnetic atoms on the particle surface, creating a disordered magnetic shell and affecting saturation magnetization. In these circumstances, the organic coating layer exerts a significant influence on MNP heat efficiency. Spherical magnetite nanoparticles with a mean diameter of 13 nm coated with different surfactants (citric acid (CA), dextran (DEX), and (3-aminopropyl) trimethoxysilane (APTES)) were subjected to MH calorimetric experiments with the RF field parameters of 15.9 kAm^{-1} and 250 kHz [102]. Furthermore, the SAR was evaluated in naked Fe_3O_4 nanoparticles. The hydrodynamic diameter was found to be 20.8 nm for the naked Fe_3O_4 nanoparticles, 46.9 nm for Fe_3O_4@CA, and 76.6 nm for Fe_3O_4@APTES. The Fe_3O_4@CA and Fe_3O_4@APTES samples exhibited relatively appropriate SAR values of 65.8 Wg^{-1} and 67.2 Wg^{-1}, respectively, compared with the uncoated Fe_3O_4 nanoparticles (63.4 Wg^{-1}). The Fe_3O_4@DEX sample demonstrated a lower SAR value (55.6 Wg^{-1}). The Fe_3O_4@CA and Fe_3O_4@APTES nanoparticles demonstrated higher SAR values in comparison to the uncoated ones due to better dispersion stability. The Fe_3O_4@DEX nanoparticles exhibited the highest hydrodynamic diameter and the thickest organic layer, which exerted a significant influence on Brownian loss and the SAR value, respectively. Furthermore, the high thickness of the DEX layer induces magnetic disorder at the particle surface, decreasing the saturation magnetization and, correspondingly, the SAR [102].

MH experiments conducted with varying field parameters in coated and uncoated iron oxide SPM-NPs demonstrated, by the Box–Lucas method, a decrease in SAR values in the case of the coated particles [103]. In the study, the dextran-coated Fe_3O_4 nanoparticles with a hydrodynamic diameter of 352 nm exhibited significantly lower SAR values (247 Wg^{-1}) compared with the naked Fe_3O_4 nanoparticles with a hydrodynamic diameter of 52 nm(997 Wg^{-1}). This reduction in the SAR is attributed to a decrease in Brownian motion. The MH heat efficiency of Fe_3O_4 nanoparticles coated with sodium citrate and carboxymethyl cellulose (CMC) was investigated through dynamic hysteresis measurements performed at 69 kHz for 13.3 and 24.8 kAm^{-1} and calorimetric measurements performed at 100 kHz and 48 kAm^{-1} [104]. The dispersion of the citrate–Fe_3O_4 nanoparticles in solution and their smallest hydrodynamic mean diameter were found to be directly related to the highest SLP values (99.8 Wg^{-1}) obtained from calorimetric measurements. In contrast, the naked Fe_3O_4 nanoparticles with a hydrodynamic diameter of 135 nm demonstrated complete sedimentation and a SAR value of 60.4 Wg^{-1}. Furthermore, the CMC-coated Fe_3O_4 nanoparticles with a hydrodynamic diameter of 152 nm exhibited a SAR value of 50.8 Wg^{-1}. The dynamic hysteresis measurements shown the largest loop area for citrate–Fe_3O_4 and the smallest for uncoated Fe_3O_4 nanoparticles. The magnetic measurements do not account the Brownian relaxation contribution.

Another important factor that influences MH performance is related to the heat transfer capacity of the coating layer. Various coating molecules and their geometrical configurations around the nanoparticle define the heat transfer behavior between the magnetic core and the environment. In [105], a study on the influence of the thermal diffusivity of different coating molecules on the SAR is performed. Magnetite nanoparticles of around 10 nm in size coated with palmitic (PA) and stearic (SA) acids were subjected to MH experiments under similar RF field conditions. The SAR values obtained for SA were 36.1%, higher than in the case of PA. Numerical simulations generated similar SAR values for setting the thermal diffusivity of SA at a 1.7 value, higher than for PA. Based on numerical simulations, another work [106] showed that both polymer layer thickness and its density influence MH performance. It was demonstrated that for a given coating thickness, the SLP has a polynomial dependence with respect to the surface layer density. The SLP slightly decreased with the increase in layer thickness, reached a minimum value and increased again with the increase in thickness. This behavior was explained by the competition between the steric forces (repulsion) and magnetic dipolar forces (attraction). This allows one to tune MH by controlling the thickness of the polymer coating.

Lazaro et al. [93] reported MH experiments using naked and different-polymer-coated magnetite/maghemite nanorods of 40 nm. The calculated SAR values for the same frequency of 144 kHz and different field amplitudes showed higher values for the naked particles than for the coated ones. The SAR values differences between the naked and coated particles increased with the field amplitude (at 23 kAm^{-1}, the SAR for the naked particles was around 150 Wg^{-1}, and that for the coated particles was two times smaller). Keeping the field amplitude at the same value of 17 kAm^{-1}, at different frequencies, the SAR values were relatively unchanged for the coated nanorods (about 15 Wg^{-1}), but the naked nanorods' SAR showed significant higher values, which dramatically changed with frequency and reached 80 Wg^{-1}.

5. SAR at Cellular Level

Nanoparticle clustering in tumor cells has been experimentally observed, particularly when the particles are functionalized for targeting specific cellular components [36]. Therefore, optimizing the injected magnetic fluid doses, particle volume fraction, and their shape and size emerges as a pivotal task to preserve the heat dispersion efficiency of the nanoparticles utilized in MH. Conversely, a specific geometrical configuration of a nanoparticle cluster might prove advantageous in facilitating optimal heat transfer to the targeted cellular component to hinder its functionality. As previously mentioned, the cellular effects of magnetic hyperthermia are predominantly influenced by the heat transfer in the proximity of the nanoparticle. A cluster of nanoparticles measuring a few nanometers is comparable in dimensions to subcellular components. From this perspective, the size and shape of the cluster may be optimized with respect to the size of the cellular component to ensure efficient heat transfer. Additionally, large particles (hundreds of nm) that can be internalized by cells and come into physical contact with the thermosensitive part of the cell are of particular interest in cellular hyperthermia [16,107]. These particles dissipate thermal energy through hysteresis loss, exhibiting, in general, a lower SLP factor compared with SPM-NPs. However, the total thermal energy transferred to a vital component of the cell may exceed the minimum necessary for it to lose its functionality. Another type of nanostructure that is of interest in the context of cellular hyperthermia is the nanoflower [83,108,109], which possess particular shapes and complex geometries, allowing it to attach to cellular components in specific ways that lead to amplifying the heat transfer. In this regard, the significance of the SAR/SLP/ILP diminishes when assessing the consequences of hyperthermia at the subcellular level. Cellular MH is a new and more focused research direction in cancer hyperthermia that requires advanced knowledge in the biology and biochemistry fields regarding the physical and chemical interactions between magnetic nanostructures and cellular components.

6. Conclusions and Remarks

Magnetic hyperthermia constitutes an alternative therapeutic approach for the treatment of cancer. Its principal mechanism of action involves the generation of heat in a tumor tissue (by embedding a system of MNPs) subsequent to the application of a radiofrequency field. The MNPs release heat by specific magnetic relaxation processes, Neel and Brownian relaxation [65], or hysteresis losses [76,94–99]. It is well established that cancer cells exhibit heightened sensitivity to temperature increases that surpass physiological thresholds, leading to a process known as apoptosis.

In the classical MH approach, the implementation at the clinical level necessitates prior knowledge of the magnetic material distribution within the tumor tissue and the magnetic field parameters that can be utilized under safe patient conditions. This can be achieved through the theoretical and experimental modeling of magnetic hyperthermia. From a theoretical standpoint, the modeling of magnetic hyperthermia can be achieved through the utilization of the BHTE equation [75], contingent upon the availability of data pertaining to the power dissipated in the tissue (q_P) by the mechanisms of magnetic hyperthermia and the bio-thermodynamic parameters of the tissue in normal physiological metabolism. To determine the q_P term, it is necessary to employ laboratory experimental methods involving calorimetric and magnetic measurements.

Calorimetric measurements face limitations in directly evaluating the q_P term due to their inability to maintain adiabatic conditions, which are required to avoid heat loss. In a real, non-adiabatic system, the most straightforward and accessible method to circumvent the impact of heat dissipation is to consider the initial heating slope (IS) solely [83] of the temperature curve recorded during magnetic hyperthermia. For higher SAR evaluation accuracy, similar methods (BL and TIS) relying on finding the maximum heating slope have been developed [84–86] Despite their versatility, the IS, BL, and TIS methods have drawbacks, particularly for slow magnetic hyperthermia processes, where the ferrofluid samples may lose considerable heat with respect to that produced. Another strategy is not to avoid the heat loss but to record it to reconstruct the experimental adiabatic heating curve. For this purpose, various simple or advanced heat loss calorimetric methods providing high accuracy results [69,75,88–92] have been developed based on the consideration of the heat loss induced by the thermal conduction between the experimental setup components, natural convective cooling, and radiative loss. If the heat slope methods give good results in the case of significant heating rates, the heat loss compensation methods could be complementarily used in the case of low heating rates (<0.1 °C).

Magnetic methods (dynamic hysteresis and susceptibility measurements) are versatile and provide accurate and rapid information on the MNP response in alternating magnetic fields [76,94–99]. The principal disadvantage of these techniques is that magnetometer devices operating in the range of magnetic hyperthermia field parameters are not commercially available, only locally constructed in the laboratory, but provide remarkable results.

Unfortunately, MH experiments have proved important variations in SAR results collected with different evaluation methods and devices [82,91,93]. Therefore, it is hard to believe that all these evaluation methods and methodologies are able to allow for a comparison with high-accuracy SAR values obtained from different laboratories around the world without involving standardized measurement equipment and protocols. The standardization of the equipment and working protocols represent the unique way to avoid the pitfalls given by a wide class of experimental factors, i.e., various geometries used in coil building, their cooling systems, ambient thermal conditions, the volume and shape of the sample's holder and its material composition, ferrofluid volume fraction and suspension stability, etc., in the case of the calorimetric approach and different electronic parts, constructive design, and calibration protocols in the case of the magnetic approach.

At the cellular level, heat dissipation during hyperthermia exerts a specific influence on the growth and multiplication capacity of tumor cells. While the underlying mechanisms remain to be fully elucidated, ongoing research endeavors seek to identify the heat-sensitive cellular components that could expedite the process of apoptosis following magnetic hyperthermia treatment. In this regard, a novel trend in the study of magnetic hyperthermia entails targeting specific subcellular structures within the cytoplasmic environment or extracellular matrix [30–41]. This approach aims to localize the effect of magnetic hyperthermia and optimize the utilization of MNPs and applied fields. This objective can be achieved by binding molecular vectors with chemical affinity for specific cell receptors to the nanoparticle surface. In addition to the effects of cellular magnetic hyperthermia, scientific interest has shifted to the study of the kinetic effect [45–47,49,50] of nanoparticles on subcellular components by applying low-frequency magnetic fields to nanostructures with high magnetic anisotropy (nanodisks, nanorods, nanoflowers, etc.).

The oscillations and vibrations of these particles have the potential to induce local cellular damage, thereby triggering apoptosis processes.

Considering the classical aspects of magnetic hyperthermia, but also the new trend to focus the effect of MNPs on cell growth and proliferation mechanisms at the cellular level, questions inevitably arise: What is the relationship between the SAR/SLP/ILP and the local effect of magnetic hyperthermia on the cellular infrastructure? Does it still make sense to consider the power dissipated by MNPs in tissue a continuous macroscopic variable? Or should we focus our scientific interest on local thermal and mechanical effects at the subcellular level, looking for the optimal way to release the heat and kinetic energy of a single nanoparticle or a cluster of nanoparticles? In this case, we would need to find ways of expressing SLP/ILP for a single particle, incorporating the kinetic term. However, it is difficult to establish a link between the power (caloric or kinetic) dissipated by a single particle and its effect on cellular metabolic mechanisms. In this way, the new approach to magnetic hyperthermia becomes highly specialized, adding destructive mechanical effects to the thermal ones. Depending on the targeted cell part, combining these effects could be the safest way to approach magnetic hyperthermia. The temperature achieved in the proximity of the nanoparticle surface has become the new variable of interest in cellular hyperthermia and innovative ways of measuring it must be found. Most of the research in the MH field is performed by chemists and physicists, but a strong involvement of biologists and biochemists is crucial to understanding the fundamental cellular mechanisms of interactions between magnetic particles and cells. Therefore, designing optimal MH approaches of high precision and efficiency without side effects is necessary. It is like in a modern war: instead of using a lot of low-precision projectiles, you can use a few with high targeting precision. High local heat waves and kinetic disrupting effects targeting the most thermally and structurally sensitive parts of the cell will probably be the magic bullet in MH. However, until there is a long and challenging process, material and design optimization steps in nanoparticle production should be carried out continuously. On the other hand, a systematic and complex review of all clinical and preclinical experimental results, including the vast number of in vitro and in vivo experimental results, is needed to provide focused research directions. This may be possible with the new AI algorithms. Therefore, a unified perspective on the SAR at the macroscopic and cellular levels is more than welcome.

Funding: This research study was funded by the Romanian Ministry of Research, Innovation and Digitalization, Core Program of the National Institute of Materials Physics under Project PN23080101.

Institutional Review Board Statement: Not applicable.

Informed Consent Statement: Not applicable.

Data Availability Statement: Data are contained within the article.

Conflicts of Interest: The author declares no conflicts of interest.

References

1. Moon, J.; Kitty, I.; Renata, K.; Qin, S.; Zhao, F.; Kim, W. DNA Damage and Its Role in Cancer Therapeutics. *Int. J. Mol. Sci.* **2023**, *24*, 4741. [CrossRef] [PubMed]
2. Kay, J.; Thadhani, E.; Samson, L.; Engelward, B. Inflammation-Induced DNA Damage, Mutations and Cancer. *DNA Repair* **2019**, *83*, 102673. [CrossRef] [PubMed]
3. Chatterjee, N.; Walker, G.C. Mechanisms of DNA Damage, Repair, and Mutagenesis. *Environ. Mol. Mutagen.* **2017**, *58*, 235–263. [CrossRef] [PubMed]
4. Greten, F.R.; Grivennikov, S.I. Inflammation and Cancer: Triggers, Mechanisms and Consequences. *Immunity* **2019**, *51*, 27–41. [CrossRef]

5. Huang, R.; Zhou, P.-K. DNA Damage Repair: Historical Perspectives, Mechanistic Pathways, and Clinical Translation for Targeted Cancer Therapy. *Signal Transduct. Target. Ther.* **2021**, *6*, 254. [CrossRef]
6. Li, L.; Guan, Y.; Chen, X.; Yang, J.; Cheng, Y. DNA Repair Pathways in Cancer Therapy and Resistance. *Front. Pharmacol.* **2021**, *11*, 629266. [CrossRef]
7. Imashiro, C.; Takeshita, H.; Morikura, T.; Miyata, S.; Takemura, K.; Komotori, J. Development of Accurate Temperature Regulation Culture System with Metallic Culture Vessel Demonstrates Different Thermal Cytotoxicity in Cancer and Normal Cells. *Sci. Rep.* **2021**, *11*, 21466. [CrossRef]
8. De Maio, A.; Alfieri, G.; Mattone, M.; Ghanouni, P.; Napoli, A. High-Intensity Focused Ultrasound Surgery for Tumor Ablation: A Review of Current Applications. *Radiol. Imaging Cancer* **2023**, *6*, e230074. [CrossRef]
9. Ji, Y.; Zhu, J.; Zhu, L.; Zhu, Y.; Zhao, H. High-Intensity Focused Ultrasound Ablation for Unresectable Primary and Metastatic Liver Cancer: Real-World Research in a Chinese Tertiary Center with 275 Cases. *Front. Oncol.* **2020**, *10*, 519164. [CrossRef]
10. Schena, E.; Saccomandi, P.; Fong, Y. Laser Ablation for Cancer: Past, Present and Future. *J. Funct. Biomater.* **2017**, *8*, 19. [CrossRef]
11. Fan, Y.; Xu, L.; Liu, S.; Li, J.; Xia, J.; Qin, X.; Li, Y.; Gao, T.; Tang, X. The State-of-the-Art and Perspectives of Laser Ablation for Tumor Treatment. *Cyborg Bionic Syst.* **2024**, *5*, 0062. [CrossRef] [PubMed]
12. Adibzadeh, F.; Sumser, K.; Curto, S.; Yeo, D.T.B.; Shishegar, A.A.; Paulides, M.M. Systematic review of pre-clinical and clinical devices for magnetic resonance-guided radiofrequency hyperthermia. *Int. J. Hyperth.* **2020**, *37*, 15–27. [CrossRef] [PubMed]
13. Chang, D.; Lim, M.; Goos, J.A.C.M.; Qiao, R.; Ng, Y.Y.; Mansfeld, F.M.; Jackson, M.; Davis, T.P.; Kavallaris, M. Biologically Targeted Magnetic Hyperthermia: Potential and Limitations. *Front. Pharmacol.* **2018**, *9*, 831. [CrossRef]
14. Dutz, S.; Hergt, R. Magnetic Particle Hyperthermia—A Promising Tumour Therapy? *Nanotechnology* **2014**, *25*, 452001. [CrossRef]
15. Feddersen, T.V.; Hernandez-Tamames, J.A.; Franckena, M.; van Rhoon, G.C.; Paulides, M.M. Clinical Performance and Future Potential of Magnetic Resonance Thermometry in Hyperthermia. *Cancers* **2021**, *13*, 31. [CrossRef]
16. Gavilán, H.; Kumar Avugadda, S.; Fernández-Cabada, T.; Soni, N.; Cassani, M.; Mai, B.T.; Chantrell, R.; Pellegrino, T. Magnetic Nanoparticles and Clusters for Magnetic Hyperthermia: Optimizing Their Heat Performance and Developing Combinatorial Therapies to Tackle Cancer. *Chem. Soc. Rev.* **2021**, *50*, 11614–11667. [CrossRef]
17. Kok, H.P.; Cressman, E.N.K.; Ceelen, W.; Brace, C.L.; Ivkov, R.; Grüll, H.; ter Haar, G.; Wust, P.; Crezee, J. Heating Technology for Malignant Tumors: A Review. *Int. J. Hyperth.* **2020**, *37*, 711–741. [CrossRef]
18. Ribeiro, T.P.; Moreira, J.A.; Monteiro, F.J.; Laranjeira, M.S. Nanomaterials in Cancer: Reviewing the Combination of Hyperthermia and Triggered Chemotherapy. *J. Control. Release* **2022**, *347*, 89–103. [CrossRef]
19. Shirvalilou, S.; Khoei, S.; Esfahani, A.J.; Kamali, M.; Shirvaliloo, M.; Sheervalilou, R.; Mirzaghavami, P. Magnetic Hyperthermia as an Adjuvant Cancer Therapy in Combination with Radiotherapy versus Radiotherapy Alone for Recurrent/Progressive Glioblastoma: A Systematic Review. *J. Neurooncol.* **2021**, *152*, 419–428. [CrossRef]
20. Price, P.M.; Mahmoud, W.E.; Al-Ghamdi, A.A.; Bronstein, L.M. Magnetic Drug Delivery: Where the Field Is Going. *Front. Chem.* **2018**, *6*, 619. [CrossRef]
21. Pusta, A.; Tertis, M.; Crăciunescu, I.; Turcu, R.; Mirel, S.; Cristea, C. Recent Advances in the Development of Drug Delivery Applications of Magnetic Nanomaterials. *Pharmaceutics* **2023**, *15*, 1872. [CrossRef] [PubMed]
22. Toderascu, L.I.; Sima, L.E.; Orobeti, S.; Florian, P.E.; Icriverzi, M.; Maraloiu, V.-A.; Comanescu, C.; Iacob, N.; Kuncser, V.; Antohe, I.; et al. Synthesis and anti-melanoma activity of L-cysteine-coated iron oxide nanoparticles loaded with doxorubicin. *Nanomaterials* **2023**, *13*, 621. [CrossRef] [PubMed]
23. Reddy, L.H.; Arias, J.L.; Nicolas, J.; Couvreur, P. Magnetic Nanoparticles: Design and Characterization, Toxicity and Biocompatibility, Pharmaceutical and Biomedical Applications. *Chem. Rev.* **2012**, *112*, 5818–5878. [CrossRef] [PubMed]
24. Kush, P.; Kumar, P.; Singh, R.; Kaushik, A. Aspects of High-Performance and Bio-Acceptable Magnetic Nanoparticles for Biomedical Applications. *Asian J. Pharm. Sci.* **2021**, *16*, 704–737. [CrossRef]
25. Bychkova, A.V.; Sorokina, O.N.; Rosenfeld, M.A.; Kovarski, A. Multifunctional biocompatible coatings on magnetic nanoparticles. *Russ. Chem. Rev.* **2012**, *81*, 1070–1083. [CrossRef]
26. Atkinson, W.J.; Brezovich, I.A.; Chakraborty, D.P. Usable frequencies in hyperthermia with thermal seeds. *IEEE Trans. Biomed. Eng.* **1984**, *31*, 70. [CrossRef]
27. Dutz, S.; Hergt, R. Magnetic Nanoparticle Heating and Heat Transfer on a Microscale: Basic Principles, Realities, and Physical Limitations of Hyperthermia for Tumour Therapy. *Int. J. Hyperth.* **2013**, *29*, 790–800. [CrossRef]
28. de la Parte, B.H.; Rodrigo, I.; Gutiérrez-Basoa, J.; Correcher, S.I.; Mar Medina, C.; Echevarría-Uraga, J.J.; Garcia, J.A.; Plazaola, F.; García-Alonso, I. Proposal of New Safety Limits for In Vivo Experiments of Magnetic Hyperthermia Antitumor Therapy. *Cancers* **2022**, *14*, 3084. [CrossRef]
29. Kwok, M.K.Y.; Maley, C.C.J.; Dworkin, A.; Hattersley, S.; Southern, P.; Pankhurst, Q.A. Nonspecific Eddy Current Heating in Magnetic Field Hyperthermia. *Appl. Phys. Lett.* **2023**, *122*, 240502. [CrossRef]

30. Gu, Y.; Piñol, R.; Moreno-Loshuertos, R.; Brites, C.D.S.; Zeler, J.; Martínez, A.; Maurin-Pasturel, G.; Fernández-Silva, P.; Marco-Brualla, J.; Téllez, P.; et al. Local Temperature Increments and Induced Cell Death in Intracellular Magnetic Hyperthermia. *ACS Nano* **2023**, *17*, 6822–6832. [CrossRef]
31. Li, J.; Tran, N.; Webster, T.J. Towards a Better Understanding of the Effects of the Magnetic Nanoparticles Size and Magnetic Field on Cellular Endocytosis. *J. Phys. D Appl. Phys.* **2020**, *53*, 175401. [CrossRef]
32. Sabourian, P.; Yazdani, G.; Ashraf, S.S.; Frounchi, M.; Mashayekhan, S.; Kiani, S.; Kakkar, A. Effect of Physico-Chemical Properties of Nanoparticles on Their Intracellular Uptake. *Int. J. Mol. Sci.* **2020**, *21*, 8019. [CrossRef] [PubMed]
33. Peng, Y.; Yang, Z.; Sun, H.; Li, J.; Lan, X.; Liu, S. Understanding Cellular Uptake, Localization, and Retention for Enhanced Disease Diagnosis and Therapy. *Aging Dis.* **2025**, *16*, 168–208. [CrossRef] [PubMed]
34. Potrč, T.; Kralj, S.; Nemec, S.; Kocbek, P.; Erdani Kreft, M. The Shape Anisotropy of Magnetic Nanoparticles: An Approach to Cell-Type Selective and Enhanced Internalization. *Nanoscale* **2023**, *15*, 8611–8618. [CrossRef]
35. Blanco-Andujar, C.; Ortega, D.; Southern, P.; Nesbitt, S.A.; Thanh, N.T.K.; Pankhurst, Q.A. Real-Time Tracking of Delayed-Onset Cellular Apoptosis Induced by Intracellular Magnetic Hyperthermia. *Nanomedicine* **2016**, *11*, 121–136. [CrossRef]
36. Calatayud, M.P.; Soler, E.; Torres, T.E.; Campos-Gonzalez, E.; Junquera, C.; Ibarra, M.R.; Goya, G.F. Cell damage produced by magnetic fluid hyperthermia on microglial BV2 cells. *Sci. Rep.* **2017**, *7*, 8627. [CrossRef]
37. Beola, L.; Asín, L.; Fratila, R.M.; Herrero, V.; de la Fuente, J.M.; Grazu, V.; Gutiérrez, L. Dual Role of Magnetic Nanoparticles as Intracellular Hotspots and Extracellular Matrix Disruptors Triggered by Magnetic Hyperthermia in 3D Cell Culture Models. *ACS Appl. Mater. Interfaces* **2018**, *10*, 44301–44313. [CrossRef]
38. Hogan, K.J.; Perez, M.R.; Mikos, A.G. Extracellular Matrix Component-Derived Nanoparticles for Drug Delivery and Tissue Engineering. *J. Control. Release* **2023**, *360*, 888–912. [CrossRef]
39. Chen, B.-W.; Chiu, G.-W.; He, Y.-C.; Huang, C.-Y.; Huang, H.-T.; Sung, S.-Y.; Hsieh, C.-L.; Chang, W.-C.; Hsu, M.-S.; Wei, Z.-H.; et al. Extracellular and Intracellular Intermittent Magnetic-Fluid Hyperthermia Treatment of SK-Hep1 Hepatocellular Carcinoma Cells Based on Magnetic Nanoparticles Coated with Polystyrene Sulfonic Acid. *PLoS ONE* **2021**, *16*, e0245286. [CrossRef]
40. Hannon, G.; Bogdanska, A.; Volkov, Y.; Prina-Mello, A. Comparing the Effects of Intracellular and Extracellular Magnetic Hyperthermia on the Viability of BxPC-3 Cells. *Nanomaterials* **2020**, *10*, 593. [CrossRef]
41. Kobayashi, S.; Ohki, A.; Tanoue, M.; Inaoka, Y.; Murase, K. Comparative Study of Extracellular and Intracellular Magnetic Hyperthermia Treatments Using Magnetic Particle Imaging. *Open J. Appl. Sci.* **2017**, *7*, 647–660. [CrossRef]
42. Idiago-López, J.; Ferreira, D.; Asín, L.; Moros, M.; Armenia, I.; Grazú, V.; Fernandes, A.R.; Baptista, P.V.; Fratila, R.M. Membrane-Localized Magnetic Hyperthermia Promotes Intracellular Delivery of Cell-Impermeant Probes. *Nanoscale* **2024**, *16*, 15176–15195. [CrossRef] [PubMed]
43. Fernández-Afonso, Y.; Ruta, S.; Páez-Rodríguez, A.; van Zanten, T.S.; Gleadhall, S.; Fratila, R.M.; Moros, M.; Morales, M.P.; Satoh, A.; Chantrell, R.W.; et al. Reversible Alignment of Nanoparticles and Intracellular Vesicles During Magnetic Hyperthermia Experiments. *Adv. Funct. Mater.* **2024**, *34*, 2405334. [CrossRef]
44. Kim, D.-H.; Rozhkova, E.A.; Ulasov, I.V.; Bader, S.D.; Rajh, T.; Lesniak, M.S.; Novosad, V. Biofunctionalized Magnetic-Vortex Microdiscs for Targeted Cancer-Cell Destruction. *Nat. Mater.* **2009**, *9*, 165–171. [CrossRef]
45. Zhang, E.; Kircher, M.F.; Koch, M.; Eliasson, L.; Goldberg, S.N.; Renström, E. Dynamic Magnetic Fields Remote-Control Apoptosis via Nanoparticle Rotation. *ACS Nano* **2014**, *8*, 3192–3201. [CrossRef]
46. Naud, C.; Thébault, C.; Carrière, M.; Hou, Y.; Morel, R.; Berger, F.; Diény, B.; Joisten, H. Cancer treatment by magneto-mechanical effect of particles, a review. *Nanoscale Adv.* **2020**, *2*, 3632–3655. [CrossRef]
47. Cheng, D.; Li, X.; Zhang, G.; Shi, H. Morphological Effect of Oscillating Magnetic Nanoparticles in Killing Tumor Cells. *Nanoscale Res. Lett.* **2014**, *9*, 195. [CrossRef]
48. Nishita, M.; Park, S.-Y.; Nishio, T.; Kamizaki, K.; Wang, Z.; Tamada, K.; Takumi, T.; Hashimoto, R.; Otani, H.; Pazour, G.J.; et al. Ror2 Signaling Regulates Golgi Structure and Transport Through IFT20 for Tumor Invasiveness. *Sci. Rep.* **2017**, *7*, 1. [CrossRef]
49. Leulmi, S.; Chauchet, X.; Morcrette, M.; Ortiz, G.; Joisten, H.; Sabon, P.; Livache, T.; Hou, Y.; Carrière, M.; Lequien, S.; et al. Triggering the Apoptosis of Targeted Human Renal Cancer Cells by the Vibration of Anisotropic Magnetic Particles Attached to the Cell Membrane. *Nanoscale* **2015**, *7*, 15904–15913. [CrossRef]
50. Dieny, B.; Morel, R.; Joisten, H.; Naud, C.; Nicolas, A.; Visonà, A.; Obeïd, P.; Belin, S.; Berger, F. Magnetism for mechanobiology and related biomedical applications. *Phys. Rev. Appl.* **2025**, *23*, 010501. [CrossRef]
51. Unni, M.; Uhl, A.M.; Savliwala, S.; Savitzky, B.H.; Dhavalikar, R.; Garraud, N.; Arnold, D.P.; Kourkoutis, L.F.; Andrew, J.S.; Rinaldi, C. Thermal Decomposition Synthesis of Iron Oxide Nanoparticles with Diminished Magnetic Dead Layer by Controlled Addition of Oxygen. *ACS Nano* **2017**, *11*, 2284–2303. [CrossRef] [PubMed]
52. Fokina, V.; Wilke, M.; Dulle, M.; Ehlert, S.; Förster, S. Size Control of Iron Oxide Nanoparticles Synthesized by Thermal Decomposition Methods. *J. Phys. Chem. C* **2022**, *126*, 21356–21367. [CrossRef]

53. Demessie, A.A.; Park, Y.; Singh, P.; Moses, A.S.; Korzun, T.; Sabei, F.Y.; Albarqi, H.A.; Campos, L.; Wyatt, C.R.; Farsad, K.; et al. An Advanced Thermal Decomposition Method to Produce Magnetic Nanoparticles with Ultrahigh Heating Efficiency for Systemic Magnetic Hyperthermia. *Small Methods* **2022**, *6*, 2200916. [CrossRef] [PubMed]
54. Gavilan, H.; Rizzo, G.M.R.; Silvestri, N.; Mai, B.T.; Pellegrino, T. Scale-up approach for the preparation of magnetic ferrite nanocubes and other shapes with benchmark performance for magnetic hyperthermia applications. *Nat. Protoc.* **2023**, *18*, 783–809. [CrossRef]
55. Niţică, S.; Fizeşan, I.; Dudric, R.; Barbu-Tudoran, L.; Pop, A.; Loghin, F.; Vedeanu, N.; Lucaciu, C.M.; Iacovita, C. A Fast, Reliable Oil-In-Water Microemulsion Procedure for Silica Coating of Ferromagnetic Zn Ferrite Nanoparticles Capable of Inducing Cancer Cell Death In Vitro. *Biomedicines* **2022**, *10*, 1647. [CrossRef]
56. Bruschi, M.L.; Toledo, L.A.S. Pharmaceutical Applications of Iron-Oxide Magnetic Nanoparticles. *Magnetochemistry* **2019**, *5*, 50. [CrossRef]
57. Arias, L.S.; Pessan, J.P.; Vieira, A.P.M.; Lima, T.M.T.; Delbem, A.C.B.; Monteiro, D.R. Iron Oxide Nanoparticles for Biomedical Applications: A Perspective on Synthesis, Drugs, Antimicrobial Activity, and Toxicity. *Antibiotics* **2018**, *7*, 46. [CrossRef]
58. Li, B.; Chen, X.; Qiu, W.; Zhao, R.; Duan, J.; Zhang, S.; Pan, Z.; Zhao, S.; Guo, Q.; Qi, Y.; et al. Synchronous Disintegration of Ferroptosis Defense Axis via Engineered Exosome-Conjugated Magnetic Nanoparticles for Glioblastoma Therapy. *Adv. Sci.* **2022**, *9*, 2105451. [CrossRef]
59. Soares, G.A.; Faria, J.V.C.; Pinto, L.A.; Prospero, A.G.; Pereira, G.M.; Stoppa, E.G.; Buranello, L.P.; Bakuzis, A.F.; Baffa, O.; Miranda, J.R.A. Long-Term Clearance and Biodistribution of Magnetic Nanoparticles Assessed by AC Biosusceptometry. *Materials* **2022**, *15*, 2121. [CrossRef]
60. Vandendriessche, S.; Mattos, M.S.; Bialek, E.L.; Schuermans, S.; Proost, P.; Marques, P.E. Complement Activation Drives the Phagocytosis of Necrotic Cell Debris and Resolution of Liver Injury. *Front. Immunol.* **2024**, *15*, 1512470. [CrossRef]
61. Coey, J.M.D. *Magnetism and Magnetic Materials*; Cambridge University Press: Cambridge, UK, 2010; ISBN 978-0-521-81614-4.
62. Kuncser, V.; Palade, P.; Kuncser, A.; Greculeasa, S.; Schinteie, G. Engineering Magnetic Properties of Nanostructures via Size Effects and Interphase Interactions. In *Size Effects in Nanostructures*; Kuncser, V., Miu, L., Eds.; Springer Series in Materials Science; Springer: Berlin/Heidelberg, Germany, 2014; pp. 169–237. [CrossRef]
63. Bedanta, S.; Kleemann, W. Supermagnetism. *J. Phys. D Appl. Phys.* **2009**, *42*, 013001. [CrossRef]
64. Rajan, A.; Sahu, N.K. Review on Magnetic Nanoparticle-Mediated Hyperthermia for Cancer Therapy. *J. Nanopart. Res.* **2020**, *22*, 319. [CrossRef]
65. Rosensweig, R.E. Heating Magnetic Fluid with Alternating Magnetic Field. *J. Magn. Magn. Mater.* **2002**, *252*, 370–374. [CrossRef]
66. Rognin, E.; Brun, P.; Sauvage, E.; Lacombe, J. Computation of Eddy Currents in Highly Conductive Particles Dispersed in a Moderately Conductive Matrix. *Int. J. Appl. Electromagn. Mech.* **2017**, *53*, 2241. [CrossRef]
67. Wheeler, H.A. Formulas for the Skin Effect. *Proc. IRE* **1942**, *30*, 412–426. [CrossRef]
68. Available online: https://www.rfcafe.com/references/electrical/cond-high-freq.htm (accessed on 14 January 2025).
69. Iacob, N.; Schinteie, G.; Palade, P.; Kuncser, V. Approach for an Improved Experimental Evaluation of the Specific Absorption Rate in Magnetic Fluid Hyperthermia. *J. Nanopart. Res.* **2015**, *17*, 190. [CrossRef]
70. Coïsson, M.; Barrera, G.; Appino, C.; Celegato, F.; Martino, L.; Safronov, A.P.; Kurlyandskaya, G.V.; Tiberto, P. Specific Loss Power Measurements by Calorimetric and Thermal Methods on γ-Fe_2O_3 Nanoparticles for Magnetic Hyperthermia. *J. Magn. Magn. Mater.* **2019**, *473*, 403–409. [CrossRef]
71. Takeuchi, M.; Sakai, T.; Andocs, G.; Takao, K.; Nagaoka, R.; Hasegawa, H. Temperature Elevation in Tissue Detected In Vivo Based on Statistical Analysis of Ultrasonic Scattered Echoes. *Sci. Rep.* **2020**, *10*, 9030. [CrossRef]
72. Sun, Y.; Kong, M.; Ke, J.; Yuan, W.; Wen, C.; Gu, Y.; Luo, S.; Feng, W. Nanothermometer for In Vivo Temperature Detection with High Spatial Resolution Based on Core–Shell Rare Earth Nanoparticles. *ACS Appl. Nano Mater.* **2023**, *6*, 23173–23183. [CrossRef]
73. Valdés, D.P.; Torres, T.E.; Maldonado, A.C.M.; Urretavizcaya, G.; Nadal, M.S.; Mansilla, M.V.; Zysler, R.D.; Goya, G.F.; Biasi, E.D.; Lima, E., Jr. Thermographical Method to Assess the Performance of Magnetic Nanoparticles in Hyperthermia Experiments Through Spatiotemporal Temperature Profiles. *Phys. Rev. Appl.* **2023**, *19*, 014042. [CrossRef]
74. Kallumadil, M.; Tada, M.; Nakagawa, T.; Abe, M.; Southern, P.; Pankhurst, Q.A. Suitability of Commercial Colloids for Magnetic Hyperthermia. *J. Magn. Magn. Mater.* **2009**, *321*, 1509–1513. [CrossRef]
75. Iacob, N.; Schinteie, G.; Palade, P.; Ticos, C.M.; Kuncser, V. Stepped Heating Procedure for Experimental SAR Evaluation of Ferrofluids. *Eur. Phys. J. E* **2015**, *38*, 57. [CrossRef] [PubMed]
76. Carrey, J.; Mehdaoui, B.; Respaud, M. Simple Models for Dynamic Hysteresis Loop Calculations of Magnetic Single-Domain Nanoparticles: Application to Magnetic Hyperthermia Optimization. *J. Appl. Phys.* **2011**, *109*, 083921. [CrossRef]
77. Beković, M.; Trlep, M.; Jesenik, M.; Gorican, V.; Hamler, A. An Experimental Study of Magnetic-Field and Temperature Dependence on Magnetic Fluid's Heating Power. *J. Magn. Magn. Mater.* **2013**, *331*, 264–268. [CrossRef]
78. Gas, P.; Kurgan, E. Cooling Effects Inside Water-Cooled Inductors for Magnetic Fluid Hyperthermia. *IEEE Trans. Magn.* **2017**, *53*, 1528–1534. [CrossRef]

79. Available online: https://www.calculator.net/dew-point-calculator.html (accessed on 15 January 2025).
80. Li, J.; Ye, Y.; Lin, Z.; Wang, Z.; Chen, Y.; Li, G.; Ouyang, Z.; Li, J. Application of a New Self-Regulating Temperature Magnetic Material $Fe_{83}Zr_{10}B_7$ in Magnetic Induction Hyperthermia. *Int. J. Hyperth.* **2023**, *40*, 2211269. [CrossRef]
81. Vicentini, M.; Ferrero, R.; Manzin, A. Influence of Coil Geometry, Supply Conditions and Nanoparticle Heating Properties on Magnetic Hyperthermia in Mouse Models. *Int. J. Therm. Sci.* **2024**, *203*, 109151. [CrossRef]
82. Makridis, A.; Curto, S.; van Rhoon, G.C.; Samaras, T.; Angelakeris, M. A Standardisation Protocol for Accurate Evaluation of Specific Loss Power in Magnetic Hyperthermia. *J. Phys. D Appl. Phys.* **2019**, *52*, 255001. [CrossRef]
83. García-Soriano, D.; Milán-Rois, P.; Lafuente-Gómez, N.; Rodríguez-Díaz, C.; Navío, C.; Somoza, Á.; Salas, G. Multicore Iron Oxide Nanoparticles for Magnetic Hyperthermia and Combination Therapy Against Cancer Cells. *J. Colloid Interface Sci.* **2024**, *670*, 73–85. [CrossRef]
84. Tithito, T.; Sillapaprayoon, S.; Chantho, V.; Pimtong, W.; Thongbunchoo, J.; Charoenphandhu, N.; Krishnamra, N.; Yong, N.; Lert-itthiporn, A.; Maneeprakorn, W.; et al. Evaluation of Magnetic Hyperthermia, Drug Delivery and Biocompatibility of Trace Element Co-Doped Hydroxyapatite Combined with Mn–Zn Ferrite for Bone Tissue Applications. *RSC Adv.* **2024**, *14*, 29242–29253. [CrossRef]
85. Zeleňáková, A.; Nagy, Ľ.; Hrubovčák, P.; Barutiak, M.; Lisnichuk, M.; Huntošová, V.; Mrakovič, A.; Gerina, M.; Zákutná, D. Cobalt-Ferrite Nano-Cubes for Magnetic Hyperthermia Applications. *J. Alloys Compd.* **2024**, *989*, 174415. [CrossRef]
86. Soetaert, F.; Kandala, S.K.; Bakuzis, A.; Ivkov, R. Experimental Estimation and Analysis of Variance of the Measured Loss Power of Magnetic Nanoparticles. *Sci. Rep.* **2017**, *7*, 6661. [CrossRef]
87. Schinteie, G.; Palade, P.; Vekas, L.; Iacob, N.; Bartha, C.; Kuncser, V. Volume Fraction Dependent Magnetic Behaviour of Ferrofluids for Rotating Seal Applications. *J. Phys. D Appl. Phys.* **2013**, *46*, 395501. [CrossRef]
88. Chalkidou, A.; Simeonidis, K.; Angelakeris, M.; Samaras, T.; Martinez-Boubeta, C.; Balcells, L.; Papazisis, K.; Dendrinou-Samara, C.; Kalogirou, O. In vitro application of Fe/MgO nanoparticles as magnetically mediated hyperthermia agents for cancer treatment. *J. Magn. Magn. Mater.* **2011**, *323*, 775–780. [CrossRef]
89. Iglesias, C.A.M.; Araújo, J.C.R.; Xavier, J.; Anders, R.L.; de Araújo, J.M.; da Silva, R.B.; Soares, J.M.; Brito, E.L.; Streck, L.; Fonseca, J.L.C.; et al. Magnetic Nanoparticles Hyperthermia in a Non-Adiabatic and Radiating Process. *Sci. Rep.* **2021**, *11*, 11867. [CrossRef]
90. Coïsson, M.; Barrera, G.; Celegato, F.; Martino, L.; Vinai, F.; Martino, P.; Ferraro, G.; Tiberto, P. Specific Absorption Rate Determination of Magnetic Nanoparticles Through Hyperthermia Measurements in Non-Adiabatic Conditions. *J. Magn. Magn. Mater.* **2016**, *415*, 2–7. [CrossRef]
91. Ruta, S.; Fernández-Afonso, Y.; Rannala, S.E.; Morales, M.P.; Veintemillas-Verdaguer, S.; Jones, C.; Gutiérrez, L.; Chantrell, R.W.; Serantes, D. Beyond Newton's Law of Cooling in Evaluating Magnetic Hyperthermia Performance: A Device-Independent Procedure. *Nanoscale Adv.* **2024**, *6*, 4207. [CrossRef]
92. Natividad, E.; Castro, M.; Mediano, A. Accurate Measurement of the Specific Absorption Rate Using a Suitable Adiabatic Magnetothermal Setup. *Appl. Phys. Lett.* **2008**, *92*, 093116. [CrossRef]
93. Lázaro, M.; Sola-Leyva, A.; Jimenez-Carretero, M.; Carrasco Jiménez, M.P.; Delgado, Á.V.; Iglesias, G.R. The Role of Biocompatible Coatings of Magnetic Nanorods on Their Thermal Response in Hyperthermia: Consequences on Tumor Cell Survival. *J. Drug Deliv. Sci. Technol.* **2024**, *95*, 105622. [CrossRef]
94. Guibert, C.; Fresnais, J.; Peyre, V.; Dupuis, V. Magnetic Fluid Hyperthermia Probed by Both Calorimetric and Dynamic Hysteresis Measurements. *J. Magn. Magn. Mater.* **2017**, *421*, 384–392. [CrossRef]
95. Coïsson, M.; Barrera, G.; Celegato, F.; Martino, L.; Kane, S.N.; Raghuvanshi, S.; Vinai, F.; Tiberto, P. Hysteresis Losses and Specific Absorption Rate Measurements in Magnetic Nanoparticles for Hyperthermia Applications. *Biochim. Biophys. Acta* **2017**, *1861*, 1545–1558. [CrossRef] [PubMed]
96. Garaio, E.; Collantes, J.M.; Garcia, J.A.; Plazaola, F.; Mornet, S.; Couillaud, F.; Sandre, O. A Wide-Frequency Range AC Magnetometer to Measure the Specific Absorption Rate in Nanoparticles for Magnetic Hyperthermia. *J. Magn. Magn. Mater.* **2014**, *368*, 432–437. [CrossRef]
97. Rodrigo, I.; Castellanos-Rubio, I.; Garaio, E.; Arriortua, O.K.; Insausti, M.; Orue, I.; Garcia, J.A.; Plazaola, F. Exploring the Potential of Dynamic Hysteresis Loops via High Field, High Frequency, and Temperature-Adjustable AC Magnetometer for Magnetic Hyperthermia Characterization. *Int. J. Hyperth.* **2020**, *37*, 976–991. [CrossRef]
98. Sámano, A.H.; Rosales, S.; Mazon, E.E.; Casillas, N.; Topete, A.; Paz, J.A.; Quintero, L.H.; Estrada, J.C.; Cano, M.E. A Dynamic Hysteresis Meter for Studying Ferrofluids Designed for Magnetic Hyperthermia. *Meas. Sci. Technol.* **2020**, *31*, 055902. [CrossRef]
99. Zapata, J.C.; Restrepo, J. Dynamic Hysteretic Properties and Specific Loss Power of Magnetic Nanoparticles in a Viscous Medium at Different Thermal Baths. *AIP Adv.* **2023**, *13*, 105110. [CrossRef]
100. Glöck, G.; Hergt, R.; Zeisberger, M.; Dutz, S.; Nagel, S.; Weitschies, W. The Effect of Field Parameters, Nanoparticle Properties, and Immobilization on the Specific Heating Power in Magnetic Particle Hyperthermia. *J. Phys. Condens. Matter* **2006**, *18*, S2935–S2949. [CrossRef]

101. Nakamura, K.; Ueda, K.; Tomitaka, A.; Yamada, T.; Takemura, Y. Self-Heating Temperature and AC Hysteresis of Magnetic Iron Oxide Nanoparticles and Their Dependence on Secondary Particle Size. *IEEE Trans. Magn.* **2013**, *49*, 240–243. [CrossRef]
102. Ognjanović, M.; Stanković, D.M.; Jaćimović, Ž.K.; Kosović-Perutović, M.; Dojčinović, B.; Antić, B. The Effect of Surface-Modifier of Magnetite Nanoparticles on Electrochemical Detection of Dopamine and Heating Efficiency in Magnetic Hyperthermia. *J. Alloys Compd.* **2021**, *884*, 161075. [CrossRef]
103. Ferreira-Filho, V.C.; Morais, B.; Vieira, B.J.C.; Waerenborgh, J.C.; Carmezim, M.J.; Tóth, C.N.; Même, S.; Lacerda, S.; Jaque, D.; Sousa, C.T.; et al. Influence of SPION Surface Coating on Magnetic Properties and Theranostic Profile. *Molecules* **2024**, *29*, 1824. [CrossRef]
104. Vassallo, M.; Martella, D.; Barrera, G.; Celegato, F.; Coïsson, M.; Ferrero, R.; Olivetti, E.S.; Troia, A.; Sözeri, H.; Parmeggiani, C.; et al. Improvement of Hyperthermia Properties of Iron Oxide Nanoparticles by Surface Coating. *ACS Omega* **2023**, *8*, 2143–2154. [CrossRef]
105. Khan, F.; Lahiri, B.B.; Srujana, M.; Vidya, R.; Philip, J. Induction Heating of Magnetic Nanoparticles: Role of Thermo-Physical Properties of the Coating Moieties. *Colloids Surf. A Physicochem. Eng. Asp.* **2024**, *692*, 133982. [CrossRef]
106. Osaci, M.; Cacciola, M. About the Influence of the Colloidal Magnetic Nanoparticles Coating on the Specific Loss Power in Magnetic Hyperthermia. *J. Magn. Magn. Mater.* **2021**, *519*, 167451. [CrossRef]
107. Rennick, J.J.; Johnston, A.P.R.; Parton, R.G. Key Principles and Methods for Studying the Endocytosis of Biological and Nanoparticle Therapeutics. *Nat. Nanotechnol.* **2021**, *16*, 266–276. [CrossRef] [PubMed]
108. Patel, H.; Upadhyay, R.V.; Parekh, K.; Reis, D.; Oliveira, C.L.P.; Neto, A.M.F. Optimized $Mn_{0.5}Zn_{0.5}Fe_2O_4$ Nanoflowers Based Magnetic Fluids for Potential Biomedical Applications. *J. Magn. Magn. Mater.* **2024**, *590*, 171656. [CrossRef]
109. Bejko, M.; Al Yaman, Y.; Keyes, A.; Bagur, A.; Rosa, P.; Gayot, M.; Weill, F.; Mornet, S.; Sandre, O. Structure-Function Relationship of Iron Oxide Nanoflowers: Optimal Sizes for Magnetic Hyperthermia Depending on Alternating Magnetic Field Conditions. *ChemPhysChem* **2024**, *25*, e202400023. [CrossRef]

coatings

Review

A Review on Preparation of Palladium Oxide Films

Petre Badica * and Adam Lőrinczi

National Institute of Materials Physics, Atomistilor 405A, 077125 Măgurele, Romania
* Correspondence: badica2003@yahoo.com or petre.badica@infim.ro

Abstract: Fabrication aspects of PdO thin films and coatings are reviewed here. The work provides and organizes the up-to-date information on the methods to obtain the films. In recent years, the interest in Pd oxide for different applications has increased. Since Pd can be converted into PdO, it is instructive to pay attention to the preparation of the pure and the alloyed Pd films, heterostructures, and nanoparticles synthesized on different substrates. The development of PdO films is presented from the early reports on coatings' formation by oxidation of Pd foils and wires to present technologies. Modern synthesis/growth routes are gathered into chemical and physical categories. Chemical methods include hydrothermal, electrochemical, electroless deposition, and coating methods, such as impregnation, precipitation, screen printing, ink jet printing, spin or dip coating, chemical vapor deposition (CVD), and atomic layer deposition (ALD), while the physical ones include sputtering and cathodic arc deposition, laser ablation, ion or electron beam-induced deposition, evaporation, and supersonic cluster beam deposition. Analysis of publications indicates that many as-deposited Pd or Pd-oxide films are granular, with a high variety of morphologies and properties targeting very different applications, and they are grown on different substrates. We note that a comparative assessment of the challenges and quality among different films for a specific application is generally missing and, in some cases, it is difficult to make a distinction between a film and a randomly oriented, powder-like (granular), thin compact material. Textured or epitaxial films of Pd or PdO are rare and, if orientation is observed, in most cases, it is obtained accidentally. Some practical details and challenges of Pd oxidation toward PdO and some specific issues concerning application of films are also presented.

Keywords: thin films; deposition routes; Pd; Pd alloys; PdO; oxidation of Pd

1. General Considerations on the Growth of PdO Films and Early Research as the Background of the Present Study

1.1. General Considerations on the Growth of PdO Films

According to the International Union of Pure and Applied Chemistry (IUPAC) [1], "A film is defined as a generic term referring to condensed matter restricted in one dimension. The term thin film is associated to a film whose thickness is of the order of a characteristic scale (i.e., a parameter which characterizes the density profile of a given physical quantity, which can have a static—equilibrium—or dynamic character that should be specified, e.g., terms out of plane and in plane refer to characteristic lengths normal or parallel to film surface, respectively) or smaller. Since a film may 'look' operationally thin or thick, according to the procedure applied, it is recommended that the measurement procedure employed be specified (e.g., ellipsometrically thin film, optically thin film, or, e.g., thick compared to the electron mean free path, thin compared to the optical wavelength, etc.)".

In this review on preparation techniques of palladium oxide films, the presentation will stick to understanding that films are 2D objects and they are obtained on a substrate, unless otherwise mentioned, as for the cases of free-standing films. Thickness below about 10–30 nm, i.e., containing up to ~100 Pd(II) oxide unit cells (tetragonal with lattice constants a = 0.3035 nm and c = 0.5323 nm [2]), is often associated in the literature with the term 'ultrathin' films. Typical thin films are usually of few hundreds of nm, while thick films exceed 1 μm in thickness.

It is emphasized that many of the palladium oxide films are granular (i.e., composed of individual, well-defined particles with simple or complex morphologies). Often, a clear delimitation between the methods and procedures to obtain particles in a powder (on a substrate) and in a thin film is not possible. In this work, mostly the technologies demonstrated to produce a conventional relatively continuous film, usually with macroscopic surfaces useful for device fabrication, are briefly introduced. These films are Pd(II) monoxide films, although palladium shows different oxidation states. The common ones are 0, +1, +2, and +4 [3]. The most important oxide forms are the monoxide PdO, dioxide PdO_2, and trioxide PdO_3, as well as a hydrated sesquioxide $Pd_2O_3 \times H_2O$ [4]. In the organic synthesis methods, Pd(0) and Pd(II) are usually considered [5].

Oxidation states +3 and +6 have been reported [6–8] in intermediate compounds proposed to occur in the palladium-catalyzed cross-coupling reactions. Authors of [9] theoretically studied the stability and activity of Pd_4O_x (x = 1–7) films on Pd metal within a CO oxidation reaction model. Other articles also present nonstoichiometric palladium oxides PdO_{1-x} [4,10]. Finally, there are films in the literature referred to as Pd/PdO that assume a Pd core region covered by a Pd-oxide surface layer. These types of particles and thin films (including materials marked with PdO/Pd), containing both metal and oxide species, are of interest for various applications, especially where oxidation–reduction processes play an important role. Since there are no generally accepted terms, on many occasions, materials described as Pd or PdO are in fact a mixture of the metal and oxide [11].

1.2. *Early Research as the Background of the Present Study*

Three groups of methods for film preparation are available:

(1) Preparation of a Pd metal coating on a substrate, followed by oxidation.
(2) Preparation of a PdO film by thermal decomposition of salts.
(3) Preparation of the PdO oxide films by reactive methods.

The proposed classification is arbitrary, and often technological routes share the same processing steps and raw materials. Considering this, in this presentation, this classification is not strictly followed.

An early approach was to use a Pd metal foil or wire as a substrate, subject to surface oxidation. The functional product would be a PdO film coated on the Pd metal. This configuration was used as a pH sensor at high temperatures in geothermal brines, corrosive liquids in contact with nuclear waste containers, pressurized steam/water lines in power generation, and pressurized chemical reactors [12]. These electrodes are also suitable for measurements of pH in biological environments, such as blood and extracellular fluids [13,14], or for neurophysiological measurements [15]. It is well known that PdO is biocompatible, hence, it can also be used in wearable devices.

A high-temperature pH electrode should provide a direct voltage response to a pH value, negligible ionic and redox interfaces, minimum hysteresis, and drift. Only the oxide should be in contact with the solution. These requirements are difficult to fulfill in terms of palladium oxide thin films' quality and thickness. An additional needed condition is miniaturization and low cost [13,16].

Three methods were attempted for fabrication of the Pd oxide film on Pd metal foils or wires:

(i) Betteridge and Rhys [17] demonstrated that, when the Pd metal is heated to 700 °C in air, a PdO film forms on its surface. The electrode was not stable, and the utility of the thermal oxidation method for these purposes was reconsidered.
(ii) Grubb and King [14] used chemical oxidation of the metal surface. The Pd metal surface, after cleaning in aqua regia for 20 s and roughening, was dipped into a 50% aqueous solution of NaOH and dried for 10 min in a stream of dry nitrogen. The coated Pd wire was heated in a flow of oxygen at 800 °C for 20 min. The portion covered with NaOH resulted in formation of a black PdO layer. The oxide layer was uniform,

nonporous, and its thickness was of about 6 μm. Kinoshita et al. [18] expanded this research and studied different parameters, such as oxidation temperature in the range of 600 and 870 °C. They found that the best results were obtained for the electrodes processed at 750 °C after having been immersed in a 50% NaOH solution.

Thermal oxidation and chemical oxidation, as presented in points (i) and (ii), need relatively high processing temperatures. This impedes the preparation of oxide films and devices on low-temperature substrates, such as flexible ones (plastics, paper, and textiles), although Pd films have been coated by other methods on flexible, elastomeric, or optical fiber substrates [19–22]. Another inconvenience is that the reproducibility and stability of the oxide electrodes produced by the two addressed routes have been found to be highly sensitive to processing conditions, electrodes' design, and sizes. Other important issues are stoichiometry control of PdO vs. oxidation temperature and decomposition of PdO. Decomposition of PdO was established both theoretically and experimentally to occur above 870–900 °C [23]. Some practical aspects regarding oxidation are addressed in the next paragraphs, while details in terms of time and spatial development of the metal–oxide interface will not be approached in detail.

Considering the presented information, the search for new methods to obtain PdO films on Pd or on other substrates has continued.

(iii) An alternative method was proposed, namely, the electrochemical oxidation. Kinoshita et al. [18] oxidized the metal electrodes by treating them in 0.2 M NaOH for 5 min at 2.7 V and for 17 h at 0.74 V. Electrodes lost their near-Nernstian response to pH after a few acid–base cycles. A longer lifetime of less than 6 days was determined for an electrode fabricated by Liu et al. [13]. The authors used a molten salt electrolyte of $NaNO_3/LiCl$ for oxidation, with a current density of 20 $mA{\cdot}mm^{-2}$ at 5.9–6.2 V for 90 s. In [24], the authors applied electrolysis waveforms containing various levels of the *ac* and *dc* components for a processing time up to 24 h. The pH sensor's response depends on the processing conditions. The authors also observed surface morphology changes.

Recent review articles summarize the work on synthesis of powders and coatings of Pd [3,25–27]. Powders and granular films of Pd or Pd-based materials are intensively investigated and used as catalysts. Rare are cases when they are oxidized to obtain palladium oxide as a product, although the methods from points (i)–(iii) are expected to also be successful in these cases. A straightforward example is thermal oxidation (in air at 300 °C for 2h) of PdCo nanoparticles and formation of a PdO-doped Co_2O_3 nano-powder [28,29].

Oxidation can be a general route to obtain Pd-based oxide, oxide compounds, or oxide composites from a large class of Pd-based alloys. Although, as already pointed out, oxidation of many reported Pd products was not performed, in the next paragraphs in this section, some details regarding the types of alloys and the methods used for their synthesis are introduced. Whenever available, the information on oxidation is also provided.

Palladium is shown to combine with other metals. Bimetallic or ternary alloys in the nano-powder form were synthesized: Pd-Pt, Pd-Au, Pd-Ag, Pd-Co, Pd-Co-Au, Pd-Cu, Pd-Ni, Pd-Rh, PdBi, Pt-Pd-Bi ([25,27] and references therein, [11,30–33]), Pd-Fe [34], and Pd-Ti [3]. Heterostructures and supported catalysts can enhance the chemical performance due to cooperative and synergetic interactions of the components [35]. In these cases, one takes advantage at the interface between Pd and the support on the charge transfer, strain effects, nanoscale surface stabilization, preferential faceting, prevention of the catalyst deactivation and regeneration, control of selective catalytic oxidation reactions, and improved gas-poisoning resistance. Supports of Pd tested in the literature are carbon black, activated carbon, carbon spheres, carbon nanofibers, single-wall (SWCNT) and multiwall (MWCNT) carbon nanotubes, graphene (G) and graphene oxide (GO), nanodiamond [36–45], Ti, TiO_2, TiO_2 nanotubes, Fe_2O_3, MnO_2, V_2O_3, ZnO ([25] and references therein, [46]), Fe_3O_4, Fe_3O_3-C core-shell microsphere, Fe_3O_4-polyaniline and Fe_3O_4-$NiSiO_3$ composites ([47] and references therein), CeO_2 [48,49], CuO [26,50], sulfur-doped WO_3 [51,52], Cr_2O_3 [53],

MgO [54], La_2O_3 [55], kaolin [56], $CuMn_2O_4$ [57], $MnSiO_x$ [58], FeCrAl fibers [59], NiO [60], Co oxide [61], In-Sn-Zn-O-sputtered thin films [62], and wool [63]. Catalysts of Pd nanoparticles in which GO particles are encapsulated are also reported [64]. We observed that many of the supports are oxides stable at oxidation temperatures of metallic Pd and, therefore, they can be useful, depending on applications, as substrates for thin-film preparation of PdO-based materials. Other supports for Pd metal or alloy clusters are micelles, dendrimers, and mesoporous materials (e.g., polycarbonate, carbon tube, alumina, porous Ni, SiO_2, and zeolite) [65,66]. Often, these supporting materials play the role of hard nanostructured templates, and the skeleton of the synthesized Pd objects follows the porous architecture of the support [3]. Materials of Pd/PdO clusters in glasses were also fabricated [67]. Apart from the use of templates, the crystal chemistry of Pd as a cubic face-centered (fcc) metal is itself highly favorable to formation of different geometrical shapes/morphologies [68] of the particles, both in the powders and in the coatings. The resulting morphology is controlled through manipulation of the thermodynamic aspects, involving high reduction rates, or of the anisotropy kinetics of the crystal growth. Size and especially morphology have a strong impact on the catalytic properties of nanocrystals. Fabrication and catalytic activity of Pd or Pd-based powders composed of particles with well-defined morphology, uniformity, and size is currently a field of significant interest both from academia and industry. This is a rich domain that is developing fast and in which many excellent articles were published in recent years. Despite this tremendous effort, as already mentioned, research and development rarely proceed toward synthesis of Pd oxide products from the as-fabricated Pd or Pd alloys.

Instead of alloys, other precursors or raw materials can be used, and the technological routes usually employ two processing steps. In the first one, a precursor film is obtained and, in the second one, the precursor is decomposed, reacted, or oxidized to obtain the Pd oxide. An example of a modern route is the growth of a nanoparticulate PdO thin film by the Langmuir–Blodgett (LB) technique, followed by thermal decomposition of the multilayer precursor film of octadecylamine (ODA)–chloropalladate complex [69]. Many other modern synthetic routes usually have several processing steps, one of which is mandatory of a chemical type: a granular metallic $Pd^{(0)}$, alloy, or organometallic film/coating is obtained in the first step, and it is further used as a precursor for preparation of the Pd oxide film final product by decomposition/reaction/oxidation. Nevertheless, the final oxide film can also be prepared by physical methods and by one-step routes, reactive or not. In the next sections, the review will consider two groups of technologies, namely, chemical and physical. This classification is not strict and should be taken as arbitrary.

Depending on the application, it is of interest to also take into account the reduction of PdO into Pd. In [70], Pd nanoparticles were decorated on PdO hollow-shell supports through a one-step treatment with $NaBH_4$. In this process, PdO is partially reduced to Pd. The H_2 gas detection limit of this material is lower than 1 ppm, and the proposed strategy can be employed to extend the lifetime of the H_2 gas sensors. Another approach is precipitation on the support particles. In [71], $PdCl_2$ was precipitated on ZnO nanoparticles and decomposed at 500 °C to Pd. Decoration of different support particles (WO_3, ZnO-Eu, ZnO, β-Bi_2O_3, In_2O_3, SnO_2, and V_2O_5) with functional Pd [71–89] or PdO [88,90–95], mainly for improved gas-sensing applications, is gaining much attention.

2. Chemical Routes of PdO and Pd Precursor Films' Growth

In the category of chemical preparation methods of PdO or Pd precursor films (Figure 1), hydrothermal, electrochemical, electroless deposition, and coating methods can be included, such as the already mentioned Langmuir–Blodgett method, impregnation, precipitation, screen printing, ink jet printing, spin or dip coating, chemical vapor deposition (CVD), including its popular variant atomic layer deposition (ALD), and green biosynthesis. Chemical routes may involve in situ oxidation or reduction processes (e.g., in reactive synthesis routes or salts' decomposition routes). As mentioned in Section 1.2, additional processing steps for reduction or oxidation (such as thermal, chemical, and electrochemical treatments) are sometimes applied on the precursor films already deposited by physical or chemical routes.

Chemical routes to obtain Pd or PdO films	Physical routes to obtain Pd or PdO films
Hydrothermal	Sputtering (*dc*, *rf*, magnetron, plasma-assisted sputtering), and pulsed filtered cathodic arc deposition
Electrochemical (cyclic voltammetry, square-wave voltammetry, chronoamperometry, chronopotentiometry, chronocoulometry, pulsed electrodeposition, electrophoretic deposition, electrospinning)	Electron beam coating and Electron and Ion beam – induced deposition
Electroless (displacement -galvanic replacement-deposition, autocatalytic deposition, microemulsion synthesis, photochemical synthesis)	Pulsed laser deposition (in gas or liquid environment)
Coating (Langmuir-Blodgett, impregnation, precipitation, screen printing, ink jet printing, spin/dip coating)	Evaporation (sublimation) of Pd
Chemical vapor deposition, CVD (aerosol-assisted CVD, atomic layer deposition)	Supersonic cluster-beam deposition
Green biosynthesis	

Figure 1. Arbitrary classification of the technological routes to obtain Pd and PdO coatings.

2.1. Hydrothermal Method

The hydrothermal method was employed in [96] to obtain PdCd nanostructures on a Ti plate. The Ti substrate was introduced in the Teflon autoclave containing 10 mL of 5 mM $PdCl_2$, 5 mM $Cd(NO_3)_2{\cdot}4H_2O$, and 1 M ammonium formate as a reducing agent. The Pd/Cd ratio was varied by the amount of $Cd(NO_3)_2{\cdot}4H_2O$ precursor. The processing temperature was 180 °C for 2 h. The coated Ti plates were further heated in a furnace with a controlled Ar-flow atmosphere at 250 °C for 2 h. The PdCd film had a dendritic microstructure.

In other examples ([25] and references therein) of hydrothermal synthesis of Pd or Pd-based bimetallic nanomaterials, a popular source of Pd is H_2PdCl_4, and formaldehyde (CH_2O) and ethylene glycol (EG) are used as reducing agents. Composite $PdO/CoNi_2S_4$ particles were obtained via the hydrothermal method in [97].

The hydrothermal method is considered a single-step green technology that can use nontoxic solvents, and it minimizes waste. On the other hand, the productivity of this method is relatively low, and synthesis of large amounts of nano-powders or of large film surfaces is challenging. Via hydrothermal synthesis, different controlled shapes, such as nanospheres, nano-cubes, nano-tetrapods, nanoporous bulks, nanowires/dendrites, and core-shell particles on different substrates, are obtained [25].

A facile sol-gel hydrothermal method to obtain PdO particles is presented in [98]. After dissolving $PdCl_2$ in C_2H_5OH (0.1 Molar solution), 1 vol.% HCl is added to enhance the solubility of $PdCl_2$. Stirring for 30 min results in a transparent brown solution. The solution is aged for 24 h, loaded into a Teflon hydrothermal capsule, and heated at 200 °C. The as-obtained brown powder is washed with D.I. water and annealed at 600 °C. The final powder is fine and black and is used in the electrophoretic deposition to obtain a PdO film covered with ITO on a glass substrate for H_2 gas detection at room temperature (Figure 2).

Wang et al. [99], via the hydrothermal method at 200 °C, obtained a ring-like PdO-NiO architecture, with a lamellar structure of 3.5 μm in diameter composed of nanosheets with a thickness of ~15 nm, for CO gas detection, showing a relatively low operating temperature (180 °C) and a rapid response/recovery (2–3 s when exposed to 50 ppm CO).

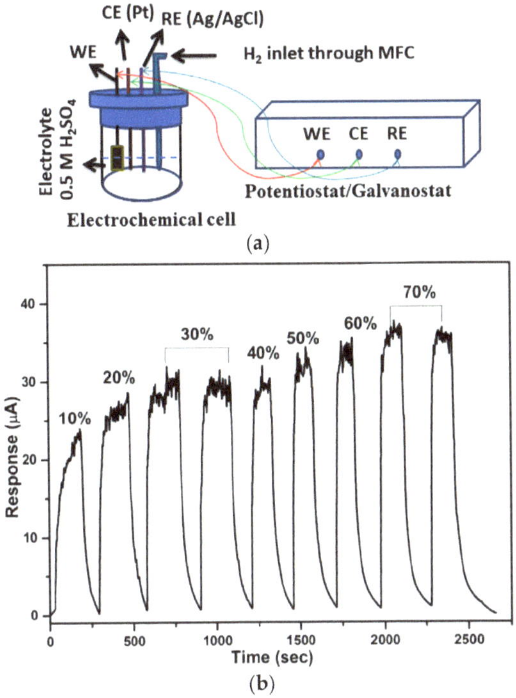

Figure 2. Adapted from [98]. (**a**) Electrochemical gas sensor arrangement (30 mL glass cell): RE—reference electrode; MFC—mass flow controller and the inlet for the gas covered with fluorinated ethylene propylene (FEP) H_2 gas-permeable membrane; WE—working electrode, PdO thin film of 1 μm thickness on ITO substrate; CE—counter electrode, Pt rod. (**b**) Room-temperature response by using the sensing arrangement from (**a**) when passing a H_2 gas (10%–70% in Ar) into the cell for 200 s and for a constant potential on electrodes of 1 V.

2.2. *Electrochemical Deposition*

In the group of electrochemical deposition methods, in an electrochemical cell with two or three electrodes (working electrode, counter electrode, and reference electrode), the electrolyte is conductive, and it is also the source of Pd. A current or potential is applied on the electrochemical cell, and ions of $Pd^{(+II)}$ in solution will deposit in the metal state ($Pd^{(0)}$) at the surface of electrodes. Depending on the current or potential features, different versions of electrochemical depositions were defined: cyclic voltammetry (CV), square-wave voltammetry (SWV), chronoamperometry (CA), chronopotentiometry (CP), chrono-coulometry (CC), and pulsed electrodeposition ([25] and references therein).

The electrochemical deposition is considered a simple, affordable, and relatively fast method. Popular Pd raw materials are $PdCl_2$ and H_2PdCl_4, but other materials can be used, e.g., K_2PdCl_6 [100] or $Pd(NH_3)_4Cl_2$ [101]. Different morphologies are obtained in the electrochemical deposition. Nanoparticles with diameters of 5–10 nm and nanorods or nanowires are the most common. Pd nano-horns or nanoparticles of PdAu tube-shaped heterostructures were electrochemically deposited in [100,102], while an AgPd coating (on an oxidized/activated stainless-steel substrate), tested as a beneficial catalyst for low-temperature fuel cells' application, was obtained in [103]. Fabrication of nanostructured bimetallic Pd-Fe thin films and their electrodechlorination activity are reported in [104]. The electrochemical deposition allows the use of template substrates, e.g., anodized aluminum oxide (AAO) or porous anodic alumina (PAA) [25]. After the directional template-assisted growth into the parallel pores, the substrate is removed, and the product is a sort of film consisting of an array of Pd nanowires [101]. By using an AAO template and subsequent etching of PdCu nanorods, Pd nano-springs were obtained in [102]. An open-pore Pd foam was electrodeposited, and a Pd-CeO_2 catalyst for environment processes was reported in [105]. It was observed that most works report deposition of metal particles and films, rather than of the Pd oxide. Additional information can be found in [25].

In [98], films of PdO nanoparticles were deposited on ITO-coated glass substrates (5 mm × 10 mm) with electrophoretic deposition (EPD). The anode is the ITO substrate, and the cathode is a Pt rod. The distance between electrodes was set to 1 cm, and the electrodes were placed in a 20 mL glass vessel. A PdO powder (0.1 g) obtained by the sol-gel hydrothermal method (see the previous paragraph) was dispersed in 10 mL of ethanol. In this electrolyte solution, as a dispersant and to enhance conductivity and mobility of the ions, magnesium nitrate ($Mg(NO_3)_2 \cdot 6H_2O$, 100 µL) was added. The deposition potential applied for 2 min was between 10 and 35 V. The authors indicated that the quality of the film depends on the deposition voltage and time. The optimum voltage would be 30 V. Films were used for H_2 gas detection at room temperature by using an amperometric gas-sensing setup (Figure 2).

It is important to emphasize that PdO oxide films can be used not only for reduction gases' detection, but also for oxidizing ones. An example of the results obtained with a chemoresistive (conductomeric) sensor sensitive to ozone is presented in Figure 3 [106].

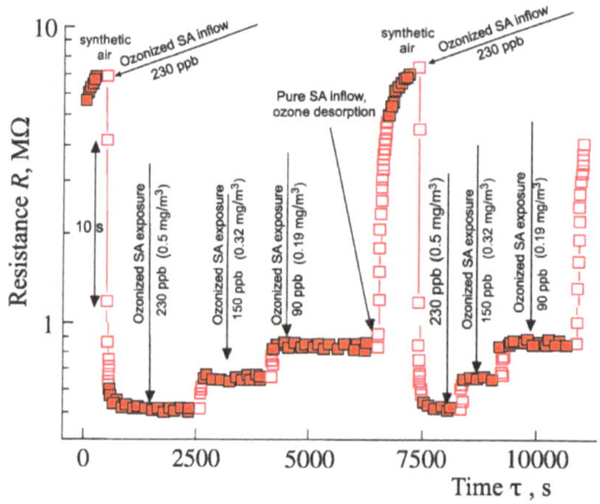

Figure 3. Reproduced with permission from [106]. PdO sensor resistance at different ozone concentrations as a function of time at an operating temperature of 220 °C. SA denotes synthetic air. Note that ozone (O_3) is harmful to human health, similar to other oxidizing gases, such as NO_x, SO_2, and Cl_2. It is a by-product of many modern technologies, and its interaction under sunlight with volatile hydrocarbons produces many toxic organic compounds.

The somehow related method of electrospinning is also a straightforward fabrication coating technique, and it was applied in [107] to obtain a decorated Pd nanolayer upon aligned polyurethane nanofibers for H_2 gas sensors. Among the coatings obtained by electrospinning of PdO, Au, and CdO on SnO_2 nanofibers, the highest sensitivity to toluene was determined for PdO-SnO_2 [108].

2.3. Electroless Deposition

The group of electroless deposition techniques has two main variants, namely, the displacement (galvanic replacement) and autocatalytic deposition [25]. These methods take advantage of a photo-reductant, a stabilizer, or a chemical reducing agent. Microemulsion and photochemical synthesis also belong to these types of methods, which are without an external current source but share a common ground with electrochemical routes.

In the displacement deposition, 25 μL of 10 mM H_2PdCl_2 was placed on Fe thin films for 30 min [104]. Reactions between $PdCl_4^{2-}/Pd$ and Fe^{2+}/Fe developed with standard redox potentials of 0.95 and −0.44 V, respectively. The final product was washed with ultrapure water and dried. The less noble metal acted as a reducing agent, and the reaction proceeded at its surface. This produced certain limitations. Other limitations were imposed by the required favorable difference in the reduction potentials of the two metals.

The autocatalytic deposition uses citric acid, ascorbic acid, ethylene glycol, sodium borohydride, and others as reducing agents. The impact of the reducing agent is very strong and can modify growth kinetics with changes in the particles' morphology and in the surface properties ([25] and references therein). In general, the reducing agents are added into aqueous solutions of $PdCl_2$, $NaPdCl_4$, or H_2PdCl_4. The method is suitable to obtain bimetallic Pd-based nanomaterials. It also has the advantage that the processes proceed at the surface and in the bulk of the metals. The polyol synthesis is a derivate method. Metal-containing compounds are obtained through the use of ethylene glycol, playing the role of both the solvent and the reducing agent. Without templates, Wang et al. [109,110] synthesized nanowires with uniform diameters of 2 nm via this technique. Synthesis proceeded from a solution of $(CF_3COO)_2Pd$ dissolved in diethylene glycol (DEG). When instead of $(CF_3COO)_2Pd$, the authors used Na_2PdCl_4, Pd polyhedrons were obtained.

Microemulsion techniques use water in oil to control the particle size. Surfactant-stabilized reverse micelles are the microreactors. Typical surfactants are dioctyl sodium sulfosuccinate (AOT) and cetyl trimethyl ammonium bromide (CTAB). In some cases, heptane is used as an oil phase, while toluene is a co-surfactant. Application of surface-confining agents, such as CO that are preferentially binding on the Pd(111) surface, produced nanosheets of Pd with a thickness of 1.8 nm [111]. Microemulsion is shown to be suitable for the synthesis of Pd-Co-Au/C tri-metallic catalysts [112]. Yuasa et al. [113] investigated preparation of composite particles of nano-sized PdO loaded on SnO_2 nanoparticles by precipitating $Pd(OH)_2$ and $Sn(OH)_4$ inside a reverse micelle. CO gas-sensing properties of the particles were highly sensitive and were maximized for a small amount of PdO (0.1 mol %), at which PdO agglomeration is avoided. This method allows loading of small amounts of PdO when compared to the conventional impregnation method.

Nanoparticles of PdAu were obtained on TiO_2 via photochemical synthesis. Tian et. al. [114] immersed a plate with TiO_2 nanotubes into a water solution of 0.04 M $Pd(NO_3)_2$ and $AuCl_3$. In the solution, 5 mL (50 vol.%) of methanol was also introduced. After 30 min of exposure to UV radiation, in the solution, PdAu particles with the size of ~20 nm formed.

2.4. Coating Techniques

The group of coating techniques, such as Langmuir–Blodgett, impregnation, precipitation, screen printing, ink jet printing, and dip and spin coating, due to their simplicity and low cost, are often employed to obtain Pd-based precursor powders or films [10,94,115–118]. The idea of these synthesis methods is to synthesize particles or films, usually of palladium chloride, nitrate, or hydroxide, which are subsequently thermally decomposed.

Jin et al. [119] used the conventional impregnation method by dipping a honeycomb cordierite substrate coated with Ce_xZr_y oxides (synthesized by co-precipitation with urea of a Ce, Zr nitrate solution, and calcination at 550 °C in air) into the aqueous solution of palladium nitrate, followed by drying and calcination in air at 500 °C for 2 h, to obtain a brown-colored catalyst complex with Pd (metal loading on the monolith of about 0.3 wt.%). This material was investigated as an alternative catalytic technology to oxygen removal of coal mine methane. To optimize CO_2 methanation pathways, coatings of $Pd/P\text{-}CeO_2\text{-}Al_2O_3$ were prepared via multistep impregnation in [120].

In [121], decorated flower-like ZnO structures with PdO obtained by precipitation in a methanol solution with dispersed ZnO particles and with dissolved $PdCl_2$ (60 mg ZnO, 3.3 mL methanol, and 6 mg $PdCl_2$, shaken for 30 min), followed by calcination at 350 °C for 1 h (heating rate of 5 °C/min), showed better sensing activity of toluene and ethanol than of ZnO (Figure 4).

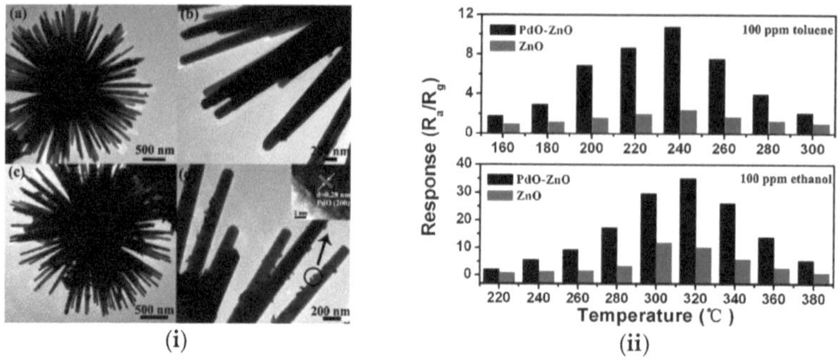

Figure 4. Reproduced with permission from [121]. (i) TEM images taken on (**a**,**b**) ZnO and (**c**,**d**) ZnO-PdO. (ii) Response to toluene and ethanol of the structures from (i) at different operating temperatures.

Use of $PdCl_2$ as an initial reagent is economically convenient since it is produced at an industrial scale and it is relatively cheap, but traces of residual chloride anions can show a strong negative influence on the sensitivity of PdO gas sensors. For this type of device, it is recommended to convert $PdCl_2$ (e.g., through a reaction with NaOH or $AgNO_3$ [10]) or to use the palladium hydroxide or nitrate directly as initial reagents. The typical media to dissolve the initial reagents is water. Balamurugan et al. [118] dissolved $Pd(NO_3)_2 \cdot H_2O$ in weak triprotic citric acid monohydrate ($C_6H_8O_7 \cdot H_2O$). Powders or films are thermally decomposed in air or in dry oxygen to obtain PdO. In [10], PdO was obtained by decomposition of Pd nitrate at 400 and 600 °C in dry oxygen. In [115], samples of $PdNO_3$ impregnated on $La_{0.6}Sr_{0.4}Co_{0.2}Fe_{0.8}O_{3-x}$ for cathodes in solid oxide fuel cells working at intermediate temperatures were calcinated at 800 °C for 2 h, in air. Nanohybrids of PdO-WO_3 were obtained by annealing at 400 °C for 24 h in air to decompose $PdCl_2$ into PdO on WO_3 nanorods [117].

Other initial reagents than those already mentioned can be used. A proposed precursor material for spin coating is the β-ketoiminato palladium(II) complex [Pd(OAc)L] (L=PhC(O)CHC(Me)$NCH_2CH_2NHCH_2CH_2NH$(Me)C=CHC(O)Ph) [122]. It decomposes at 200–500 °C and forms Pd in argon and PdO in oxygen atmospheres. Kabcum et al. [117] impregnated 0.25–2.0 wt.% Pd onto WO_3 powder using palladium acetylacetonate ($Pd(C_5H_7O_2)_2$ dissolved in ethanol. After drying at 80 °C for 24 h, a heat treatment was conducted at 300 °C for 2 h. The powder (60 mg) was mixed (30 min) with a homogeneous binder solution (0.28 mL) of ethyl cellulose dissolved (at 80 °C for 12 h) in α-terpineol solvent to obtain a paste. The paste was spin-coated at 3000 rpm for 30 s on Al_2O_3 substrates. Annealing was performed at 150 °C for 1 h and then at 450 °C for 2 h (heating rate of 1 °C/min) for binder removal. The advantage of using organometallic reagents as a precursor is that it is convenient for decreasing the decomposition temperature.

Thick films in the SnO_2–CeO_2–PdO_x system were obtained by sol-gel screen printing on alumina substrates in [123]. A solution of 20 wt.% of $SnCl_4$ in water was mixed with solutions of ceric ammonium nitrate and palladium chloride. Ammonium hydroxide solution (25% in water) was added dropwise to the mixture to obtain a pH of 9.0, followed by stirring for 20 min. The resulting solution was aged for one day at room temperature. The volatiles were removed under reduced pressure at 70 °C to obtain powders. Powders were used to obtain a paste that was screen-printed on ultrasonically cleaned alumina substrates with Au electrodes and with a heater on the backside. The printed films were heated at 700 °C for 2 h. The as-fabricated sensor had enhanced sensitivity for CO gas detection.

In [116], the precursor solution of a Pd-organoamine was spin-coated (500 rpm for 10 s, followed by 3000 rpm for 60 s, using an acceleration of 800 rpm/s in both stages) on glass substrates and annealed at 200 and 250 °C (4 min, 24 h, and 48 h). This approach allows ink jet printing, as demonstrated in [124]. Amine-stabilized Pd clusters were obtained after decomposition of the precursor. Subsequent thermolysis of the clusters under air yielded smooth films. Under nitrogen atmosphere, the films consisted of large micron aggregates, while in low vacuum, Pd/PdO bilayer films were obtained (Figure 5).

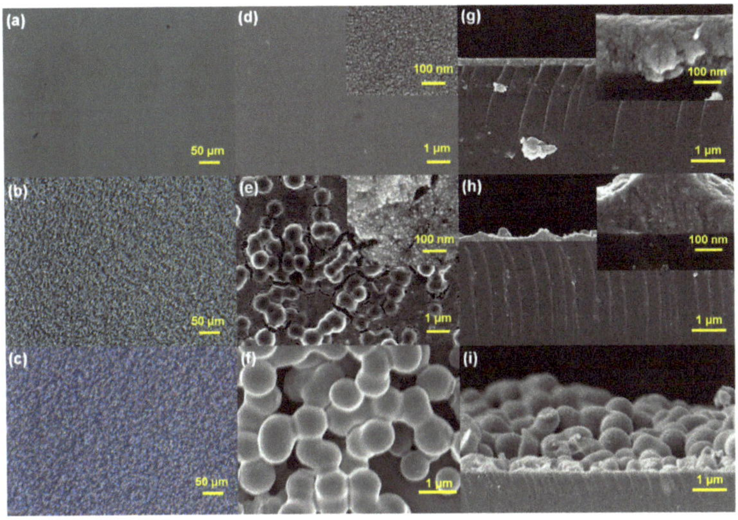

Figure 5. Reproduced with permission from [124]. (**a–c**) SEM images of Pd/PdO films obtained by thermolysis in air, low vacuum, and N_2. (**d–f**) SEM images of films from (**a–c**) were taken at higher magnification. (**g–i**) SEM images on cross-sections of the films from (**a–c**).

2.5. *Chemical Vapor Deposition*

A review on the growth via chemical vapor deposition (CVD) of iridium, platinum, rhodium, and palladium thin films is presented in [125]. The work indicated that for the deposition of Pd thin films, with the exception of $PdCl_2$, the precursors were of organic type: metal β-diketonates ($Pd(acac)_2$, $Pd(hfac)_2$), allyl complexes ($Pd(allyl)_2$, $Pd(Me\ allyl)_2$, CpPd(allyl), Pd(allyl)(acac), Pd(allyl)(hfac), Pd(Me allyl)(hfac), Pd(Me allyl)(acac), alkyl complexes ($PdMe_2(PMe_3)_2$, $PdMe_2(PEt_3)_2$, and $PdMe_2$(tmeda)), metal carboxylate ($Pd(acetate)_2$), and ethylene complexes ($Pd(C_2H_4)_2$), where acac = acetylacetonate = $C_5H_7O_2$, hfac = hexafluoroacetylacetonate = $C_5HO_2F_6$, allyl = η^3-C_3H_5, Cp = cyclopentadienyl = η^5-C_5H_5, tmeda = tetramethylethylenediamine = $C_6H_{12}N_2H_2$, acetate = $C_2H_3O_2$, and vinyl = C_2H_4. Other organic precursors proposed and reported in recent years will be mentioned in the next paragraphs. The Pd thin films were also grown via CVD in [126–129]. The PdCu alloy films on Si or SiO_2 substrates were deposited by CVD from a single-source precursor (β-diketonate complex) at 250–400 °C [129].

Among the CVD group techniques, the aerosol-assisted chemical vapor deposition (AACVD) is considered a promising variant, being simple and easy [130]. The precursor material was the commercially available palladium acetylacetonate, $Pd(acac)_2$ ($Pd(C_5H_7O_2)_2$, 100 mg, 0.328 mM), dissolved in toluene (10 mL). The aerosol mist from $Pd(acac)_2$ solution was generated by a piezoelectric ultrasonic humidifier. The aerosol was injected into the reactor tube with the help of a nitrogen (N_2) gas flow (150 cm^3/min). No hydrogen gas or any other reducing agents were used. Deposition for 30 min was performed on glass substrates heated to 475 °C.

Via AACVD, synthesis of palladium-nanoparticle-decorated WO_3-nanoneedles was accomplished in a single step [131,132]. Two source materials were tested: palladium acetylacetonate and ammonium hexachloropalladate. The first one led to better quality: the response of the PdO-decorated WO_3 toward hydrogen gas sensing was about 680–750 times higher than that of bare WO_3 nanoneedles. The operation temperature was at relatively safe and low values of 50–150 °C, and the addition of PdO minimized the effect of humidity on the sensor's response.

The AACVD was successfully demonstrated for fabrication of Pd-decorated WO_3 H_2 gas sensors on a flexible kapton substrate [133].

Pd nanoparticles were deposited by AACVD on an array of WO_3 nanorods [134]. The source material was $(NH_4)_2PdCl_4$ in a methanol solvent, the deposition temperature was 350 °C, and the N_2 gas flow rate was 300 sccm. Oxidation to PdO was performed in air at 500 °C for 2 h.

In [135], films of Pd (20–120 nm) were prepared on glass substrates via plasma-enhanced metalorganic chemical vapor deposition (PEMOCVD). The source material was $Pd(acac)_2$ (vapor pressure 0.1 hPa at 433 K) and the carrier gas was Ar (0.71 cm^3 min^{-1}). The total pressure in the deposition chamber was 4.5 Pa. The glow discharge power was 40 W. The sublimation temperature of $Pd(acac)_2$ was found to influence the films' composition relative to the Pd and C ratio in the as-deposited films. The optimum sublimation temperature was determined to be 413 K. Films were thermally decomposed at 623 K for 900 s, and they were composed of Pd nanoparticles in the range of 5–10 nm, which is considered suitable for catalytic activity.

In the CVD group methods, precursors are evaporated or sublimated and often show problems of thermal instability, resulting in poor quality and reproducibility of the as-deposited Pd film. To avoid formation of oxides is challenging, and the typically slow deposition rates in CVD are not convenient for high productivity and commercial fabrication. Some of the most popular gas-phase precursors are η^3-allyl and β-diketonato palladium complexes [136–138]. Among those of the first type, $[Pd(\eta^3\text{-}C_3H_7)_2]$ [139] and $[Pd(\eta^3\text{-2-Me-}C_3H_4)_2]$ [140] show thermal instability during evaporation and storage. They are also highly sensitive to air and moisture [137]. Another problem is that without the aid of a reactive gas [140], the deposited Pd films contain a low amount of carbon impurity (1 wt.%). The precursors of the second type, bis (β-diketonato) palladium (II) complexes, such as $[Pd(acac)_2]$ (acac = acetylacetonate), are thermally stable, but their volatility is low, and they require higher decomposition temperatures [138]. A combination of the allyl and β-diketonate groups within one complex was proposed in [141] to overcome the deficiencies of each type. Other synthesized CVD precursors are β-ketoiminate [142], β-ketoiminato palladium (II) complexes, such as $[Pd(CF_3C(O)CHC(CF_3)N^nBu)_2]$ [143,144], $[Pd(CH_3C(NH)CHC(O)CH_3)_2]$ [145,146], $[Pd(CH_3(O)CCHCN(CH_3)(CH_2)_3)_2)]$ [147], and $[Pd(2\text{-}NC_5H_4)NHNCH\text{-1-(O)-4-OMe-Ph)}(PPh_3)]$ [148], and palladium(II) allyl (β-ketoiminato) complexes [149], e.g., $[Pd(\eta^3\text{-2-Me-}C_3H_4)(MeC(O)CHCMeNPh)]$. The oxygen content in some of these precursors is minimized, leading to a lower oxidation of the substrate during formation of the metal film [145,146]. The complexes have a higher stability to air and moisture [150]. The search for new MOCVD source materials [151] continues, and it is one of the key priorities for the development of CVD technology.

Via atomic layer deposition (ALD), a variant of CVD, Pd thin films were successfully grown in [144,152–157]. In ALD, precursors are used that react with the surface of the substrate in a sequential way, one at a time. The chemisorbed metalorganic compound (e.g., β-ketoiminate [154] or $Pd^{(II)}(hfac)_2$ precursors [152]) decomposes under the influence of the second precursor, and the organic ligand is removed to form a metal film. The second precursor can be an organic (e.g., glyoxylic acid [152]) or a gas, such as oxygen [144]. In general, ALD is recognized as a technique that allows better control of the thickness and density. A recent review [157] on ALD of Pd pointed out the enhanced possibilities of this technique. Precise control of Pd deposited on challenging surfaces with different precursor chemistries and by using customizable processing conditions can generate low-dimensional nanostructures (single atoms, nanoclusters or nanoparticles, core-shell particles, and ultrathin films). The disadvantages of ALD are the slow deposition rate (in the range of nm/min) and that it allows growth of films with rather small surface areas.

2.6. Green Biosynthesis

The need for clean technologies led to a search of green biosynthesis routes ([5] and references therein). Natural, non-toxic, and eco-friendly precursor substances can be used to obtain usually metal particles. Examples of such substances are *Aspalathus linearis* (Rooibos) and *Annona squamosa* peel extracts, banana peel extract, Cinnamon zeylanicum bark, Curcuma longa tuber, broth of *Cinnamon camphora* leaf, *Gum acacia*, *Solanum trilobatum* leaf, *Soybean* leaf, *Zanthan gum*, or, more recently, *Crocus sativus* [158], coconut coir [159], *Trigonella* foenum-graecum (fenugreek) seed [159], *Rosmarinus officinalis* L. (rosemary) [160], and *Zaleya decandra* [161]. Among the most bioactive substances, aspalathin, nothofagin, and aspalalinin are considered. However, the choice of the plant and processing method and conditions are critical regarding the quality and quantity of the extracted compound. *Aspalathus linearis* leaves (0.03 g) were added to 300 mL of deionized water at room temperature for ~30 min [5]. The orange extract (pH~5) was filtered three times to remove solid residuals and was slightly fermented. In the as-prepared extract, $PdCl_2$ (3 g) was added and fully dissolved under stirring at room temperature. After 30 min of reaction time necessary to complete the oxidation–reduction reaction, the color of the solution changed. By drying in air at 90 °C for 1 h, an amorphous product was obtained, which was further annealed for 2 h at 100 °C or at a higher temperature. The product annealed at 100 °C was composed of non-agglomerated crystalline nanoparticles with different morphologies and with an average mean size diameter of 12.7 nm. Via TEM, particles were ascribed to fcc cubic Pd. At higher annealing temperatures (200–600 °C), the product was oxidized. At intermediate temperatures, X-ray diffraction indicated co-existence of Pd and tetragonal PdO, while at 600 °C, only PdO was detected. Time and/or temperature to obtain PdO were considered [5,162] lower than for other processing routes, such as heating of $PdCl_2$ in a $NaNO_3$ melt, thermal decomposition of $Pd(NO_3)_2{\cdot}2H_2O$, oxidation of Pd in air or in oxygen, or direct carbon arc vaporization of Pd.

In [37], citric acid was used as a mild and environmentally friendly stabilizing and reducing agent to obtain nanohybrids of graphene oxide decorated with Pd. Graphene oxide was treated with aliphatic amines 1-octadecylamine and 1,8-diaminooctane, followed by addition of a palladium chloride in a liquid medium and citric acid.

3. Physical Routes of Pd and PdO Thin Films' Growth

Through physical technological methods, either the Pd metal precursor or the Pd oxide films can be prepared in a single processing step. Among the physical methods are sputtering and cathodic arc deposition, laser ablation, ion or electron beam-induced deposition, evaporation, and supersonic cluster beam deposition (Figure 1).

3.1. Sputtering and Cathodic Arc Deposition

Sputtering a target of a metal through the bombardment with energetic ions of an inert gas (e.g., argon or helium) produces collision of these energetic ions with the target and ejects target metal atoms into space. These metal atoms are deposited on a substrate material, forming a metallic film. Sputtering is recognized as a fast-rate vacuum coating technique that allows creation of strongly adhesive coatings on substrates with complex geometries. Magnetron sputtering is popular for deposition of metals and oxides. The energetic ionic species and electrons are confined in a magnetically enhanced glow discharge. Confinement produces a relatively high plasma density and promotes acceleration of the ions, which will vaporize the source material (target). The ion or molecular ion-sputtered species from the target, i.e., Pd in this case, fly and deposit on the substrate. If oxygen, as a reactive gas, is introduced in the deposition chamber, one can obtain in situ Pd oxide films. In practice, to avoid oxygen contamination of the sputtering chamber, thermal oxidation of the as-prepared Pd films is preferred to direct reactive sputtering. Simultaneous co-sputtering from different targets is applied to obtain bi- or multi-component metal alloys. This is performed by tuning the power applied on each sputtering gun/target. The drawback of sputtering is considered its limitation in deposition of films into deep structures. The literature presents different solutions of directional sputtering approaches, such as collimated sputter deposition, elevated temperature/reflow deposition, long-throw sputter deposition, and ionized magnetron sputtering [163,164].

Thin films of PdO on (111) Si-wafers and alumina circuit-board substrates were deposited by magnetron sputtering from a Pd target in [12]. The deposition atmosphere was O_2:Ar = 1:1 for a total pressure of 0.35 Pa (starting base pressure was (2–4) $\times 10^{-4}$ Pa). The as-grown film consisted of Pd oxide, according to XRD, but an unidentified impurity peak was detected at $2\theta = 56°$. The lattice parameters of PdO increased, and the XRD line of the impurity disappeared during post-annealing at 420 °C in air.

Yoon at et al. [165] deposited, for hydrogen gas sensors, Pd thin films (50 nm) by magnetron sputtering in Ar gas at a pressure of 1.2×10^{-2} Torr (initial vacuum pressure was 1.2×10^{-6} Torr) and for an input power of 1000 W.

Via *dc* magnetron sputtering, Pd films were obtained in [166] on oxidized Si (SiO_2 thickness was 100 nm) and for a source power of 60 W. The distance between the target and substrate was 10 cm, and an Ar atmosphere was used (initial vacuum was 10^{-6} Torr). When the deposition pressure was increased from 22 to 122 mTorr, the average grain size in the films changed from 10 to 30 nm.

Hao et al., via *dc* magnetron sputtering, obtained porous films of Pd on porous anodic aluminum oxide (AAO) templates for room-temperature chemoresistive hydrogen sensors [167] (Figure 6). After degassing the AAO substrate in a vacuum (1×10^{-3} Pa) by heating in the sputtering chamber at 100 °C, Pd films were deposited at ~50 °C in an Ar (99.999%) atmosphere (made from a vacuum of 4×10^{-4} Pa) at a pressure of 0.7 Pa. As-prepared Pd films (45 nm) were annealed for 30 min in the deposition chamber at 150, 200, 300, and 400 °C. It was shown that annealing decreased the response time (time delay to reach 90% of resistance), and the drift of the baseline for the annealed sensor was not observed.

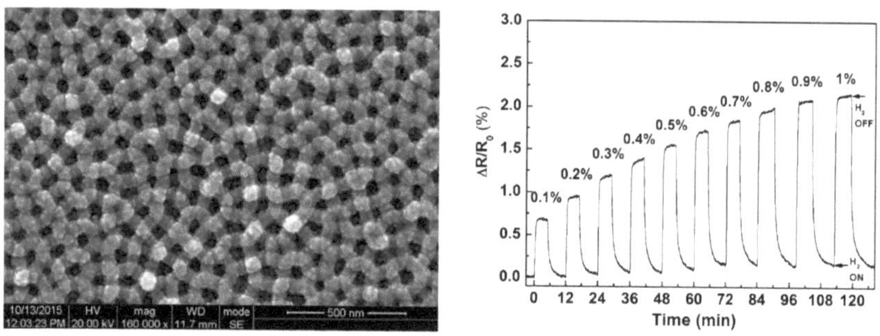

(a) (b)

Figure 6. Reproduced with permission from [167]. (**a**) SEM image of the porous Pd thin film on AAO substrate prepared by *dc* magnetron sputtering and post-annealed at 200 °C, and (**b**) room-temperature response at various hydrogen concentrations in nitrogen carrier gas. On the Pd film, Au electrodes (10 mm × 3 mm) were deposited by thermal evaporation.

Magnetron sputtering was also used in [168] to grow Pd thin films on Ti–6Al–7Nb plates. After a heat treatment in air at 450, 550, and 650 °C for 1 h, their mechanical, corrosive, and wettability properties and in vitro bioactivity were studied. These materials are considered for implant applications.

Other examples of PdO-sputtered films were reported in [169–172]. Reactive sputtering was employed to obtain PdO nanoflake thin films for CO gas sensing at low temperatures in [170,173]. Layers of Pd were incorporated by *rf* sputtering into an electrode-functional layer of a thin-film-based nickel–yttria-stabilized zirconia (Ni–YSZ) solid oxide cell (TF-SOC) [174]. Selective doping (Gd or Tb) of Pd films on (001) MgO for hydrogen gas sensing was performed by *dc* magnetron sputtering at 550 °C, and they showed the (002) preferred orientation [175]. PdO thin films [176] with nano-sized cracks (Figure 7), obtained in a vacuum atmosphere (base pressure of 4×10^{-8} Torr with 5, 10, 15, 20, 25, and 30% oxygen) by reactive sputtering at room temperature on oxidized (100) Si, demonstrated an ultrahigh sensitivity (~4.5×10^{3}%) and a fast response time at room temperature to H_2 gas in nitrogen.

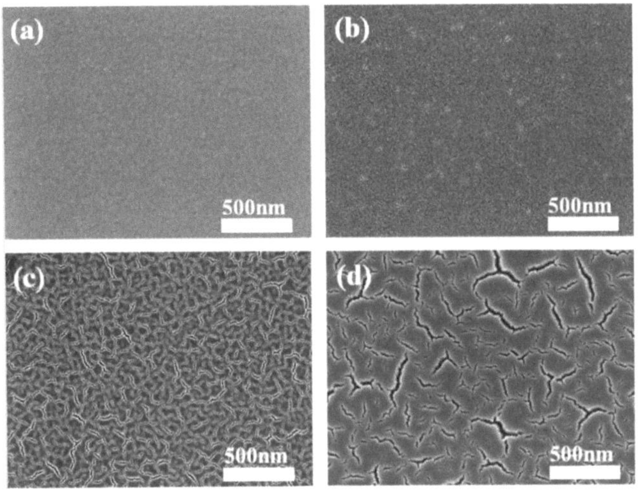

Figure 7. Reproduced with permission from [176]. SEM images of reactively sputtered films in different oxygen atmospheres: (**a**) 15%, (**b**) 20%, (**c**) 25%, and (**d**) 30%.

A variant of sputtering, namely, the plasma-assisted sputtering, was performed in [177] for an ionization current of 3 mA under 1.0×10^{-1} Torr. The thickness of the Pd nanostructures deposited on (001) GaN was in the range of 1–30 nm. Samples were annealed for 450 s at 650, 700, and 750 °C in a vacuum of 1.0×10^{-4} Torr, and evolution of the morphology was studied.

Thin films in the Pd-Zr-Y-O and Pd-Pt-O systems were grown via reactive magnetron sputtering and the pulsed filtered cathodic arc deposition technique in [178] and [179], respectively.

3.2. Electron Beam Coating and Electron and Ion Beam-Induced Deposition

For the growth of Pd films, electron and ion beam-induced deposition (EBID and IBID) were demonstrated. These methods were successful in fabrication of 3D structures/patterns [180]. In these vacuum-based techniques, a precursor material (usually of a metal-organic type) is introduced (injected) in the deposition chamber, and it is absorbed on a substrate. A focused electron or ion beam is used for processing the precursor and through its decomposition to obtain the film. Via IBID, and using Pd-bis(hexafluoroacetylaceacetonate) as the precursor film, a film composed of nanowires of Pd was obtained in [181]. EBID was applied in [182] on a spin-coated precursor film of Pd-hexadecathiolate on a Si substrate. These methods share similar features with two-step chemical methods that use a precursor Pd-based compound deposited in the first step and decomposed in the second one. A typical conventional electron beam coating to obtain a Pd film (with a thickness of 12.3 nm) was applied in [183]. The film was deposited on glass substrates. The base pressure before deposition was 3×10^{-5} mbar. The substrate temperature was 32 °C, cathode voltage was 8.5 kV, and coating rate was 12 $A \cdot s^{-1}$. Annealing was performed in air at 350, 450, 550, and 650 °C for 1 h. According to XRD, the films consisted of Pd phase. In [184], the electron beam gun was used to evaporate a Pd ingot target and deposit Pd thin films (17–100 nm) on a 2-inch Si wafer substrate with a sputtered buffer layer of SiO_2 of 400 nm in thickness. The substrate was positioned at about 35 cm above the target. The film deposition was performed at 8×10^{-5} Pa in a vacuum chamber. XRD indicated that the Pd film was (111) oriented. The sheet resistance of the Pd film (50 nm) annealed at 100, 150, and 200 °C for 3 h was not affected up to 150 °C. In [185], Pd films were deposited by e-beam evaporation of palladium pellets (99.95%). The substrate was thick fused silica (1 mm) and, during deposition in the vacuum (base pressure of 10^{-4} Pa), it was heated at 90 °C. The growth rate of the film was set at 0.5 nm/s. The Pd films with thicknesses of 7.2 (S1), 12.6 (S2), 17.5 (S3), and 27.3 (S4) nm were oxidized by annealing at 500 °C in a vacuum at 150 Pa. The annealing time was selected to be 1 h for each 10 nm of the film's thickness, ensuring the full oxidation of the films. The oxidized films were exposed for 10 min to H_2 atmosphere (5 vol.% in nitrogen) at room temperature. This procedure reduced PdO to r-Pd, and the thickness of the samples decreased without restoring the initial one, while porosity increased (density decreased by 1.68–2.6 times). Films were used in room-temperature optical H_2 gas sensors (Figure 8). A higher film thickness increased the response and recovery times of the as-prepared Pd films, and it was almost constant for r-Pd-treated films. Palladium films were deposited by electron beam evaporation on $Si/SiO_2/Pt$ substrates, and after annealing in oxygen at 500 °C for one hour to form the palladium oxide (PdO), films were tested for hydrogen ion (pH) sensing [186]. The authors found that the sensitivity of the device (extended-gate field-effect transistor (EGFET)) was 42.36 mV/pH.

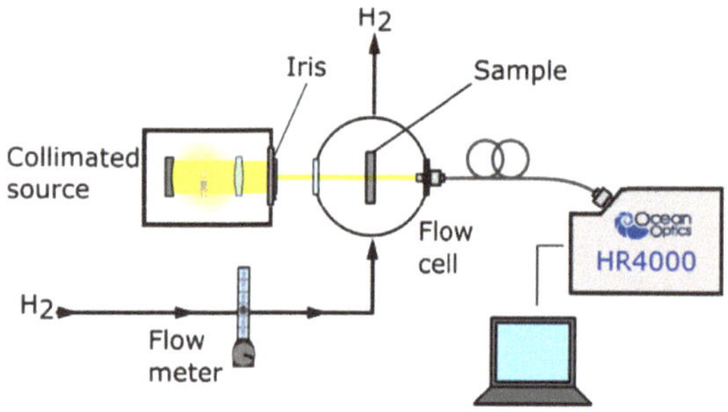

(a)

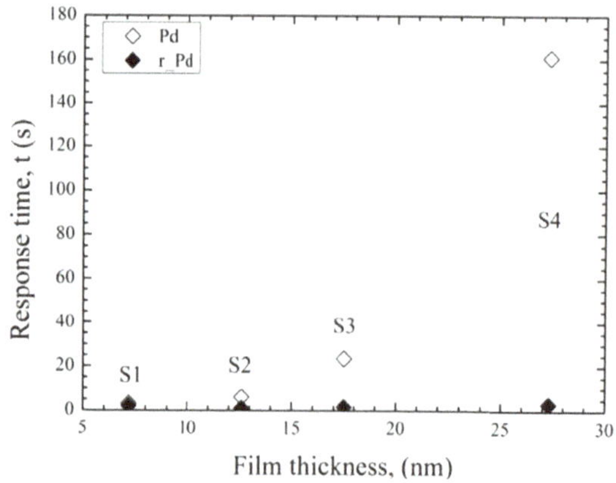

(b)

Figure 8. Reproduced with permission from [185]. (**a**) Optical sensor arrangement based on optical absorbance of the sample when irradiated from a source of a halogen lamp in the spectral range of 400–800 nm. (**b**) Response time (calculated as the average time to change from 5% to 95% of the absorbance) at room temperature of the samples with different thicknesses to 5 vol.% H_2 gas in nitrogen.

3.3. Pulsed Laser Deposition

In the pulsed laser deposition (PLD), a pulsed laser is focused on a Pd or PdO [187–191] bulk target. Through ablation, the material from the target evaporates/sublimates or converts to a plasma, and it is directed on a substrate to form a film. Laser type (pulsed or continuous), wavelength, and its fluence, gas pressure in the deposition chamber, target–substrate distance, and substrate temperature are among the parameters to control the process and the film quality. Ablation of the Pd bulk targets can also be realized in a liquid instead of a gas environment, and the product is usually a nano-powder. The type of liquid was shown to influence not only the size and morphology, but also the oxidation and hydrogenation level of the resulting Pd particles [192]. Via this approach, ultra-small Pd nanoparticles were deposited on CdS nanorods for the photocatalytic hydrogen production [193]. The morphology and porosity of the bilayer films of Pd/TiO_2 deposited by PLD (Nd-YAG laser, emission wavelength of 355 nm, 5 ns pulse duration, repetition rate of 10 Hz, energy per pulse of 71 mJ, energy density on the target of 25 J/cm^2, target–substrate distance of 40 mm, and substrate temperature is room temperature) were modified by changing the oxygen and argon pressures. The films were deposited on the quartz substrate of surface acoustic wave (SAW) room-temperature H_2 gas sensors [188]. The quality of Pd deposits versus the ablation process and Al_2O_3 substrate characteristics were reported in [187]. Pd was deposited by PLD on wood for Ni electroless plating in [194]. Different helium background gas pressures (50, 350, and 800 mTorr) during Pd thin films' PLD deposition influence the crystallite size and the tendency of their orientation in the <111> direction [189]. Pd thin films of 0.4–4 monolayers were deposited on (001) Cu at room temperature by PLD. The Pd–Cu interface formation is characterized by an alloying–dealloying mechanism. Up to two monolayers, Pd atoms are incorporated into the Cu substrate for less than half-filled layers but expelled if the Pd coverage is close to a complete layer. In this latter case, the top layer is composed of Pd [195]. PLD was used for doping the surface of polycrystalline thin-film SnO_2 or SnO_2(Cu) gas sensors for the enhancement of the sensitivity to 1% H_2 in N_2 (at 200–380 °C), with two orders of magnitude [196]. Pd-doped SnO_2 thin films for gas sensors were grown on (100) Si substrates using PLD at room temperature in [197]. The microstructure of the films depends on the O_2 background pressure (10–100 Pa) and on the target–substrate distance. When the target–substrate distance was close to the plume length, textured thin films were obtained, and the authors explained this phenomenon considering an adiabatic expansion. Nanostructured Pd films obtained by PLD (KrF excimer laser, 248 nm, pulse time of 20 ns, repetition rate of 100 Hz, background pressure in the deposition chamber of 10^{-5} mbar, laser fluence of 5 ± 0.1 J/cm^2 or 250 MW/cm^2, the laser beam was focused at a 45° angle on the target with a spot area of 1 mm^2, and Pd metal target with purity of 99.9%) have very different electrochemical characteristics from those of common coarse-grained films [198]. Stability was good under hydrogen charge/discharge cycling, and this was related to the lack of an abrupt α-to-β phase transition. Upon oxidation of Pd films with controlled thickness and grain size deposited by PLD, spontaneous formation of uniformly distributed arrays of conical PdO "tips" useful for field emission applications was reported in [191,199].

Chemoresistive acetone sensors of ZnO were decorated with Pd particles by PLD [200]. Sensors operated in the temperature range 159–200 °C, and they showed an enhancement of the response factor between 2 and 7 when compared with pure ZnO sensors. The detection limit was 26 ppm at 200 °C.

3.4. Evaporation (Sublimation) of Pd

Evaporation (sublimation) of Pd (e.g., a metal foil resistively heated by a V-shaped tungsten filament) and deposition is a simple and popular technique [106,201–205]. Evaporation was realized in a high-vacuum chamber (e.g., at 6.65 × 10^{-4} Pa in [84] or 18–57 × 10^4 Pa in [87]). These films were deposited on (11$\underline{2}$0) α-Al_2O_3 [86], polished polycrystalline Al_2O_3 [87], SiO_2/(100)Si [83,84], Pyrex glass [82], (100)Si [84], optical quality quartz [106,202], and KCl with a buffer layer of amorphous carbon [106,205]. The palladium vapor pressure, *P*, in the deposition chamber was set according to Equation (1) [206]:

$$\log_{10}P(\mathrm{Pa}) = -20{,}150T^{-1} - 0.419\log_{10}T - 0.302 \times 10^{-3}T + 13.670 \quad (1)$$

In [164], the palladium vapor pressure was:

$$\log_{10}P(\mathrm{Pa}) = 8.749 - 18{,}655T^{-1} \quad (2)$$

in the temperature range 927–1427 °C. When the rate of evaporation in the vacuum was compared among the platinum-group metals (Pt, Ir, Os, Pd, Rh, and Ru), the values at lower temperatures were the highest for Pd. The heat of vaporization was the lowest (H_{292K} = 89.2 ± 0.8 kcal/mol). The presented information promoted Pd as the most convenient element among the platinum-group metals for thin-film deposition through physical evaporation: it is considered that Pd has an anomalously high metal vapor pressure. Substrate temperature during Pd thin films' deposition was selected to be at room temperature [106,202], but in some experiments, a low temperature was used, e.g., 77 K in [201]. The physical evaporation allows deposition of Pd films with different thicknesses. Attempts to grow Pd ultrathin films are presented in [202], and films with thicknesses of 1, 5, 10, and 80 nm are compared. The study pointed out the challenging limitations for ultrathin films regarding their tendency for formation of an island-type microstructure featuring a high possibility of discontinuity, accompanied by the lack of a percolation path for electrical conduction. It also pointed out the strong substrate influence, implying occurrence of a high crack density and stoichiometry changes due to the substrate–film interdiffusion. In [106,205], PdO ultrathin and thin films with thickness of 5–40 nm were obtained. The thermal annealing of the Pd precursor films was performed under dry oxygen. Films with low thickness of 5–15 nm were heated for 1 h, while thicker ones for 2 h. Annealing temperatures were 237, 297, 397, 497, 597, and 797 °C. According to X-ray diffraction measurements, oxidation of the Pd precursor film with partial or full formation of PdO took place for annealing temperatures of 297–797 °C. Heras et al. [201] found that for Pd films deposited on cold substrates (glass), during subsequent annealing by heating from −196 °C to 400 °C, they changed their morphology and absorbed oxygen without formation of a bulk PdO oxide. A fiber morphology and texture with the axis (111) normal to the substrate developed, and the process was stronger for temperatures closer to the higher values of the indicated interval. The penetration of oxygen into the bulk depends on the surface defect density, and significant oxygen incorporation into polycrystalline palladium without formation of PdO was also reported in [207,208] at 247 °C.

3.5. Supersonic Cluster-Beam Deposition

In the supersonic cluster-beam deposition [209], a Pd target rod is subject to a He plasma ignited by a pulsed electric discharge between the Pd rod (cathode) and anode, producing the ablation of the target. The ablated Pd atoms thermalize inside the cavity via collision with the inert gas and condense into clusters. The mixture of the cluster and He is expanded in a vacuum chamber through a nozzle and forms a supersonic beam. Substrates intercept the supersonic beam in a second chamber separated from the expansion chamber by an electro-skimmer. The deposition rate is about 1 mm/min, and the kinetic energy is small enough (0.5 eV order) to avoid cluster fragmentation.

In [210], deposition was performed on glass substrates. The authors adopted two strategies for oxidation of Pd clusters. In the two-step ex situ route, Pd was deposited with He carrier gas, followed by oxidation annealing in air at 200 and 400 °C. In the one-step in situ route, the carrier gas was a mixture of 80 at.% He and 20 at.% O_2. Films were nanostructured with a particle size below 20 nm, with the smallest particles below 5 nm. They were porous and very soft and could be easily detached from the substrate [211].

4. Some Practical Aspects of Pd Films' Oxidation Toward PdO and Specific Issues Concerning Applications of the Films

The above-presented information indicates that oxidation of Pd is a complex process, and some details are not completely understood. The next paragraphs will briefly screen some practical aspects, but details of the oxidation mechanisms/models will be neglected.

When a Pd foil is thermally oxidized in ambient conditions at 650 °C, a bulk PdO layer develops, and after 1.52 min [212] of oxidation, it is detected by XRD. The thickness of the PdO layer increases with time, and the oxidation process is governed by a parabolic law directly related to temperature and oxidation time parameters [212]. The PdO film on Pd presents a preferential growth corresponding to the (002) PdO direction, since the Pd foil substrate is polycrystalline, with grains showing a preferred (002)-normal direction. The lattice constants of the Pd oxide film are a = 0.3035 nm and c = 0.5323 nm, as expected for the bulk PdO [2]. The authors noted that the oxidation rate is high as a result of lower surface energy activation for the (002) direction than for the (111) one. A model for oxidation of Pd with formation of PdO based on motion and creation of cation vacancies was proposed, and within it the necessary energy (E_a) for movement or migration of cationic vacancies depends on the thickness of the oxide layer. A lower E_a of 0.034528 eV is found for an ultrathin film, compared to E_a of 0.115883 eV for a thick one. This explains the faster diffusion process in a thin oxide layer than in a thick one and the already mentioned parabolic dependence. The activation energy for oxygen diffusion in Pd bulk has been estimated to be in the range of 84–98 kJ·mol^{-1} [201], while the free energy of formation of PdO (at 727–875 °C) is [202]:

$$\Delta G^0{}_T(\mathrm{PdO}) = -27460 + 23.9T \pm 280\ (\mathrm{cal/mol}) \quad (3)$$

It is worthy to note that the solid oxide formation results in a significant increase in weight. Other features of interest are:

(i) Kinetics of the Pd oxidation (in the mbar pressure range) depends on the crystal surface: oxidation of Pd(110) proceeds at ~100 K lower temperatures than Pd(111) [213].
(ii) The PdO layer formed as a skin on Pd is green when relatively thin [23]. When thicker (annealed in air at 800 °C), it turns gray.
(iii) Depending on the heating rate and annealing temperature, the PdO layer decomposes in air at 825–900 °C. A decomposition temperature of 870 °C was estimated for annealing under 1 atm of oxygen [23,214–216].

Samoylov et al. [214] found that the thermal stability of the PdO film in an oxygen atmosphere enhanced from 810 ± 5 to 860 ± 5 °C when the thickness of the initial Pd layer increased from 10 to 95 nm, respectively. The authors also noted that the resulting Pd film, unlike the continuous Pd film used as a precursor to obtain the oxygen-decomposed PdO film, was composed of isolated hemispherical Pd nanocrystalline particles with well-defined faceting and preferential (111) or (100) alignment.

In the ultrahigh vacuum, PdO was reported to be unstable at temperatures above 147 °C [215].

Okamoto and Aso [203] claimed decomposition of PdO at room temperature under H_2 atmosphere from the rapid increase in electrical conductivity.

The results of more recent first principles studies indicated that the thermodynamic stability of PdO depends on the crystal surfaces [216]. These aspects significantly impact the catalytic efficiency. Although for methane oxidation, Pd is one of the best catalysts, the relation between the presence of oxides and the catalytic activity is complex. In methane oxidation over a Pd(100) single crystal, it was found [217] that the metallic surface was active, but it was unstable under reaction conditions. Growth of the oxide promoted low activity, but for a thicker oxide, the activity increased, and above a certain level, it decreased, even if the thickness of the oxide continued to expand. In these processes, depending on the thickness, the exposed surface of the oxide changed, being (101) or (100), with the second one showing low activity due to lack of coordinatively unsaturated (CUS) sites (i.e., Pd atoms on the surface) where the reactants can adsorb. The first layer was of (101) PdO. For very thick films, the (100) oxide was stable, and it was no longer coupled to the substrate, so the surface became inactive. For a few layers of the oxide, a ligand effect [218] with the Pd layer below occurred, and it contributed high activity for the intermediate thickness of the oxide.

To overcome the instability of the catalytic activity, Pd alloys were obtained and tested, e.g., PdAu in [11]. Composites, such as conductive polymers (polyaniline)/PdO (PANI/PdO), are another promising solution. In [219], a CH_4 gas sensor of the quartz crystal microbalance (QMS—a mass sensitive device) type was obtained by a layer-by-layer self-assembly method that is based on electrostatic force and in which an in situ chemical oxidation polymerization approach in the presence of PdO nanoparticles at room temperature is applied. The method uses dip coating on hydrophilic glass substrates. Substrates were treated sequentially and ultrasonically with $CHCl_3$, C_2H_5OH, and deionized H_2O, each for 20 min, and dried in nitrogen gas. Treated substrates were coated by immersing in 1.0 wt.% PDDA (poly(diallyldimethylammonium chloride), molecular weight 200,000–350,000, polycation) aqueous solution for 5 min and washing with deionized water.

Washing prevents cross-contamination with the second coating solution of 0.2 wt.% PSS (poly(sodium-pstyrenesulfonate), molecular weight 70,000, polyanion), with the opposite charge to PDDA. The PSS polyanion layer provides the charges for the adsorption of the first-layer polycation in the formation of a polyaniline/PdO thin film. The film PANI/PdO was deposited on a glass/PDDA/PSS substrate. Dip coating of the substrate was performed in a solution prepared as follows: (a) Aniline monomer was added into the 1.0 M camphor sulfonic acid (CSA) solution with PdO nanoparticles at room temperature. The ratio of aniline to PdO was 5 to 1. (b) The mixture was reacted with ammonium persulfate for 20 min. The progress in this research direction can have high importance and impact because the greenhouse effect of methane (CH_4) gas is about 30–80 times higher than for carbon dioxide (CO_2) [220]. Methane is also a colorless and odorless gas, and the key component of the natural gas. It is flammable and explosive when its concentration is 5–14% in air.

(iv) At rather high annealing temperatures (~1000 °C) of Pd, a slight weight increase was detected because of higher oxygen solubility at higher temperatures [221,222].

The questions to be addressed in the future studies will also have to deal with possibilities and novel techniques of fast films' characterization, e.g., in situ local mapping of the degree of oxidation on the surface and in the volume of the film. In [11,172], the authors used advanced characterization methods, such as grazing-incidence X-ray diffraction (GIXRD) and photoemission spectroscopy, respectively, to observe the details of oxidation. Evolution during oxidation of Pd particles (2.3 nm) was investigated with in situ time-resolved X-ray absorption spectroscopy (XAS), supported by theoretical simulations [223]. Obtaining coatings with organized surfaces, e.g., with ordered and periodic patterns of hillocks of Pd thin films, was studied by XRD and AFM in [224]. The authors observed the variation in compressive stress and the total hillock surface area versus the thickness of the Pd film. They introduced formation of a PdO layer at the interface of Pd as one relaxation mechanism and found that this mechanism was the dominant relaxation process in the films with a thickness higher than 40 nm. Hence, to generate spontaneously ordered hillocks over the Pd surface, a thickness of over 40 nm is required since it is expected to lead to a constant value of the total hillock surface area. The aspects of oxygenation and morphology formation will help in designing the films and finding their optimum deposition route to achieve the desired and controlled functional characteristics that are needed for the targeted application. Ultimately, it will enhance the performance (response time, recovery time, response percent, sensitivity, selectivity, and reliability [225]) of the sensor and will provide its reproducible and stable operation. Notably, the effect of the preparation process parameters on the responses of the gas sensors of SnO_2 loaded with Pt, Pd, and Au was investigated for prediction purposes with artificial neural networks in [226].

Since H_2 gas is highly flammable and explosive when its concentration is 4–75% in air and it cannot be detected by human senses, in the emerging clean economy of hydrogen, detection of this gas in a very sensitive, highly stable, and selective manner and in a wide range of temperatures will be crucial. This is challenging, and the current typical conductometric and electrochemical (amperometric and potentiometric) sensors show limitations.

Review articles [225,227] indicated eight types of commercial hydrogen gas sensors based on electrical conductivity, thermal conductivity, acoustic, mechanical, optical, electrochemical, catalytic, and work functions (Figure 9). Metal oxide sensors of conductometric-type work optimally only at high temperatures, in low relative humidity, and their sensitivity for hydrogen is quite low. On the other hand, Pd is a highly electroactive element, but undesirable processes occur, and we mention hydrogen embrittlement due to formation of the palladium hydride phase (accompanied by lattice expansion and resistance increase [167]) and of the irreversible contamination with the other gases from the atmosphere. These processes decrease the sensing efficiency of a sensor made of a pure metal. A sensor made of PdO will not be poisoned due to its higher working temperature. Hence, these sensors are thought of as candidates with a higher potential than those made of Pd but, ultimately, this will also depend on the application needs. In addition, PdO electrochemical H_2 gas sensors show high sensitivity and selectivity and might be a convenient solution for sensing at room temperature [228,229].

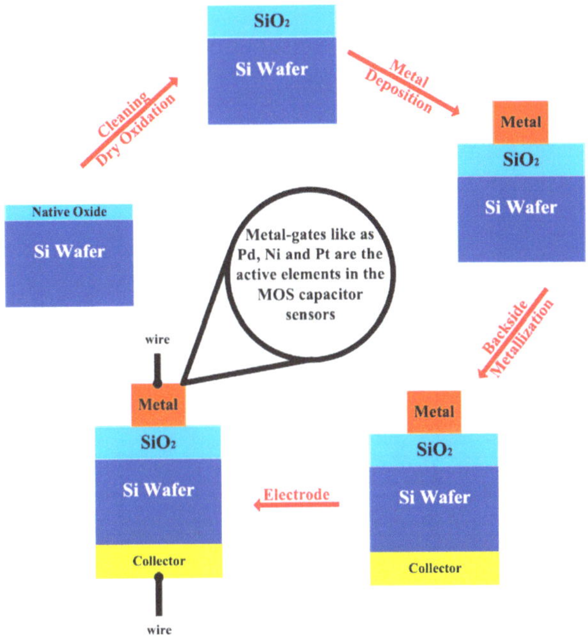

Figure 9. Reproduced with permission from [225]. Fabrication of a typical Pd/MOS (MOS—metal oxide semiconductor) capacitor hydrogen sensor. The hydrogen diffuses from the metal Pd gate (active element) and creates a dipole layer at the (Pd/SiO_2) interface that changes the work function of the active element. The response, R (%) = $(C_H - C_N)/C_N \times 100$, where C_H and C_N are the capacitance of the sensor in hydrogen gas and pure nitrogen, respectively. The carrier gas is nitrogen, argon, and air.

The present strategies for improvement of gas sensors are to use composite and/or alloyed films and to modify/control their morphology, e.g., particle size, shape, distribution, porosity, nano-structuring, and defects (e.g., cracks) [20,230–242], and all these features depend on preparation and post-processing technology. A low detection limit of 300 ppb, a fast response time of 3 s toward 500 ppb, excellent selectivity at 200 °C, long-term stability, and outstanding tolerance to humidity were determined for a H_2 gas sensor built with Fe_2O_3-core/Pd/PdO-shell nanoparticles [243]. Excellent sensing properties were discussed within the frame of the Pd/PdO 'spillover effect', where the electron transfer and charge accumulation related to adsorbed oxygen are promoted by formation of PdO/Fe_2O_3 p-n heterojunctions.

The indicated strategies also promote new practical opportunities. For example, indium oxide-palladium (In_2O_3-Pd) sensors have a good sensitivity toward nitric oxide; hence, it can be used for detection of nitroaromatic explosives [244]. In another example, binary metal oxide solid solution ($Ir_{(1-x)}M_xO_y$, where M = Pd, Rh, and Ru) thin films obtained by *dc* reactive magnetron sputtering (each of 2 Ir targets were supplied 25 W power at a frequency of 144 kHz, while for the Pd target, power was 50 W, and pressure in the chamber was ~4 Pa, with 20% oxygen partial pressure) on 316 stainless-steel or SiO_2/Si substrates were assessed, targeting implantable neural interfacing applications [245]. Undesirable nanoflake growth of IrO_x is suppressed by introduction of Pd (Figure 10). The presence of Pd also activates higher oxidation states, including a +5 oxidation state. Both effects are believed to be the reason for better performance of the binary metal oxide films when compared to single metal oxides.

Salagare et al. [246], in a selective three-electrode electrochemical sensor, modified a graphite electrode with PdO-rGO composite flakes and demonstrated a high degree of consistency of electrode sensitivity measurements of nitrite at room temperature. In a recent article [247], palladium nanoparticles were decorated on TiO_2 nanotubes (TNT) by gamma-ray irradiation. They degraded nitrogen monoxide (NO) in the photocatalysis process, achieving 53.30% NO removal efficiency, which is 1.6 times greater than that of pure TNTs. A layered nanoelectrode based on GO and rGO decorated with n/p nanoparticles of palladium oxide and cadmium sulfide has shown high efficiency for supercapacitor applications [248]. Nanoparticles of PdO (n-type) and SrO (p-type) on GO and rGO were tested [249] for decomposition of organic pollutants, such as mixed dyes of Rhodamine B and methylene blue (RhB/MB), insecticides, such as imidacloprid, and the removal of heavy metals, such as chromium ions. Results were considered promising for water cleaning.

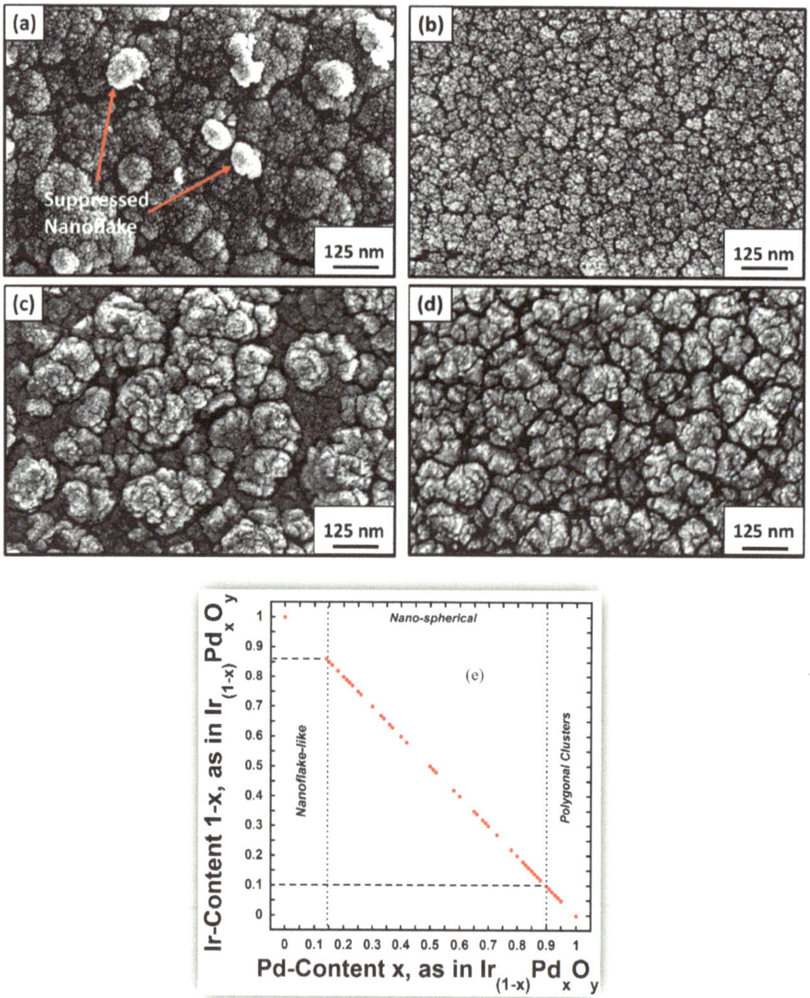

Figure 10. Reproduced with permission from [245]. SEM images showing the morphology of $Ir_{(1-x)}Pd_xO_y$ films deposited on 316 SS substrates for: (**a**) x = 0.14, (**b**) x = 0.50, (**c**) x = 0.90, and (**d**) x = 0.95. Map of morphology summarizing results from (**a**–**d**) depending on the composition of the $Ir_{(1-x)}Pd_xO_y$ films (**e**).

5. Conclusions

This review presented different deposition routes of PdO and Pd films and heterostructures. The largest contribution in the literature is on Pd films' fabrication, without a further investigation toward PdO formation as a product. This situation is due to the dominant interest from the industry in catalytic processes where Pd rather than PdO has a key role. However, the development of different applications, such as gas sensors, especially of ozone, NO_x, H_2, CH_4, etc., has become important and promoted a higher motivation, expanded attention, and research on fabrication of PdO films. Gas sensors are used in different industries, such as medical, aerospace, petrochemical, mining, and automotive.

In the literature, there are very few papers to compare films for a specific application obtained by different technological routes. An isolated example is in [250], in which the electrochemically grown Pd nanoparticle films were observed to show better hydrogen sensing responses than the sputtered Pd thin films.

Some of the reviewed deposition routes are highly complex or sophisticated, require expensive and specific raw materials, and are using or generating harmful, undesirable substances. This situation leads to fabrication cost increases and environmental and integration problems. Recent approaches tried looking at green synthesis or biosynthesis routes, and some progress was achieved in recent years ([5] and references therein, [37,251]), taking advantage of non-toxic, clean, eco-friendly, natural substances. Although these developments are promising, in general, green synthesis produced particles rather than films.

Films of Pd or PdO are usually composed of randomly oriented crystallites, although in some reports, the authors observed a certain degree of texture [175,184,252,253], which was accidentally obtained in most cases. As we presented above, the activity of different crystal planes of the palladium oxide is very different. Therefore, preparation and exploration of textured or epitaxial, single-crystal or polycrystalline palladium-oxide-based films might provide interesting new opportunities for applications and may afford a better understanding of the complex catalytic/sensing processes. Despite the huge effort of the international scientific community, revealing the details of these processes requires further research—the physical–chemical mechanisms governing the interaction with the medium or taking place inside the material containing Pd are insufficiently understood. In this regard, there is an urgent need for new, fast, in situ, and reliable investigation techniques that are effective at different scales.

The number of devices reported and fabricated on PdO films is significantly lower than those on Pd, although, as addressed above, there are advantages in using the oxide. Gas and pH sensors are the main ones to use PdO films. These sensors are expanding their domains of application due to rapid developments linked to wellbeing devices and gadgets, clean and safe environments, and sustainable economies. Therefore, these devices based on PdO films can contribute to improvements of the quality of life. Palladium oxide has p-type conductivity and, due to this, it can successfully compete with other oxides, such as SnO_2, ZnO, or In_2O_3, especially in detection of oxidizing gases [214].

Author Contributions: Conceptualization, P.B.; methodology, P.B. and A.L.; formal analysis, P.B. and A.L.; writing—original draft preparation, P.B.; writing—review and editing, A.L.; visualization, P.B and A.L.; funding acquisition, P.B. and A.L. All authors have read and agreed to the published version of the manuscript.

Funding: This research was funded by MCI-UEFISCDI Romania through Core Programs PC2-PN23080202 and PC3-PN23080303.

Institutional Review Board Statement: Not applicable.

Informed Consent Statement: Not applicable.

Data Availability Statement: This review analyzed results reported in the literature, and all information is cited in the References Section.

Conflicts of Interest: The authors declare no conflicts of interest. The funders had no role in the design of the study; in the collection, analyses, or interpretation of data; in the writing of the manuscript, or in the decision to publish the results.

References

1. IUPAC. *Compendium of Chemical Terminology (The "Gold Book")*, 2nd ed.; McNaught, A.D., Wilkinson, A., Eds.; Blackwell Scientific Publications: Oxford, UK, 1997; ISBN 0-9678550-9-8. [CrossRef]
2. Swanson, H.E.; Tatge, E. Standard X-ray Diffraction Powder Patterns. *Natl. Bur. Stand. Circ.* **1955**, *539*, 27.
3. Saldan, I.; Semenyuk, Y.; Marchuk, I.; Reshetnyak, O. Chemical synthesis and application of palladium nanoparticles. *J. Mater. Sci.* **2015**, *50*, 2337–2354. [CrossRef]
4. Yamaguchi, S. Zur bildung von nichtstöchiometrischem palladiumoxyd. *Mater. Chem.* **1980**, *5*, 257–266. [CrossRef]
5. Ismail, E.; Khenfouch, M.; Dhlamini, M.; Dube, S.; Maaza, M. Green palladium and palladium oxide nanoparticles synthesized via Aspalathus linearis natural extract. *J. Alloys Compd.* **2017**, *695*, 3632–3638. [CrossRef]
6. Powers, D.C.; Ritter, T. *Palladium (III) in Synthesis and Catalysis in Higher Oxidation State Organopalladium and Platinum Chemistry*; Canty, A.J., Ed.; Springer: Berlin/Heidelberg, Germany, 2011; pp. 129–156.
7. Chen, W.Z.; Shimada, S.; Tanaka, M. Synthesis and structure of formally hexavalent palladium complexes. *Science* **2002**, *295*, 308–310. [CrossRef] [PubMed]
8. Crabtree, R.H. A new oxidation state for Pd? *Science* **2002**, *295*, 288–289. [CrossRef] [PubMed]
9. Wang, Z.; Hu, P. Identifying the general trend of activity of non-stoichiometric metal oxide phase for CO oxidation on Pd(111). *Sci. China Chem.* **2019**, *62*, 784–789. [CrossRef]
10. Samoylov, A.M.; Gvarishvili, L.J.; Ivkov, S.A.; Pelipenko, D.I.; Badica, P. Two-stage synthesis of palladium (II) oxide nanocrystalline powders for gas sensor application. *Res. Dev. Mater. Sci.* **2018**, *8*, 857–863. [CrossRef]
11. Edström, H.; Schaefer, A.; Jacobse, L.; von Allmen, K.; Hagman, B.; Carlsson, P.-A.; Gustafson, J. Alloying and oxidation of PdAu thin films. *Thin Solid Film.* **2024**, *790*, 140212. [CrossRef]
12. Kreider, K.G.; Tarlov, M.J.; Cline, J.P. Sputtered thin-film pH electrodes of platinum. palladium, ruthenium, and iridium oxides. *Sens. Actuators* **1995**, *28*, 167–172. [CrossRef]
13. Liu, C.-C.; Bocchicchio, D.B.; Overmyer, P.A.; Neuman, M.R. A Palladium–Palladium Oxide Miniature pH electrode. *Science* **1980**, *207*, 188–189. [CrossRef] [PubMed]
14. Grubb, W.T.; King, L.H. Palladium-Palladium oxide pH electrodes. *Anal. Chem.* **1980**, *52*, 270–273. [CrossRef]
15. Bergveld, P. Development of an ion-sensitive solid-state device for neurophysiological measurements. *IEEE Trans. Biomed. Eng.* **1970**, *1*, 70–71. [CrossRef] [PubMed]
16. Kim, J.Y.; Lee, Y.-H. Pd-PdO pH microprobe for local pH measurement. *Biotechnol. Bioeng.* **1989**, *34*, 131–136. [CrossRef]
17. Betteridge, W.; Rhys, D.W. *First International Congress on Metallic Corrosion*; Butterworths: London, UK; Woburn, MA, USA, 1962; pp. 186–192.
18. Kinoshita, E.; Ingman, F.; Edwall, G. An examination of the Palladium/Palladium oxide system and its utility for pH-sensing electrodes. *Electrochim. Acta* **1986**, *31*, 29–38. [CrossRef]
19. Eryürek, M.; Karadag, Y.; Taşaltın, N.; Kılınç, N.; Kiraz, A. Optical sensor for hydrogen gas based on a palladium-coated polymer microresonator. *Sens. Actuators B Chem.* **2015**, *212*, 78–83. [CrossRef]
20. Lee, J.; Noh, J.S.; Lee, S.H.; Song, B.; Jung, H.; Kim, W.; Lee, W. Cracked palladium films on an elastomeric substrate for use as hydrogen sensors. *Int. J. Hydrogen Energy* **2012**, *37*, 7934–7939. [CrossRef]
21. Si, L.; Yu, P.; Huang, J.; Zhao, Z.; Huang, M.; He, S.; Liu, H.; Wang, X.; Liu, W. Advances in gas-sensitive materials based on polyurethane film, foam, and fiber. *Mater. Today Comm.* **2024**, *38*, 108528. [CrossRef]
22. Zhang, Y.-N.; Peng, H.; Qian, X.; Zhang, Y.; An, G.; Zhao, Y. Recent advancements in optical fiber hydrogen sensors. *Sens. Actuators B* **2017**, *244*, 393–416. [CrossRef]
23. Coughlin, J.P. Contributions to the data on theoretical metallurgy XIII. *Heats and free energies of formation of inorganic oxides., US Bur. Mines Bull.* **1954**, *542*, 35.
24. Bloor, L.J.; Malcolme-Lawes, D.J. An electrochemical preparation of palladium oxide pH sensors. *J. Electroanal. Chem. Interfacial Electrochem.* **1990**, *278*, 161–173. [CrossRef]
25. Chen, A.; Ostrom, C. Palladium-based materials: Synthesis and electrochemical applications. *Chem. Rev.* **2018**, *115*, 11999–12044. [CrossRef] [PubMed]
26. Tri, P.N.; Ouellet-Plamondon, C.; Rtimi, S.; Assadi, A.A.; Nguyen, T.A. Methods for synthesis of hybrid nanoparticles. In *Noble Metal—Metal Oxide Hybrid nanoparticles: Fundamentals and Applications*; Mohapatra, S., Nguyen, T.A., Nguyen-Tri, P., Eds.; Elsevier: Amsterdam, The Netherlands, 2019; Chapter 3; pp. 51–63.
27. Sarkar, S.; Pal, T. Theoretical aspects of synthesis for controlled morphological Nanostructures. In *Noble Metal—Metal Oxide Hybrid Nanoparticles: Fundamentals and Applications*; Mohapatra, S., Nguyen, T.A., Nguyen-Tri, P., Eds.; Elsevier: Amsterdam, The Netherlands, 2019; Chapter 2; pp. 7–50.
28. Hu, T.; Wang, Y.; Liu, Q.; Zhang, L.; Wang, H.; Tang, T.; Chen, W.; Zhao, M.; Jia, J. In-situ synthesis of palladium-base binary metal oxide nanoparticles with enhanced electrocatalytic activity for ethylene glycol and glycerol oxidation. *Int. J. Hydrogen Energy* **2017**, *42*, 25951–25959. [CrossRef]

29. Hu, T.; Wang, Y.; Liu, Q.; Zhang, L.; Tang, T.; Xiao, H.; Chen, W.; Zhao, M.; Jia, J.; Zhu, H. Facile synthesis of PdO-doped Co_3O_4 nanoparticles as an efficient bifunctional oxygen electrocatalyst. *Appl. Catal. B Environ.* **2019**, *243*, 175–182. [CrossRef]
30. Chung, M.; Maalouf, J.H.; Adams, J.S.; Jiang, C.; Román-Leshkov, Y.; Manthiram, K. Direct propylene epoxidation via water activation over Pd-Pt electrocatalysts. *Science* **2024**, *383*, 49–55. [CrossRef]
31. Zhang, N.; He, C.; Jing, Y.; Qian, Y.; Toyao, T.; Shimizu, K.-I. Enhanced N_2O decomposition on Rh/ZrO_2 catalysts through the promotional effect of palladium. *Surf. Interfaces* **2024**, *46*, 104120. [CrossRef]
32. Konrad, M.; Popescu-Pelin, G.; Socol, G.; Mardare, A.I.; Hassel, A.W. Combinatorial Analysis of Silver-Palladium Alloy Thin Film Libraries. *Phys. Status Solidi A* **2024**, 2400355. [CrossRef]
33. Wang, T.-J.; Sun, L.-B.; Ai, X.; Chen, P.; Chen, Y.; Wang, X. Boosting Formate Electrooxidation by Heterostructured PtPd Alloy and Oxides Nanowires. *Adv. Mater.* **2024**, *36*, 2403664. [CrossRef]
34. Lokteva, E.S.; Pesotskiy, M.D.; Golubina, E.V.; Maslakov, K.I.; Kharlanov, A.N.; Shishova, V.V.; Kaplin, I.Y. Effect of Iron Content in Alumina-Supported Palladium Catalysts and Their Reduction Conditions on Diclofenac Hydrodechlorination in an Aqueous Medium. *Kinet. Catal.* **2024**, *65*, 133–154. [CrossRef]
35. Huang, J.; Klahn, M.; Tian, X.; Bartling, S.; Zimina, A.; Radtke, M.; Rockstroh, N.; Naliwajko, P.; Steinfeldt, N.; Peppel, T.; et al. Fundamental Structural and Electronic Understanding of Palladium Catalysts on Nitride and Oxide Supports. *Angew. Chem. Int. Ed.* **2024**, *63*, e202400174. [CrossRef]
36. Karczmarska, A.; Adamek, M.; El Houbbadi, S.; Kowalczyk, P.; Laskowska, M. Carbon-supported noble-metal nanoparticles for catalytic applications—A review. *Crystals* **2022**, *12*, 584. [CrossRef]
37. Rodríguez-Otamendi, D.I.; Bizarro, M.; Meza-Laguna, V.; Álvarez-Zauco, E.; Rudolf, P.; Basiuk, V.A.; Basiuk, E.V. Eco-friendly synthesis of graphene oxide–palladium nanohybrids. *Mater. Today Commun.* **2023**, *35*, 106007. [CrossRef]
38. Leve, Z.; Ross, N.; Pokpas, K.; Carleschi, E.; Doyle, B.P.; Sanga, N.A.; Mokwebo, K.V.; Iwuoha, E. Synthesis and Characterization of Palladium/Silver Modified Reduced Graphene Oxide–Nanocomposite Platform for Electrochemical Sensors. *ChemistrySelect* **2024**, *9*, e202400606. [CrossRef]
39. Li, Z.; Xing, X.; Feng, D.; Du, L.; Tian, Y.; Chen, X.; Yang, D. Nitrogen-doped carbon microfibers decorated with palladium and palladium oxide nanoparticles for high-concentration hydrogen sensing. *Ceram. Int.* **2024**, *50*, 21519–21525. [CrossRef]
40. Renjini, S.; Pillai, A.M.; Abraham, P.; Pavitha, P.A. Electrochemical synthesis of graphene oxide/palladium composite for the detection of norepinephrine in the presence of ascorbic acid and uric acid. *Ionics* **2024**. [CrossRef]
41. Feng, Y.; Cheng, G.; Wang, Z.; Wu, K.; Deng, A.; Li, J. Electrochemiluminescence immunosensor based on tin dioxide quantum dots and palladium-modified graphene oxide for the detection of zearalenone. *Talanta* **2024**, *271*, 125740. [CrossRef]
42. Ndlovu, L.; Ndlwana, L.; Mishra, A.K.; Nxumalo, E.; Mishra, S.B. Imobilizing palladium nanoparticles in beta-cyclodextrin-grafted graphene oxide modified polyvinylidene fluoride mixed matrix membranes for the removal of anionic azo dyes. *Chem. Eng. Res. Des.* **2024**, *203*, 149–164. [CrossRef]
43. Brusko, V.V.; Prytkova, A.; Kirsanova, M.; Vakhitov, I.; Sabirova, A.; Tayurskii, D.; Kadirov, M.; Dimiev, A.M. A copper–palladium/reduced graphene oxide composite as a catalyst for the oxygen reduction reaction. *New J. Chem.* **2024**, *48*, 4126–4136.
44. Chauhan, A.S.; Kumar, A.; Bains, R.; Kumar, M.; Das, P. A comprehensive study of palladium-based catalysts on different supports for the hydrogenolysis of 5-hydroxymethylfurfural (HMF) to 2,5-dimethylfuran (DMF) biofuel. *Biomass Bioenergy* **2024**, *185*, 107209. [CrossRef]
45. Pocklanová, R.; Warkad, I.R.; Prucek, R.; Balzerová, A.; Panácek, A.; Kadam, R.G.; Kvítek, L.; Gawande, M.B. Nanodiamond Supported Ultra-Small Palladium Nanoparticles as an Efficient Catalyst for Suzuki Cross-Coupling Reactions. *Catalysts* **2024**, *14*, 53. [CrossRef]
46. van Oossanena, R.; Maierc, A.; Godarta, J.; Pignola, J.-P.; Denkovab, A.G.; van Rhoon, G.C.; Djanashvili, K. Magnetic hybrid Pd/Fe-oxide nanoparticles meet the demands for ablative thermo-brachytherapy. *Int. J. Hyperth.* **2024**, *41*, 2299480. [CrossRef] [PubMed]
47. Leung, K.C.-F.; Xuan, S. Noble metal-iron oxide hybrid nanomaterials: Emerging applications. *Chem. Rec.* **2016**, *16*, 458–472. [CrossRef] [PubMed]
48. Zhang, L.; Ding, L.-X.; Luo, Y.; Zeng, Y.; Wang, S.; Wang, H. PdO/Pd-CeO_2 holow spheres with fresh Pd surface for enhancing formic acid oxidation. *Chem. Eng. J.* **2018**, *347*, 193–201. [CrossRef]
49. Wu, Y.; Yang, W.; Zhao, M.; Xu, H.; Wang, J.; Chen, Y. Silicate-induced high-temperature-resistant small-crystallite ceria support enhancing palladium-catalyzed low-concentration methane combustion. *Sep. Purif. Technol.* **2025**, *353*, 128385. [CrossRef]
50. Sivasankaran, S.; Kumar, M.J.K. Sonochemical synthesis of Pd-metal oxide hybrid nanoparticles. In *Noble Metal—Metal Oxide Hybrid Nanoparticles: Fundamentals and Applications*; Mohapatra, S., Nguyen, T.A., Nguyen-Tri, P., Eds.; Elsevier: Amsterdam, The Netherlands, 2019; Chapter 10; pp. 189–194.
51. Veerakumar, P.; Sangili, A.; Chen, S.-M.; Kumar, R.S.; Arivalagan, G.; Firdhouse, M.J.; Hameed, K.S.; Sivakumar, S. Photocatalytic degradation of phenolic pollutants over palladium-tungsten trioxide nanocomposite. *Chem. Eng. J.* **2024**, *489*, 151127. [CrossRef]
52. Yang, Z.; Jiang, N.; Bei, S.; Bao, K.; Xiang, M.; Yu, C.; Dong, S.; Qin, H. Single-atom palladium on nonstoichiometric tungsten oxide as bifunctional electrocatalyst for zinc-air battery. *Electrochim. Acta* **2024**, *476*, 143768. [CrossRef]
53. Rani, R.; Sharma, M.; Sharma, S.; Chandra, R.; Malik, V.K. Palladium enhanced electrochemical supercapacitive performance of chromium oxide thin films synthesized by sputtering process. *Thin Solid Film.* **2024**, *799*, 140379. [CrossRef]

54. Chen, Y.; Rana, R.; Zhang, Y.; Hoffman, A.S.; Huang, Z.; Vila, F.D.; Perez-Aguilar, J.E.; Hong, J.; Li, X.; Zeng, J.; et al. Dynamic structural evolution of MgO-supported palladium catalysts: From metal to metal oxide nanoparticles to surface then subsurface atomically dispersed cations. *Chem. Sci.* **2024**, *15*, 6454–6464. [CrossRef]
55. Rüzgar, A.; Şener, L.; Karataş, Y.; Gülcan, M. Palladium (0) nanoparticles distributed on lanthanum (III) oxide as an effective catalyst for the methanolysis of hydrazine-borane to produce hydrogen. *Turk. J. Chem.* **2024**, *48*, 137–151. [CrossRef]
56. Bekmezci, M.; Cibo, M.C.; Akin, M.; Poyraz, H.B.; Kaya, G.; Sen, F. Kaolin and zinc oxide supported PdCu catalysts as a superior catalyst in methanol oxidation reaction. *Int. J. Hydrogen Energy* **2024**, *82*, 456–463. [CrossRef]
57. Hua, Y.; Vikrant, K.; Kim, K.-H.; Heynderickx, P.M.; Boukhvalov, D.W. Room temperature thermocatalytic removal of formaldehyde in air using a copper manganite spinel-supported palladium catalyst with ultralow noble metal content. *Sep. Purif. Technol.* **2025**, *354*, 128863. [CrossRef]
58. Feng, M.; Wang, M.-Y.; Wang, F.; Xu, J.; Xue, B. Soft-templating synthesis of mesoporous MnSiOx composites as catalytic supports for Pd nanoparticles towards solvent-free oxidation of benzyl alcohol under atmospheric pressure O_2. *Appl. Catal. A Gen.* **2024**, *683*, 119829. [CrossRef]
59. Ibrahim, O.M. A Comparative Study of Platinum- Versus Palladium-Based Catalysts on FeCrAl-Sintered Metal Fiber Filter Substrate for Reducing Gaseous Diesel Engine Emissions. *Emiss. Control. Sci. Technol.* **2024**. [CrossRef]
60. Choi, J.; Nguyen, Q.T.; Park, S.; Ghule, B.G.; Park, J.H.; Park, J.R.; Nakate, U.T.; Jang, J.-H.; Kim, D.-W.; Park, S. Interfacially engineered palladium nanoparticle-decorated nickel oxide nanostructured electrocatalysts for high-performance hydrogen evolution reaction. *Chem. Eng. J.* **2024**, *497*, 154407. [CrossRef]
61. Bhalothia, D.; Yan, C.; Hiraoka, N.; Ishii, H.; Liao, Y.-F.; Dai, S.; Chen, P.-C.; Chen, T.-Y. Iridium Single Atoms to Nanoparticles: Nurturing the Local Synergy with Cobalt-Oxide Supported Palladium Nanoparticles for Oxygen Reduction Reaction. *Adv. Sci.* **2024**, *11*, 2404076. [CrossRef]
62. Yao, P.-C. Study of a New Hydrogen Sensor Based on the Synthesis of a Sputtered In–Sn–Zn-O Thin Film and Evaporated Palladium Nanoparticles. *IEEE Trans. Electron Devices* **2024**, *71*, 2612–2617. [CrossRef]
63. Alrashdi, K.S.; Munshi, A.M.; Alrefaee, S.H.; Alalawy, A.I.; Abumelha, H.M.; Alamoudi, W.M. Miniaturization of Palladium and Gold in Nanosize to Hold Immense Potentiality for Application in Wool Functional Finishing. *Fibers Polym.* **2024**, *25*, 2555–2568. [CrossRef]
64. Formenti, M.; Casaletto, M.P.; Barone, G.; Pagliaro, M.; Della Pina, C.; Butera, V.; Ciriminna, R. GrafeoPlad Palladium: Insight on Structure and Activity of a New Catalyst Series of Broad Scope. *Adv. Sustain. Syst.* **2024**, *8*, 2300643. [CrossRef]
65. Zhang, Y.; Yan, Z.; Xiao, M.; Zhang, C.; Ruan, L.; Zhang, Y.; Zhong, Y.; Yan, Y.; Yu, Y.; He, H. Catalytic performance of Pd catalyst supported on CeO_2 or ZrO_2 modified beta zeolite for methane oxidation. *J. Environ. Sci.* **2025**, *152*, 248–261. [CrossRef]
66. Ni, F.; Lewis, R.J.; López-Martín, Á.; Smith, L.R.; Morgan, D.J.; Davies, T.E.; Taylor, S.H.; Hutchings, G.J. The direct synthesis of H_2O_2 and in situ oxidation of methane: An investigation into the role of the support. *Catal. Today* **2024**, *442*, 114910. [CrossRef]
67. Catalano, M.; Carlino, E.; Tagliente, M.A.; Licciulli, A.; Tapfer, L. Microstructural and Microanalytical Characterization of Pd Clusters in ORMOCER Matrix. *Microsc. Microanal. Microstruct.* **1995**, *6*, 611–619. [CrossRef]
68. Xiong, Y.; Xia, Y. Shape-controlled synthesis of metal nanostructures: The case of Palladium. *Adv. Mater.* **2007**, *19*, 3385–3391. [CrossRef]
69. Choudhury, S.; Betty, C.A.; Bhattacharyya, K.; Saxena, V.; Bhattacharya, D. Nanostructured PdO Thin Film from Langmuir–Blodgett Precursor for Room-Temperature H_2 Gas Sensing. *ACS Appl. Mater. Interfaces* **2016**, *8*, 16997–17003. [CrossRef] [PubMed]
70. Gao, Z.; Wang, T.; Li, X.; Li, Q.; Zhang, X.; Cao, T.; Li, Y.; Zhang, L.; Guo, L.; Fu, Y. Pd-Decorated PdO Hollow Shells: A H_2-Sensing System in Which Catalyst Nanoparticle and Semiconductor Support are Interconvertible. *ACS Appl. Mater. Interfaces* **2020**, *12*, 42971–42981. [CrossRef] [PubMed]
71. Mhlongo, G.H.; Motaung, D.E.; Cummings, F.R.; Swart, H.C.; Ray, S.S. A highly responsive NH_3 sensor based on Pd-loaded ZnO nanoparticles prepared via a chemical precipitation approach. *Sci. Rep.* **2019**, *9*, 9881. [CrossRef]
72. Lupan, C.; Khaledialidusti, R.; Mishra, A.K.; Postica, V.; Terasa, M.-I.; Magariu, N.; Pauporté, T.; Viana, B.; Drewes, J.; Vahl, A.; et al. Pd-Functionalized ZnO:Eu Columnar Films for Room-Temperature Hydrogen Gas Sensing: A Combined Experimental and Computational Approach. *ACS Appl. Mater. Interfaces* **2020**, *12*, 24951–24964. [CrossRef]
73. Xuan, J.; Zhao, G.; Sun, M.; Jia, F.; Wang, X.; Zhou, T.; Yin, G.; Liu, B. Low-temperature operating ZnO-based NO_2 sensors: A review. *RSC Adv.* **2020**, *10*, 39786–39807. [CrossRef]
74. Goodarzi, M.T.; Ranjbar, M. Atmospheric flame vapor deposition of WO_3 thin films for hydrogen detection with enhanced sensing characteristics. *Ceram. Int.* **2020**, *46*, 21248–21255. [CrossRef]
75. Castillo, C.; Cabello, G.; Chornik, B.; Huentupil, Y.; Buono-Core, G.E. Characterization of photochemically grown Pd loaded WO3 thin films and its evaluation as ammonia gas sensor. *J. Alloys Compd.* **2020**, *825*, 154166. [CrossRef]
76. Malik, R.; Tomer, V.K.; Mishra, Y.K.; Lin, L. Functional gas sensing nanomaterials: A panoramic view. *Appl. Phys. Rev.* **2020**, *7*, 021301. [CrossRef]
77. Liu, X.; Zhao, K.; Sun, X.; Duan, X.; Zhang, C.; Xu, X. Electrochemical sensor to environmental pollutant of acetone based on Pd-loaded on mesoporous In_2O_3 architecture. *Sens. Actuators B Chem.* **2019**, *290*, 217–225. [CrossRef]
78. Zappa, D.; Galstyan, V.; Kaur, N.; Munasinghe Arachchige, H.M.M.; Sisman, O.; Comini, E. Metal oxide—Based heterostructures for gas sensors—A review. *Anal. Chim. Acta* **2018**, *1039*, 1–23. [CrossRef] [PubMed]

79. Wang, Z.; Huang, S.; Men, G.; Han, D.; Gu, F. Sensitization of Pd loading for remarkably enhanced hydrogen sensing performance of 3DOM WO_3. *Sens. Actuators B Chem.* **2018**, *262*, 577–587. [CrossRef]
80. Tian, J.; Wang, J.; Hao, Y.; Du, H.; Li, X. Toluene sensing properties of porous Pd-loaded flower-like SnO_2 microspheres. *Sens. Actuators B Chem.* **2014**, *202*, 795–802. [CrossRef]
81. Van Toan, N.; Chien, N.V.; Van Duy, N.; Hong, H.S.; Nguyen, H.; Hoa, N.D.; Van Hieu, N. Fabrication of highly sensitive and selective H_2 gas sensor based on SnO_2 thin film sensitized with microsized Pd islands. *J. Hazard. Mater.* **2016**, *301*, 433–442. [CrossRef]
82. Chang, Y.; Xu, J.; Zhang, Y.; Ma, S.; Xin, L.; Zhu, L.; Xu, C. Optical properties and photocatalytic performance of Pd modified ZnO samples. *J. Phys. Chem. C* **2009**, *113*, 18761–18767. [CrossRef]
83. Hu, J.; Gao, F.; Sang, S.; Li, P.; Deng, X.; Zhang, W.; Chen, Y.; Lian, K. Optimization of Pd content in ZnO microstructures for high-perfomance gas detection. *J. Mater. Sci.* **2015**, *50*, 1935–1942. [CrossRef]
84. Khan, T.M. Into the nature of Pd-dopant induced local phonon modes and associated disorders in ZnO based on spatial correlation model. *Mater. Chem. Phys.* **2015**, *153*, 248–255. [CrossRef]
85. Mane, A.; Suryawanshi, M.; Kim, J.; Moholkar, A. Superior selectivity and enhanced response characteristics of Palladium sensitized Vanadium Pentoxide nanorods for detection of Nitrogen Dioxide gas. *J. Colloid. Interf. Sci.* **2017**, *495*, 53–60. [CrossRef]
86. Trung, D.D.; Hoa, N.D.; Van Tong, P.; Van Duy, N.; Dao, T.; Chung, H.; Nagao, T.; Van Hieu, N. Effective decoration of Pd nanoparticles on the surface of SnO_2 nanowires for enhancement of CO gas-sensing performance. *Hazard. Mater.* **2014**, *265*, 124–132. [CrossRef]
87. Yi, Z.; Lou, Z.; Wang, L.; Zou, B. Enhanced Ammonia sensing performances of Pd-sensitized flowerlike ZnO nanostructure. *Sens. Actuators B Chem.* **2011**, *156*, 395–400.
88. Jiao, M.; Van Duy, N.; Chien, N.V.; Hoa, N.D.; Van Hieu, N.; Hjort, K.; Nguyen, H. On-chip growth of patterned ZnO nanorod sensors with PdO decoration for enhancement of Hydrogen-sensing performance. *Int. J. Hydrogen Energy* **2017**, *42*, 16294–16304. [CrossRef]
89. Ma, N.; Suematsu, K.; Yuasa, M.; Shimanoe, K. Pd Size effect on the gas sensing properties of Pd-loaded SnO_2 in humid atmosphere. *ACS Appl. Mater. Interfaces* **2015**, *7*, 15618–15625. [CrossRef] [PubMed]
90. Cai, L.; Zhu, S.; Wu, G.; Jiao, F.; Li, W.; Wang, X.; An, Y.; Hu, Y.; Sun, J.; Dong, X.; et al. Highly sensitive H_2 sensor based on PdO-decorated WO_3 nanospindle p-n heterostructure. *Int. J. Hydrogen Energy* **2020**, *45*, 31327–31340. [CrossRef]
91. Liu, M.; Cui, F.; Ma, Q.; Xu, L.; Zhang, J.; Zhang, R.; Cui, T. Janus coordination polymer derived PdO/ZnO nanoribbons for efficient 4-nitrophenol reduction. *New J. Chem.* **2020**, *44*, 4042–4048. [CrossRef]
92. Rao, F.; Zhu, G.; Wang, M.; Zubairu, S.M.; Peng, J.; Gao, J.; Hojamberdiev, M. Constructing the Pd/PdO/β-Bi_2O_3 microspheres with enhanced photocatalytic activity for Bisphenol A degradation and NO removal. *J. Chem. Technol. Biotechnol.* **2020**, *95*, 862–874. [CrossRef]
93. Lupan, O.; Postica, V.; Hoppe, M.; Wolff, N.; Polonskyi, O.; Pauporté, T.; Viana, B.; Majérus, O.; Kienle, L.; Faupel, F.; et al. PdO/PdO_2 functionalized ZnO:Pd films for lower operating temperature H_2 gas sensing. *Nanoscale* **2018**, *10*, 14107–14127. [CrossRef]
94. Geng, X.; Luo, Y.; Zheng, B.; Zhang, C. Photon assisted room-temperature hydrogen sensors using PdO loaded WO_3 nanohybrids. *Int. J. Hydrogen Energy* **2017**, *42*, 6425–6434. [CrossRef]
95. Guo, P.; Wei, Z.; Ye, W.; Qin, W.; Wang, Q.; Guo, X.; Lu, C.; Zhao, X.S. Preparation and characterization of nanostructured Pd with high electrocatalytic activity. *Colloids Surf. A* **2012**, *395*, 75–81. [CrossRef]
96. Adams, B.D.; Wu, G.; Nigro, S.; Chen, A. Facile Synthesis of Pd−Cd Nanostructures with High Capacity for Hydrogen Storage. *J. Am. Chem. Soc.* **2009**, *131*, 6930–6931. [CrossRef]
97. Tahira, A.; Aftab, U.; Solangi, M.Y.; Gradone, A.; Morandi, V.; Medany, S.S.; Kasry, A.; Infantes-Molina, A.; Nafady, A.; Ibupoto, Z.H. Facile deposition of palladium oxide (PdO) nanoparticles on $CoNi_2S_4$ microstructures towards enhanced oxygen evolution reaction. *Nanotechnology* **2022**, *33*, 275402. [CrossRef] [PubMed]
98. Arora, K.; Puri, N.K. Electrophoretically deposited nanostructured PdO thin film for room temperature amperometric H_2 sensing. *Vacuum* **2018**, *154*, 302–308. [CrossRef]
99. Wang, L.; Lou, Z.; Wang, R.; Fei, T.; Zhang, T. Ring-like PdO–NiO with lamellar structure for gas sensor application. *J. Mater. Chem.* **2012**, *22*, 12453. [CrossRef]
100. Wang, H.; Xu, C.; Cheng, F.; Jiang, S. Pd nanowire arrays as electrocatalysts for ethanol electrooxidation. *Electrochem. Commun.* **2007**, *9*, 1212–1216. [CrossRef]
101. Cui, C.-H.; Yu, J.-W.; Li, H.-H.; Gao, M.-R.; Liang, H.-W.; Yu, S.-H. Remarkable enhancement of electrocatalytic activity tuning the interface of Pd- Au bimetallic nanoparticle tubes. *ACS Nano* **2011**, *5*, 4211–4218. [CrossRef] [PubMed]
102. Liu, L.; Yoo, S.-H.; Lee, S.A.; Park, S. Wet-chemical synthesis of Palladium nanosprings. *Nano Lett.* **2011**, *11*, 3979–3982. [CrossRef]
103. Elezovic, N.R.; Lovic, J.D.; Jovic, B.M.; Zabinski, P.; Włoch, G.; Jovic, V.D. Synthesis and characterization of AgPd alloy coatings as beneficial catalysts for low temperature fuel cells application. *Electrochim. Acta* **2019**, *307*, 360–368. [CrossRef]
104. Qiu, C.; Dong, X.; Huang, M.; Wang, S.; Ma, H. Facile fabrication of nanostructured Pd-Fe bimetallic thin films and their electrodechlorination activity. *J. Mol. Catal. A Chem.* **2011**, *350*, 56–63. [CrossRef]

105. Ho, P.H.; Ambrosetti, M.; Groppi, G.; Tronconi, E.; Jaroszewicz, J.; Ospitali, F.; Rodríguez-Castellón, E.; Fornasari, G.; Vaccari, A.; Benito, P. One-step electrodeposition of Pd-CeO_2 on high pore density foams for environmental catalytic processes. *Catal. Sci. Technol.* **2018**, *8*, 4678–4689. [CrossRef]
106. Ryabtsev, S.V.; Ievlev, V.M.; Samoylov, A.M.; Kuschev, S.B.; Soldatenko, S.A. Microstructure and electrical properties of palladium oxide thin films for oxidizing gases detection. *Thin Solid Film.* **2017**, *636*, 751–759. [CrossRef]
107. Chen, R.; Ruan, X.; Liu, W.; Stefanini, C. A reliable and fast hydrogen gas leakage detector based on irreversible cracking of decorated palladium nanolayer upon aligned polymer fibers. *Int. J. Hydrogen Energy* **2015**, *40*, 746–751. [CrossRef]
108. Hu, R.-J.; Wang, J.; Zhu, H.-C. Preparation and Gas Sensing Properties of PdO. *Au, CdO Coatings on SnO2 Nanofibers, Acta Phys.-Chim. Sin.* **2015**, *31*, 1997–2004. [CrossRef]
109. Wang, Y.; Peng, H.-C.; Liu, J.; Huang, C.Z.; Xia, Y. Use of reduction rate as a quantitative knob for controlling the twin structure and shape of Palladium nanocrystals. *Nano Lett.* **2015**, *15*, 1445–1450. [CrossRef] [PubMed]
110. Wang, Y.; Choi, S.-L.; Zhao, X.; Xie, S.; Peng, H.-C.; Chi, M.; Huang, C.Z.; Xia, Y. Polyol synthesis of ultrathin Pd nanowires via attachment-based growth and their enhanced activity towards formic acid oxidation. *Adv. Funct. Mater.* **2014**, *24*, 131–139. [CrossRef]
111. Huang, X.; Tang, S.; Yang, J.; Tan, Y.; Zheng, N. Etching growth under surface confinement: An effective strategy to prepare mesocrystalline Pd nanocorolla. *J. Am. Chem. Soc.* **2011**, *133*, 15946–15949. [CrossRef]
112. Raghuveer, V.; Ferreira, P.; Manthiram, A. Comparison of Pd-Co-Au electrocatalysts prepared by conventional borohydrideand microemulsion methods for oxygen reduction in fuel cells. *Electrochem. Commun.* **2006**, *8*, 807–814. [CrossRef]
113. Yuasa, M.; Masaki, T.; Kida, T.; Shimanoe, K.; Yamazoe, N. Nano-sized PdO loaded SnO_2 nanoparticles by reverse micelle method for highly sensitive CO gas sensor. *Sens. Actuators B Chem.* **2009**, *136*, 99–104. [CrossRef]
114. Tian, M.; Malig, M.; Chen, S.; Chen, A. Synthesis and electrochemical study of TiO_2-supported PdAu nanoparticles. *Electrochem. Commun.* **2011**, *13*, 370–373. [CrossRef]
115. Wei, M.; Li, H.; Guo, G.; Liu, Y.; Zhang, D. Effects of PdO modification on the performance of $La_{0.6}Sr_{0.4}Co_{0.2}Fe_{0.8}O_{3-\delta}$ cathodes for solid oxide fuel cells: A first principle study. *Int. J. Hydrogen Energy* **2017**, *42*, 23180–23188. [CrossRef]
116. Qin, Y.; Alam, A.U.; Pan, S.; Howlander, M.M.R.; Ghosh, R.; Selvaganapathy, P.R.; Wu, Y.; Deen, M.J. Low-temperature solution processing of palladium/palladium oxide films and their pH sensing performance. *Talanta* **2016**, *146*, 517–524. [CrossRef]
117. Kabcum, S.; Channel, D.; Tuantranont, A.; Wisitsoraat, A.; Liewhiran, C.; Phanichphant, S. Ultra-responsive hydrogen gas sensors based on PdO nanoparticle-decorated WO_3 nanorods synthesized via precipitation and impregnation methods. *Sens. Actuators B Chem.* **2016**, *226*, 76–89. [CrossRef]
118. Balamurugan, C.; Jeong, Y.J.; Lee, D.W. Enhanced H_2S sensing performance of a p-type semiconducting PdO-NiO nanoscale heteromixture. *Appl. Surf. Sci.* **2017**, *420*, 638–650. [CrossRef]
119. Jin, J.; Li, C.; Tsang, C.-W.; Xu, B.; Liang, C. Catalytic combustion of methane over Pd/Ce–Zr oxides washcoated monolithic catalysts under oxygen lean conditions. *RSC Adv.* **2015**, *5*, 102147–102156. [CrossRef]
120. Wei, Y.; Ji, J.; Liang, F.; Ma, D.; Du, Y.; Pang, Z.; Wang, H.; Li, Q.; Shi, G.; Wang, Z. Adjusting active sites and metal-support interactions of ceramic-loaded Pd/P-CeO_2-Al_2O_3 coating to optimize CO_2 methanation pathways. *J. Environ. Chem. Eng.* **2023**, *11*, 110773. [CrossRef]
121. Lou, Z.; Deng, J.; Wang, L.; Wang, L.; Fei, T.; Zhang, T. Toluene and ethanol sensing performances of pristine and PdO-decorated flower-like ZnO structures. *Sens. Actuators B Chem.* **2013**, *176*, 323–329. [CrossRef]
122. Preus, A.; Korb, M.; Ruffer, T.; Bankwitz, J.; Geogi, C.; Jakob, A.; Schulz, S.E.; Lang, H. A β-ketoiminato palladium(II) complex for palladium deposition. *Z. Für Naturforschung B* **2019**, *74*, 901–912. [CrossRef]
123. Kim, I.J.; Han, S.D.; Singh, I.; Lee, H.D.; Wang, J.S. Sensitivity enhancement for CO gas detection using a SnO_2–CeO_2–PdO_x system. *Sens. Actuators B Chem.* **2005**, *107*, 825–830. [CrossRef]
124. Qin, Y.; Alam, A.U.; Howlader, M.; Hu, N.-X.; Deen, M.J. Morphology and electrical properties of inkjet-printed palladium/palladium oxide. *J. Mater. Chem. C* **2017**, *5*, 1893–1902. [CrossRef]
125. Garcia, J.R.V.; Goto, T. Chemical vapor deposition of iridium, platinum, rhodium, and palladium. *Mater. Trans.* **2003**, *44*, 1717. [CrossRef]
126. Wang, L.; Griffin, G.L. Batch CVD Process for Depositing Pd Activation Layers. *J. Electrochem. Soc.* **2007**, *154*, D151. [CrossRef]
127. Guerrero, R.M.; Hernández-Gordillo, A.; Santes, V.; García, J.R.V.; Escobar, J.; Díaz-García, L.; Arceo, L.D.B.; Febles, V.G. Monometallic Pd and Pt and Bimetallic Pd-Pt/Al_2O_3-TiO_2 for the HDS of DBT: Effect of the Pd and Pt Incorporation Method. *J. Chem.* **2014**, *2014*, 679281.
128. Belousov, O.V.; Tarabanko, V.E.; Borisov, R.V.; Simakova, I.L.; Zhyzhaev, A.M.; Tarabanko, N.; Isakova, V.G.; Parfenov, V.V.; Ponomarenko, I.V. Synthesis and catalytic hydrogenation activity of Pd and bimetallic Au–Pd nanoparticles supported on high-porosity carbon materials. *React. Kinet. Catal. Lett.* **2019**, *127*, 25–39. [CrossRef]
129. Krisyuk, V.V.; Turgambaeva, A.E.; Mirzaeva, I.V.; Urkasymkyzy, S.; Koretskaya, T.P.; Trubin, S.V.; Sysoev, S.V.; Shubin, Y.V.; Maksimovskiy, E.A.; Petrova, N.I. MOCVD Pd–Cu alloy films from single source heterometallic precursors. *Vacuum* **2019**, *166*, 248–254. [CrossRef]
130. Ehsan, M.A.; Sohail, M.; Jamil, R.; Hakeem, A.S. Single-step fabrication of nanostructured Palladium thin films via aerosol-assisted chemical vapor deposition (AACVD) for the electrochemical detection of hydrazine. *Electrocatalysis* **2019**, *10*, 214–221. [CrossRef]

131. Annanouch, F.E.; Roso, S.; Haddi, Z.; Vallejos, S.; Umek, P.; Bittencourt, C.; Blackman, C.; Vilic, T.; Llobet, E. p-Type PdO nanoparticles supported on n-type WO3 nanoneedles for hydrogen sensing. *Thin Solid Film.* **2016**, *618*, 238–245. [CrossRef]
132. Annanouch, F.E.; Haddi, Z.; Ling, M.; Di Maggio, F.; Vallejos, S.; Vilic, T.; Zhu, Y.; Shujah, T.; Umek, P.; Bittencourt, C.; et al. Aerosol-Assisted CVD-Grown PdO Nanoparticle-Decorated Tungsten Oxide Nanoneedles Extremely Sensitive and Selective to Hydrogen. *ACS Appl. Mater. Interfaces* **2016**, *8*, 10413–10421. [CrossRef]
133. Alvarado, M.; De La Flor, S.; Llobet, E.; Romero, A.; Ramírez, J.L. Performance of Flexible Chemoresistive Gas Sensors after Having Undergone Automated Bending Tests. *Sensors* **2019**, *19*, 5190. [CrossRef]
134. Ling, M.; Blackman, C.S. Gas-phase synthesis of hybrid nanostructured materials. *Nanoscale* **2018**, *10*, 22981–22989. [CrossRef]
135. Kapica, R.; Redzynia, W.; Tyczkowski, J. Characterization of Palladium-based Thin Films Prepared by Plasma-enhanced Metalorganic Chemical Vapor Deposition. *Mater. Sci.* **2012**, *18*, 128–131. [CrossRef]
136. Lang, H.; Dietrich, S. *Comprehensive Inorganic Chemistry II: From Elements to Applications*, 2nd ed.; Reedijk, J., Poeppelmeier, K., Eds.; Elsevier Ltd.: Amsterdam, The Netherlands, 2013; Volume 4, pp. 211–269.
137. Hierso, J.-C.; Feurer, R.; Kalck, P. Platinum, palladium and rhodium complexes as volatile precursors for depositing materials. *Coord. Chem. Rev.* **1998**, *178–180*, 1811–1834. [CrossRef]
138. Assim, K.; Melzer, M.; Korb, M.; Rüffer, T.; Jakob, A.; Noll, J.; Georgi, C.; Schulz, S.E.; Lang, H. Bis (β-diketonato)-and allyl-(β-diketonato)-palladium (II) complexes: Synthesis, characterization and MOCVD application. *RSC Adv.* **2016**, *6*, 102557–102569. [CrossRef]
139. Gozum, J.E.; Pollina, D.M.; Jensen, J.A.; Girolami, G.S. "Tailored" organometallics as precursors for the chemical vapor deposition of high-purity palladium and platinum thin films. *J. Am. Chem. Soc.* **1988**, *110*, 2688–2689. [CrossRef]
140. Henc, B.; Jolly, P.W.; Salz, R.; Stobbe, S.; Wilke, G.; Benn, R.; Mynott, R.; Seevogel, K.; Goddard, R.; Krüger, C. Transition metal allyls: IV. The (η3-allyl)2M complexes of nickel, palladium and platinum: Reaction with tertiary phosphines. *J. Organomet. Chem.* **1980**, *191*, 449–475. [CrossRef]
141. Yuan, Z.; Puddephatt, R.J. Allyl(β-diketonato)palladium(II) complexes as precursors for palladium films. *Adv. Mater.* **1994**, *6*, 51–54. [CrossRef]
142. Nikolaeva, N.S.; Kuratieva, N.V.; Vikulova, E.S.; Stabnikov, P.A.; Morozova, N.B. Volatile asymmetric fluorinated (O^N)-chelated palladium complexes: From ligand sources to MOCVD application. *Polyhedron* **2019**, *171*, 455–463. [CrossRef]
143. Aaltonen, T.; Ritala, M.; Tung, Y.L.; Chi, Y.; Arstila, K.; Meinander, K.; Leskelä, M. Atomic layer deposition of noble metals: Exploration of the low limit of the deposition temperature. *J. Mater. Res.* **2004**, *19*, 3353–3358. [CrossRef]
144. Liu, Y.H.; Cheng, Y.C.; Tung, Y.L.; Chi, Y.; Chen, Y.L.; Liu, C.S.; Peng, S.M.; Lee, G.H. Synthesis and characterization of fluorinated β-ketoiminate and imino-alcoholate Pd complexes: Pre-cursors for palladium chemical vapor deposition. *J. Mater. Chem.* **2003**, *13*, 135–142. [CrossRef]
145. Zharkova, G.I.; Stabnikov, P.A.; Baidina, I.A.; Smolentsev, A.I.; Tkachev, S.V. Synthesis, properties and crystal structures of volatile β-ketoiminate Pd complexes, precursors for palladium chemical vapor deposition. *Polyhedron* **2009**, *28*, 2307–2312. [CrossRef]
146. Zharkova, G.I.; Sysoev, S.V.; Stabnikov, P.A.; Logvinenko, V.A.; Igumenov, I.K. Vapor pressure and crystal lattice energy of volatile palladium(II) β-iminoketonates. *J. Therm. Anal. Calorim.* **2011**, *103*, 381–385. [CrossRef]
147. Vikulova, E.S.; Cherkasov, S.A.; Nikolaeva, N.S.; Smolentsev, A.I.; Sysoev, S.V.; Morozova, N.B. Thermal behavior of volatile palladium(II) complexes with tetradentate Schiff bases containing propylene-diimine bridge. *J. Therm. Anal. Calorim.* **2019**, *135*, 2573–2582. [CrossRef]
148. Manjunatha, K.B.; Dileep, R.; Umesh, G.; Bhat, B.R. Nonlinear optical and all-optical switching studies of palladium(II) complex. *Mater. Lett.* **2013**, *105*, 173–176. [CrossRef]
149. Tung, Y.-L.; Tseng, W.-C.; Lee, C.-Y.; Hsu, P.-F.; Chi, Y.; Peng, S.-M.; Lee, G.-H. Synthesis and Characterization of Allyl(β-ketoiminato)palladium(II) Complexes: New Precursors for Chemical Vapor Deposition of Palladium Thin Films. *Organometallics* **1999**, *18*, 864–869. [CrossRef]
150. Giebelhaus, I.; Müller, R.; Tyrra, W.; Pantenburg, I.; Fischer, T.; Mathur, S. First air stable tin(II) β-heteroarylalkenolate: Synthesis. characterization and application in chemical vapor deposition. *Inorg. Chim. Acta* **2011**, *372*, 340–346. [CrossRef]
151. Kuratieva, N.V.; Vikulova, E.S.; Shushanyan, A.D.; Nikolaeva, N.S.; Dorovskikh, S.I.; Mikhaleva, N.S.; Morozova, N.B. Structure of Cu(II) and Pd(II) complexes with 2-(2,2-dimethylhydrazone)pentanone-4. *J. Struct. Chem.* **2017**, *58*, 1004. [CrossRef]
152. Senkevich, J.J.; Tang, F.; Rogers, D.; Drotar, J.T.; Jezewski, C.; Lanford, W.A.; Wang, G.-C.; Lu, T.-M. Substrate-independent palladium atomic layer deposition. *Chem. Vap. Depos.* **2003**, *9*, 258–264. [CrossRef]
153. Goldstein, D.N.; George, S.M. Enhancing the nucleation of palladium atomic layer deposition on Al_2O_3 using trimethylaluminum to prevent surface poisoning by reaction products. *Appl. Phys. Lett.* **2009**, *95*, 143106. [CrossRef]
154. Achari, I.; Ambrozik, S.; Dimitrov, N. Electrochemical Atomic Layer Deposition of Pd Ultrathin Films by Surface Limited Redox Replacement of Underpotentially Deposited H in a Single Cell. *J. Phys. Chem. C* **2017**, *121*, 4404–4411. [CrossRef]
155. Nallan, H.C.; Yang, X.; Coffey, B.M.; Dolocan, A.; Ekerdt, J.G. Area-selective atomic layer deposition of palladium. *J. Vac. Sci. Technol. A* **2024**, *42*, 022401. [CrossRef]
156. Cao, T.Y.; Wang, C.Y.; Shan, K.; Vohs, J.M.; Gorte, R.J. Investigation into support effects for Pt and Pd on $LaMnO_3$. *Appl. Catal. A-Gen.* **2022**, *646*, 118873. [CrossRef]
157. Lausecker, C.; Munoz-Rojas, D.; Webber, M. Atomic layer deposition (ALD) of palladium: From processes to applications. *Crit. Rev. Solid State Mater. Sci.* **2024**, *49*, 908–930. [CrossRef]

158. Feng, J.; He, L.; Hui, J.Q.; Kavithaa, K. Synthesis of Bimetallic Palladium/Zinc Oxide Nanocomposites Using Crocus sativus and Its Anticancer Activity via the Induction of Apoptosis in Cervical Cancer. *Appl. Biochem. Biotechnol.* **2024**. [CrossRef] [PubMed]
159. Chandrashekharan, B.; Sampatkumar, H.G.; George, D.; Antony, A.M.; Doddamani, S.V.; Sasidhar, B.S.; Balakrishna, R.G.; Patil, S.A. Palladium nanoparticle immobilized on coconut coir extract coated boron carbon nitride: A green and sustainable nanocatalyst for cross-coupling reactions and HER studies. *Diam. Relat. Mater.* **2024**, *147*, 111261. [CrossRef]
160. Tiri, R.N.E.; Aygun, A.; Bekmezci, M.; Gonca, S.; Ozdemir, S.; Kaymak, G.; Karimi-Maleh, H.; Sen, F. Environmental Energy Production and Wastewater Treatment Using Synthesized Pd Nanoparticles with Biological and Photocatalytic Activity. *Top. Catal.* **2024**, *67*, 714–724. [CrossRef]
161. Naveenkumar, S.; Kamaraj, C.; Prem, P.; Raja, R.K.; Priyadharsan, A.; Alrefaei, A.F.; Govindarajan, R.-V.K.; Thamarai, R.; Subramaniyan, V. Eco-friendly synthesis of palladium nanoparticles using Zaleya decandra: Assessing mosquito larvicidal activity, zebrafish embryo developmental toxicity, and impacts on freshwater sludge worm Tubifex tubifex. *J. Environ. Chem. Eng.* **2024**, *12*, 111912. [CrossRef]
162. Griffith, W.P.; Robinson, S.D.; Swars, K. Palladium and Oxygen. In *Pd Palladium. Gmelin Handbook of Inorganic Chemistry/Gmelin Handbuch der Anorganischen Chemie*; Griffith, W.P., Swars, K., Eds.; Springer: Berlin/Heidelberg, Germany, 1989; Volume P-d/B/2. [CrossRef]
163. Matsushima, T. *Micro-Fabrication by Sputtering in Handbook of Sputtering Technology*, 2nd ed.; Wasa, K., Kanno, I., Kotera, H., Eds.; Elsevier: Amsterdam, The Netherlands, 2012; pp. 597–622.
164. Rossnagel, S.M.; Hopwood, J. Magnetron sputter deposition with high levels of metal ionization. *Appl. Phys. Lett.* **1993**, *63*, 3285–3287. [CrossRef]
165. Yoon, J.-H.; Kim, B.-J.; Kim, J.-S. Design and fabrication of micro hydrogen gas sensors using palladium thin film. *Mater. Chem. Phys.* **2012**, *133*, 987–991. [CrossRef]
166. Joshi, R.K.; Krishnan, S.; Yoshimura, M.; Kumar, A. Pd nanoparticles and thin films for room temperature hydrogen sensor. *Nanoscale Res. Lett.* **2009**, *4*, 1191–1196. [CrossRef]
167. Hao, M.; Wu, S.; Zhou, H.; Ye, W.; Wei, X.; Wang, X.; Chen, Z.; Li, S. Room-temperature and fast response hydrogen sensor based on annealed nanoporous palladium film. *J. Mater. Sci.* **2016**, *51*, 2420–2426. [CrossRef]
168. Rafieerad, A.R.; Bushroa, A.R.; Nasiri-Tabrizi, B.; Vadivelu, J.; Baradaran, S.; Mesbah, M.; Zavareh, M.A. Mechanical properties, corrosion behavior and in-vitro bioactivity of nanostructured Pd/PdO coating on Ti–6Al–7Nb implant. *Mater. Des.* **2016**, *103*, 10–24. [CrossRef]
169. Hassan, E.S.; Abd, A.N.; Habubi, N.F.; Mansour, H.L. Sensing properties controlled by thickness variable of palladium oxide synthesized by RF-reactive sputtering. *Optik* **2018**, *174*, 481–488. [CrossRef]
170. Chiang, Y.-J.; Pan, F.-M. PdO nanoflake thin films for CO gas sensing at low temperatures. *J. Phys. Chem. C* **2013**, *117*, 15593–15601. [CrossRef]
171. Arai, T.; Shima, T.; Nakano, T.; Tominaga, J. Thermally-induced optical property changes of sputtered PdOx films. *Thin Solid Film.* **2007**, *515*, 4774–4777. [CrossRef]
172. Ketteler, G.; Ogletree, D.F.; Bluhm, H.; Liu, H.; Hebenstreit, E.L.; Salmeron, M. In Situ Spectroscopic Study of the Oxidation and Reduction of Pd(111). *J. Am. Chem. Soc.* **2005**, *127*, 18269–18273. [CrossRef] [PubMed]
173. Chiang, Y.-J.; Li, K.-C.; Lin, Y.-C.; Pan, F.-M. A mechanistic study of hydrogen gas sensing by PdO nanoflake thin films at temperatures below 250 °C. *Phys. Chem. Chem. Phys.* **2015**, *17*, 3039–3049. [CrossRef]
174. Thieu, C.-A.; Hong, J.; Kim, H.; Yoon, K.J.; Lee, J.-H.; Kim, B.-K.; Son, J.-W. Incorporation of a Pd catalyst at the fuel electrode of a thin-film-based solid oxide cell by multi-layer deposition and its impact on low-temperature co-electrolysis. *J. Mater. Chem. A* **2017**, *5*, 7433–7444. [CrossRef]
175. Nguyen, D.-C.; Chu, C.-C.; Lee, C.-H.; Hsu, T.; Chang, C.-S. Fabrication and tailoring of the nano-scale textures of Pd films by selective doping for hydrogen gas sensing. *Thin Solid Film.* **2016**, *616*, 722–727. [CrossRef]
176. Lee, Y.T.; Lee, J.M.; Kim, Y.J.; Joe, J.H.; Lee, W. Hydrogen gas sensing properties of PdO thin films with nano-sized cracks. *Nanotechnology* **2010**, *21*, 165503. [CrossRef]
177. Sui, M.; Kunwar, S.; Pandey, P.; Zhang, Q.; Li, M.-Y. Fabrication and determination of growth regimes of various Pd NPs based on the control of deposition amount and temperature on c-plane GaN. *J. Mater. Res.* **2017**, *32*, 3593–3604. [CrossRef]
178. Horwat, D.; Zakharov, D.I.; Endrino, J.L.; Soldera, F.; Anders, A.; Migot, S.; Karoum, R.; Vernoux, P.; Pierson, J.F. Chemistry, phase formation, and catalytic activity of thin palladium-containing oxide films synthesized by plasma-assisted physical vapor deposition. *Surf. Coat. Technol.* **2011**, *205*, S171–S177. [CrossRef]
179. Anders, A. A review comparing cathodic arcs and high power impulse magnetron sput-tering (HiPIMS). *Surf. Coat. Technol.* **2014**, *257*, 308–325. [CrossRef]
180. Amin-Ahmadi, B.; Idrissi, H.; Galceran, M.; Colla, M.S.; Raskin, J.P.; Pardoen, T.; Godet, S.; Schyvers, D. Effect of deposition rate on the microstructure of electron beam evaporated nanocrystalline palladium thin films. *Thin Solid Film.* **2013**, *539*, 145–150. [CrossRef]
181. Barzola-Quiquia, J.; Schulze, S.; Esquinazi, P. Transport properties and atomic structure of ion-beam deposited W. Pd, and Pt nanostructures. *Nanotechnology* **2009**, *20*, 165704–165711. [CrossRef] [PubMed]
182. Bhuvana, T.; Kulkarni, G.U. Highly conducting patterned Pd nanowires by direct-write electron-beam lithography. *ACS Nano* **2008**, *2*, 457–462. [CrossRef] [PubMed]

183. Samovat, F.; Mahmoodi, F.; Ahmad, P.T.; Samavat, M.F.; Tavakoli, M.H.; Hadidchi, S. Effect of annealing temperature on the optical properties of Palladium thin film. *Open J. Phys. Chem.* **2012**, *2*, 103–106. [CrossRef]
184. Nakagawa, H.; Aoyagi, M.; Kurosawa, I.; Takada, S. Palladium thin-film resistors for Josephson LSI circuits. *Jpn. J. Appl. Phys.* **1992**, *31*, 2550–2553. [CrossRef]
185. Corso, A.J.; Tessarolo, E.; Guidolin, M.; Gaspera, E.D.; Martucci, A.; Angiola, M.; Donazzan, A.; Pelizzo, M.G. Room-temperature optical detection of hydrogen gas using palladium nano-islands. *Int. J. Hydrogen Energy* **2018**, *43*, 5783–5792. [CrossRef]
186. Sharma, P.; Singh, R.; Sharma, R.; Mukhiya, R.; Awasthi, K.; Kumar, M. Palladium-oxide extended gate field effect transistor as pH sensor. *Mater. Lett. X* **2021**, *12*, 100102. [CrossRef]
187. Guidoni, A.G.; Di Palma, T.M.; Teghil, R.; Marotta, V.; Ambrico, M.; Piccirillo, S.; Orlando, S. Pulsed lased deposition of Pd on amorphous alumina substrate. *Surf. Coat. Technol.* **1996**, *80*, 216–220. [CrossRef]
188. Constantinoiu, I.; Viespe, C. Development of Pd/TiO_2 Porous Layers by Pulsed Laser Deposition for Surface Acoustic Wave H_2 Gas Sensor. *Nanomaterials* **2020**, *10*, 760. [CrossRef]
189. Bouhtiyya, S.; Roué, L. On the characteristics of Pd thin films prepared by pulsed laser deposition under different helium pressures. *Int. J. Hydrogen Energy* **2008**, *33*, 2912–2920. [CrossRef]
190. Scarisoreanu, N.D.; Nicolae, I.; Grigoriu, C.; Dinescu, M.; Hirai, M.; Suzuki, T.; Yatsui, K. Pt, Pd, Ni metallic thin films deposited by pulsed laser ablation. In Proceedings of the SPIE 5581, ROMOPTO 2003: Seventh Conference on Optics, Constanta, Romania, 8–11 September 2004.
191. Aggarwal, S.; Monga, A.P.; Perusse, S.R.; Ramesh, R.; Ballarotto, V.; Williams, E.D.; Chalamala, B.R.; Wei, Y.; Reuss, R.H. Spontaneous Ordering of Oxide Nanostructures. *Science* **2000**, *287*, 2235–2237. [CrossRef] [PubMed]
192. Semaltianos, N.G.; Petkov, P.; Scholz, S.; Guetaz, L. Palladium or Palladium hydride nanoparticles synthesized by laser ablation od a bulk Palladium target in liquids. *J. Colloid Interface Sci.* **2013**, *402*, 307–311. [CrossRef] [PubMed]
193. Park, H.; Reddy, D.A.; Kim, Y.; Lee, S.; Ma, R.; Kim, T.K. Synthesis of Ultra-Small Palladium Nanoparticles Deposited on CdS Nanorods by Pulsed Laser Ablation in Liquid: Role of Metal Nanocrystal Size in the Photocatalytic Hydrogen Production. *Chem. A Eur. J.* **2017**, *23*, 13112–13119. [CrossRef] [PubMed]
194. Amer, J.; Alkhawwam, A.; Jazmati, A.K. Activation of wood surface by Pd pulsed laser deposition for Ni electroless plating: Effects of wood morphology on coated films. *Int. J. Struct. Integr.* **2020**, *12*, 165–176. [CrossRef]
195. Meyerheim, H.L.; Soyka, E.; Kirschner, J. Alloying and dealloying in pulsed laser deposited Pd films on Cu(100). *Phys. Rev. B* **2006**, *74*, 085405. [CrossRef]
196. Shatokhin, A.N.; Putilin, F.N.; Safonova, O.V.; Rumyantseva, M.N. Sensor Properties of Pd-Doped SnO_2 Films Deposited by Laser Ablation. *Inorg. Mater.* **2002**, *38*, 374–379. [CrossRef]
197. Pereira, A.; Cultrera, L.; Dima, A.; Susu, M.; Perrone, A.; Du, H.L.; Volkov, A.O.; Cutting, R.; Datta, P.K. Pulsed laser deposition and characterization of textured Pd-doped-SnO_2 thin films for gas sensing applications. *Thin Solid Film.* **2006**, *497*, 142–148. [CrossRef]
198. Paillier, J.; Roue, L. Nanostructured Palladium Thin Films Prepared by Pulsed Laser Deposition. Structural Characterizations and Hydrogen Electrosorption Properties. *J. Electrochem. Soc.* **2005**, *152*, E1–E8. [CrossRef]
199. Aggarwal, S.; Ogale, S.B.; Ganpule, C.S.; Shinde, S.R.; Novikov, V.A.; Monga, A.P.; Burr, M.R.; Ramesh, R. Oxide nanostructures through self-assembly. *Appl. Phys. Lett.* **2001**, *78*, 1442–1444. [CrossRef]
200. Alexiadou, M. Pulsed laser deposition of ZnO thin films decorated with Au and Pd nanoparticles with enhanced acetone sensing performance. *Appl. Phys. A* **2017**, *123*, 262. [CrossRef]
201. Heras, J.M.; Estiu, G.; Viscido, L. Aneealing behavior of clean and oxygen covered polycrystalline palladium films: A work function and electrical resistance study. *Thin Solid Film.* **1990**, *188*, 165–172. [CrossRef]
202. Kalli, K.; Othonos, A.; Christofides, C. Temperature-induced reflectivity changes and activation of hydrogen sensitive optically thin palladium films on silicone oxide. *Rev. Sci. Instrum.* **1998**, *69*, 3331–3338. [CrossRef]
203. Okamoto, H.; Aso, T. Formation of thin films of PdO and their electric properties. *Jpn. J. Appl. Phys.* **1967**, *6*, 779. [CrossRef]
204. Li, Y.; Cheng, Y.-T. Hydrogen diffusion and solubility in Pd thin films. *Int. J. Hydrogen Energy* **1996**, *21*, 281–291. [CrossRef]
205. Ievlev, V.M.; Ryabtsev, S.V.; Samoylov, A.M.; Shaposhninik, A.V.; Kuschev, S.B. Thin and ultrathin films of palladium oxide for oxidizing gases detection. *Sens. Actuators B Chem.* **2018**, *255*, 1335–1342. [CrossRef]
206. Efimov, A.I.; Belokurova, L.P.; Vasilkova, I.V.; Chechev, V.P. *Properties of inorganic compounds/Handbook*; Khimiya: Leningrad, Russia, 1983; p. 392. (In Russian)
207. Darling, A.S. Some properties and applications of the platinum group metals. *Int. Metall. Rev.* **1973**, *18*, 91–122. [CrossRef]
208. Campbell, C.T.; Foyt, D.C.; White, J.M. Oxygen Penetration into the Bulk of Palladium. *J. Phys. Chem.* **1977**, *81*, 491–494. [CrossRef]
209. Tafreshi, H.V.; Piseri, P.; Benedek, G.; Milani, P. The role of gas dynamics in operation conditions of a pulsed microplasma cluster source for nanostructured thin films deposition. *J. Nanosci. Nanotechnol.* **2006**, *6*, 1140–1149. [CrossRef]
210. Cassina, V.; Gerosa, L.; Podesta, A.; Ferrari, G.; Sampietro, M.; Fiorentini, F.; Mazza, T.; Lenardi, C.; Milani, P. Nanoscale electrical properties of cluster-assembled palladium oxide thin films. *Phys. Rev. B* **2009**, *79*, 115422. [CrossRef]
211. Garcia-Serrano, O.; Andraca-Adame, A.; Baca-Arroyo, R.; Pena-Sierra, R.; Romero-Peredes, R.G. Thermal oxidation of ultrathin palladium (Pd) foils at room conditions. In Proceedings of the CCE 2011—2011 8th International Conference on Electrical Engineering, Computing Science and Automatic Control, Program and Abstract Book, Merida City, Mexico, 26–28 October 2011.
212. Kleykamp, H. Freie Bildungsenthalpie von Palladiumoxid. *Z. Phys. Chemie. Neue Folge* **1970**, *71*, 142–148. [CrossRef]

213. Zemlyanov, D.; Klötzer, B.; Gabasch, H.; Smeltz, A.; Ribeiro, F.H.; Zafeiratos, S.; Teschner, D.; Schnörch, P.; Vass, E.; Hävecker, M.; et al. Kinetics of Palladium oxidation in the mbar pressure range: Ambient pressure XPS study. *Top. Catal.* **2013**, *56*, 885–895. [CrossRef]
214. Samoylov, A.M.; Pelipenko, D.I.; Ivkov, S.A.; Tyulyanova, E.S.; Agapov, B. Thermal stability limit of of thin Palladium(II) oxide films. *Inorg. Mater.* **2022**, *58*, 48–55. (In Russian) [CrossRef]
215. Peuckert, M. XPS study on surface and bulk palladium oxide, its thermal stability and a comparison with other noble metal oxides. *J. Phys. Chem.* **1985**, *89*, 2481–2486. [CrossRef]
216. Rogal, J.; Reuter, K.; Scheffler, M. Thermodynamic stability of PdO surfaces. *Phys. Rev. B* **2004**, *69*, 075421. [CrossRef]
217. Hellman, A.; Resta, A.; Martin, N.M.; Gustafson, J.; Trinchero, A.; Carlsson, P.A.; Balmes, O.; Felici, R.; Van Rijn, R.; Frenken, J.W.; et al. The active phase of palladium during methane oxidation. *J. Phys. Chem. Lett.* **2012**, *3*, 678–682. [CrossRef] [PubMed]
218. Martin, N.M.; Van Den Bossche, M.; Hellman, A.; Grönbeck, H.; Hakanoglu, C.; Gustafson, J.; Blomberg, S.; Johansson, N.; Liu, Z.; Axnanda, S.; et al. Intrinsic ligand effect governing the catalytic activity of Pd oxide thin films. *ACS Catal.* **2014**, *4*, 3330–3334. [CrossRef]
219. Xie, G.; Sun, P.; Yan, X.; Du, X.; Jiang, Y. Fabrication of methane gas sensor by layer-by-layer self-assembly of polyaniline/PdO ultra-thin films on quartz crystal microbalance. *Sens. Actuators B Chem.* **2010**, *145*, 373–377. [CrossRef]
220. Myhre, G.; Shindell, D.; Bréon, F.-M.; Collins, W.; Fuglestvedt, J.; Huang, J.; Koch, D.; Lamarque, J.-F.; Lee, D.; Mendoza, B.; et al. Anthropogenic and natural radiative forcing. In *Climate Change 2013: The Physical Science Basis*; Stocker, T.F., Qin, D., Plattner, G.-K., Tignor, M., Allen, S.K., Boschung, J., Nauels, A., Xia, Y., Bex, V., Midgle, P.M., Eds.; Contribution of Working Group I to the Fifth Assessment Report of the Intergovernmental Panel on Climate Change; Cambridge University Press: Cambridge, UK; New York, NY, USA, 2013; pp. 659–740.
221. Raub, E.; Plate, W. Über das Verhalten der Edelmetalle und ihrer Legierungen zu Sauerstoff bei hoher Temperatur im festen Zustand. *Int. J. Mater. Res.* **1957**, *48*, 529–539. [CrossRef]
222. Jehn, H. High temperature behaviour of platinum group metals in oxidizing atmospheres. *J. Less Common Met.* **1984**, *100*, 321–339. [CrossRef]
223. Usoltsev, O.; Stoian, D.; Skorynina, A.; Kozyr, E.; Njoroge, P.N.; Pellegrini, R.; Groppo, E.; van Bokhoven, J.A.; Bugaev, A. Restructuring of Palladium Nanoparticles during Oxidation by Molecular Oxygen. *Small* **2024**, 2401184. [CrossRef]
224. Nazarpour, S.; Chaker, M. Fractal analysis of Palladium hillocks generated due to oxide formation. *Surf. Coat. Technol.* **2012**, *206*, 2991–2997. [CrossRef]
225. Pour, G.B.; Aval, L.F.; Sarvi, M.N.; Aval, S.F.; Fard, H.N. Hydrogen sensors: Palladium-based electrode. *J. Mater. Sci. Mater. Electron.* **2019**, *30*, 8145–8153. [CrossRef]
226. Zou, Q.; Itoh, T.; Choi, P.G.; Masuda, Y.; Shin, W. Prediction of the effects of process informatics parameters on platinum, palladium, and gold-loaded tin oxide sensors with an artificial neural network. *Sens. Actuators B Chem.* **2024**, *410*, 135704. [CrossRef]
227. Kafil, V.; Sreenan, B.; Hadj-Nacer, M.; Wang, Y.; Yoon, J.; Greiner, M.; Chu, P.; Wang, X.; Fadali, M.S.; Zhu, X. Review of noble metal and metal-oxide-semiconductor based chemiresistive hydrogen sensors. *Sens. Actuators A Phys.* **2024**, *373*, 115440. [CrossRef]
228. Stetter, J.R.; Li, J. Amperometric gas sensors a review. *Chem. Rev.* **2008**, *108*, 352–366. [CrossRef] [PubMed]
229. Buttner, W.J.; Post, M.B.; Burgess, R.; Rivkin, C. An overview of hydrogen safety sensors and requirements. *Int. J. Hydrogen Energy* **2011**, *36*, 2462–2470. [CrossRef]
230. Hübert, T.; Boon-Brett, L.; Black, G.; Banach, U. Hydrogen sensors—A review. *Sens. Actuators B Chem.* **2011**, *157*, 329–352. [CrossRef]
231. Korotcenkov, G.; Han, S.D.; Stetter, J.R. Review of electrochemical hydrogen sensors. *Chem. Rev.* **2009**, *109*, 1402–1433. [CrossRef]
232. Zhao, Z.; Carpenter, M.A.; Xia, H.; Welch, D. All-optical hydrogen sensor based on a high alloy content palladium thin film. *Sens. Actuators B Chem.* **2006**, *113*, 532–538. [CrossRef]
233. Hughes, R.C.; Schubert, W.K. Thin films of Pd/Ni alloys for detection of high hydrogen concentrations. *J. Appl. Phys.* **1992**, *71*, 542–544. [CrossRef]
234. Ding, D.; Chen, Z.; Lu, C. Hydrogen sensing of nanoporous palladium films supported by anodic aluminum oxides. *Sens. Actuators B Chem.* **2006**, *120*, 182–186. [CrossRef]
235. Ding, D.; Chen, Z.; Pyrolytic, A. Carbon-stabilized, nanoporous Pd film for wide-range H_2 sensing. *Adv. Mater.* **2007**, *19*, 1996–1999. [CrossRef]
236. Ding, D.; Chen, Z.; Rajaputra, S.; Singh, V. Hydrogen sensors based on aligned carbon nanotubes in an anodic aluminum oxide template with palladium as a top electrode. *Sens. Actuators B Chem.* **2007**, *124*, 12–17. [CrossRef]
237. Zeng, X.Q.; Wang, Y.L.; Deng, H.; Latimer, M.L.; Xiao, Z.L.; Pearson, J.; Xu, T.; Wang, H.H.; Welp, U.; Crabtree, G.W.; et al. Networks of ultrasmall Pd/Cr nanowires as high-performance hydrogen sensors. *ACS Nano* **2011**, *4*, 7443–7452. [CrossRef] [PubMed]
238. Xu, T.; Zach, M.P.; Xiao, Z.L.; Rosenmann, D.; Welp, U.; Kwok, W.K.; Crabtree, G.W. Self-assembled monolayer-enhanced hydrogen sensing with ultrathin palladium films. *Appl. Phys. Lett.* **2005**, *86*, 203104. [CrossRef]
239. Favier, F.; Walter, E.C.; Zach, M.P.; Benter, T.; Penner, R.M. Hydrogen sensors and switches from electrodeposited palladium mesowire arrays. *Science* **2001**, *293*, 2227–2231. [CrossRef]

240. Liu, N.; Tang, M.L.; Hentschel, M.; Giessen, H.; Alivisatos, A.P. Nanoantenna-enhanced gas sensing in a single tailored nanofocus. *Nat. Mater.* **2011**, *10*, 631–636. [CrossRef]
241. Xie, B.; Zhang, S.; Liu, F.; Peng, X.; Song, F.; Wang, G.; Han, M. Response behavior of a palladium nanoparticle array based hydrogen sensor in hydrogen–nitrogen mixture. *Sens. Actuators A Phys.* **2012**, *181*, 20–24. [CrossRef]
242. Lee, J.; Shim, W.; Noh, J.-S.; Lee, W. Design Rules for Nanogap-Based Hydrogen Gas Sensors. *ChemPhysChem* **2012**, *13*, 1395–1403. [CrossRef]
243. Zhao, X.; Du, L.; Xing, X.; Tian, Y.; Li, Z.; Wang, C.; Feng, D.; Liu, H.; Yang, D. Palladium and palladium oxide enwrapped iron oxide shell/core nanoparticles for stable detection of ppb-level hydrogen. *Chem. Eng. J.* **2023**, *457*, 141258. [CrossRef]
244. Guthe, S.; Mudhalwadkar, R. Thin film sensors for nitro aromatic explosive detection. *Mater. Today Proc.* **2017**, *4*, 10324–10327. [CrossRef]
245. Taylor, G.; Shallenberger, J.; Tint, S.; Fones, A.; Hamilton, H.; Yu, L.; Amini, S.; Hettinger, J. Investigation of iridium; ruthenium; rhodium, and palladium binary metal oxide solid solution thin films for implantable neural interfacing applications. *Surf. Coat. Technol.* **2021**, *426*, 127803. [CrossRef]
246. Salagare, S.; Adarakatti, P.S.; Venkataramanappa, Y. Electrochemical nitrite sensing employing palladium oxide–reduced graphene oxide (PdO-RGO) nanocomposites: Application to food and environmental samples. *Ionics* **2022**, *28*, 927–938. [CrossRef]
247. Nguyen, P.H.; Pham, M.-T.; Nguyen, H.Q.; Cao, T.M.; Van Pham, V. Boosting visible-light-driven photocatalysis of nitrogen oxide degradation by Mott–Schottky Pd/TiO_2 heterojunctions. *Sep. Purif. Technol.* **2025**, *354*, 129012. [CrossRef]
248. El-Shafai, N.M.; Mostafa, Y.S.; Alamri, S.A.; El-Mehasseb, I.M. A nanoelectrode of hybrid nanomaterials of palladium oxide with cadmium sulfide based on 2D-carbon nanosheets for developing electron transfer efficiency for supercapacitor applications. *New J. Chem.* **2024**, *48*, 11932–11948. [CrossRef]
249. El-Gohary, R.M.; El-Shafai, N.M.; El-Mehasseb, I.M.; Mostafa, Y.S.; Alamri, S.A.; Beltagi, A.M. Removal of pollutants through photocatalysis. adsorption, and electrochemical sensing by a unique plasmonic structure of palladium and strontium oxide nanoparticles sandwiched between 2D nanolayers. *J. Environ. Manag.* **2024**, *363*, 121257. [CrossRef] [PubMed]
250. Karagonis, V.A.; Liu, C.C.; Neuman, M.R.; Romankiw, L.T.; Leary, P.W.; Cuomo, J.J. A Pd-Rd film potentiometric pH sensor. *IEEE Trans. Biomed. Eng. BME* **1986**, *33*, 113–116. [CrossRef]
251. Tellez, V.C.; Portillo, M.C.; Santiesteban, H.J.; Castillo, M.P. Green synthesis of palladium mixed with PdO nanoparticles by chemical bath deposition. *Opt. Mater.* **2021**, *112*, 110747. [CrossRef]
252. Brodyn, M.S.; Volkov, V.I.; Rudenko, V.I.; Liakhovetskyi, V.R.; Borshch, A.O. Large third-order optical nonlinearity in PdO thin films. *J. Nonlinear Opt. Phys. Mater.* **2017**, *26*, 1750037. [CrossRef]
253. Liakhovetskyi, V.; Brodin, A.; Rudenko, V.; Brodyn, M.; Styopkin, V. High refractive nonlinearity of PdO films under femtosecond 800 nm laser pulses. *J. Appl. Phys.* **2020**, *128*, 013108. [CrossRef]

coatings

Review

Advances in Organic Multiferroic Junctions

Bogdana Borca

National Institute of Materials Physics, Atomistilor 405A, 077125 Magurele, Romania; bogdana.borca@infim.ro

Abstract: Typically, organic multiferroic junctions (OMFJs) are formed of an organic ferroelectric layer sandwiched between two ferromagnetic electrodes. The main scientific interest in OMFJs focuses on the magnetoresistive properties of the magnetic spin valve combined with the electroresistive properties associated with the ferroelectric junction. In consequence, memristive properties that couple magnetoelectric functionalities, which are one of the most active fields of research in material sciences, are opening a large spectrum of technological applications from nonvolatile memory to elements in logic circuits, sensing devices, energy harvesting and biological synapsis models in the emerging area of neuromorphic computing. The realization of these multifunctional electronic elements using organic materials is presenting various advantages related to their low-cost, versatile synthesis and low power consumption functioning for sustainable electronics; green disintegration for transient electronics; and flexibility, light weight and/or biocompatibility for flexible electronics. The purpose of this review is to address the advancement of all OMFJs including not only the achievements in the charge and spin transport through OMFJs together with the effects of electroresistance and magnetoresistance but also the challenges and ways to overcome them for the most used materials for OMFJs.

Keywords: organic ferroelectrics; organic spin valve; multiferroic junction; magnetoresistance; electroresistance

1. Introduction

Multiferroic tunnel junctions consisting of a ferroelectric barrier in magnetic spin valves were first proposed about 15 years ago [1–3] and are capturing considerable scientific interest, especially in the field of electric-field-controlled spintronics and information technologies. As magnetic tunnel junctions, these systems allow for the tunneling spin-polarized current to be efficiently filtered according to the orientation of the magnetization of the ferromagnetic electrodes, which generate high resistive and low resistive states for their antiparallel and parallel relative orientation, respectively, based on the tunnel magnetoresistance (MR) effect [4,5]. The activity of the ferroelectric barrier in a tunnel junction allows for switching between different resistive states defined by the orientation of its electrical polarization that alters the induced charge densities and the electrostatic potential at the barrier–electrode interface and thus the charge transport characteristics of the tunneling current, based on the electroresistance (ER) effect [6–8]. In addition, the piezoelectric effect of the ferroelectric that affects the strain at the interface may couple with the magnetostrictive effects of the ferromagnetic layers (Figure 1). This affects the magnetic anisotropy and the magnetization, leading to drastic changes in the magnetoresistive properties, such as the reversal of the magnetoresistance hysteresis induced, for instance, by the electrical polarization [3].

Furthermore, advanced spin and charge transport properties may be added to the multilevel-resistive or memristive states of the multiferroic junctions when the separating layer between the ferromagnetic electrodes of the junctions is an organic material. Organic electronics and spintronics constitute developing fields for the next generation of applications. For organic spintronics, the light molecular elements present weak spin–orbit and hyperfine interactions that lead to a long spin relaxation time and thus long spin transport distances even at room temperature [9,10] that open routes for their inclusion in electronic devices, such as memory and computing elements to store and process information, neuromorphic computing and models of synaptic and neuronal operations, etc. Moreover, important assets for technological applications of organic systems are the low-cost and versatile fabrication, sustainability, light weight, mechanical flexibility and the possibility of a controlled degradation for electronic waste management [11,12].

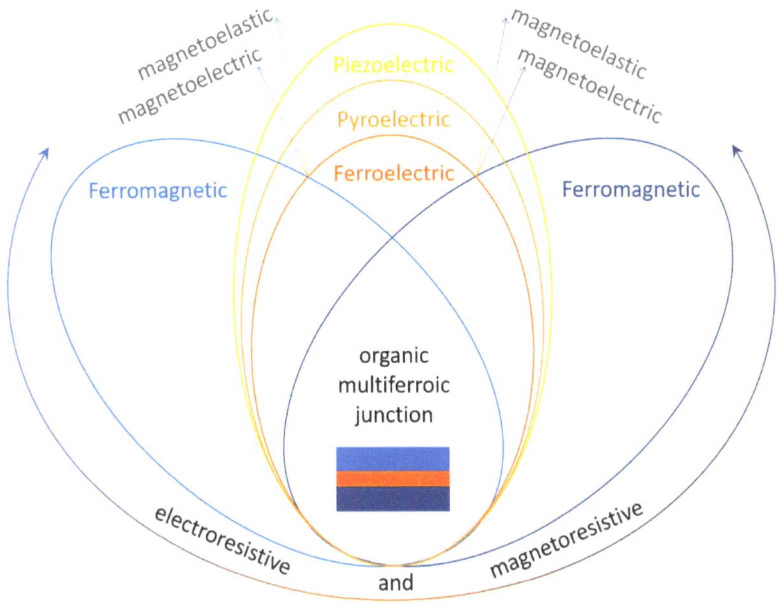

Figure 1. A schematic representation of an organic multiferroic junction composed of an organic ferroelectric film sandwiched between two different ferromagnetic electrodes. The representation highlights the pyroelectric and piezoelectric properties of a ferroelectric material and the coupling of properties at the ferromagnetic–ferroelectric interfaces. The junction has memristive properties based on the electroresistive and magnetoresistive effects.

The purpose of this article is to review the major advances in OMFJs through the analysis of mechanisms, functioning and applications. It is envisaged not only to merge current scientific knowledge but also to identify new paths for research and innovation in this field. This report contains a detailed description of the main characteristics of OMFJs in Section 2 that comprise the properties of organic spin valves and organic ferroelectric junctions, respectively. This section includes the origins of the functional properties of OMFJs, i.e., the magnetoresistance and the electroresistance effects and the control of the multilevel-resistive properties with external electric and magnetic fields. Section 2.1 describes the mechanism of the spin and charge transport through organic spin valves and summarizes magnetic and organic materials that are mostly used in spin valves and can be potentially used in OMFJ junctions. Furthermore, Section 2.2 summarizes the organic ferroelectric junctions and their properties, together with a description of organic ferroelectrics. Section 3 describes the reported OMFJs, their properties that are based on the coupling of magnetoresistive and electroresistive effects and their optimal configuration for adjusted functional devices. This review also includes challenges regarding the operational processes of OMFJs, together with proposed solutions to overcome these difficulties. Finally, concluding remarks on the progress strategies in OMFJs and a general perspective are presented in Section 4.

2. Characteristics of Organic Spin Valves and Organic Ferroelectric Junctions

2.1. Magnetic Junctions and Organic Spin Valves

Magnetic junctions are originally formed of two ferromagnetic films, preferably with different magnetic coercive fields, separated by a non-magnetic spacer that can be a tunneling barrier or a spin transport medium. In magnetic tunnel junctions, the tunneling current between metallic ferromagnetic films is spin-dependent and generates a tunneling MR that is based on the relative orientation of the magnetization of the two ferromagnetic electrodes [4]. The MR ratio is defined by the Julliere expression as follows:

$$\mathrm{MR} = \frac{R_{ap} - R_p}{R_p} 100\ (\%) = \frac{2P_1P_2}{1 + P_1P_2} 100(\%) \tag{1}$$

where R_{ap} and R_p are resistances for the antiparallel and parallel orientations of the magnetizations, respectively, and P_1, P_2 are the spin polarizability of the two ferromagnetic electrodes. Equivalently, the MR is defined for spin valves where spin-polarized currents are transported through the separating medium. Usually, the MR ratio is the maximum for an antiparallel magnetic configuration and the minimum for the parallel configuration (Figure 2a). However, these values can appear reversed, leading to negative MR values, observed for different junctions (Figure 2b).

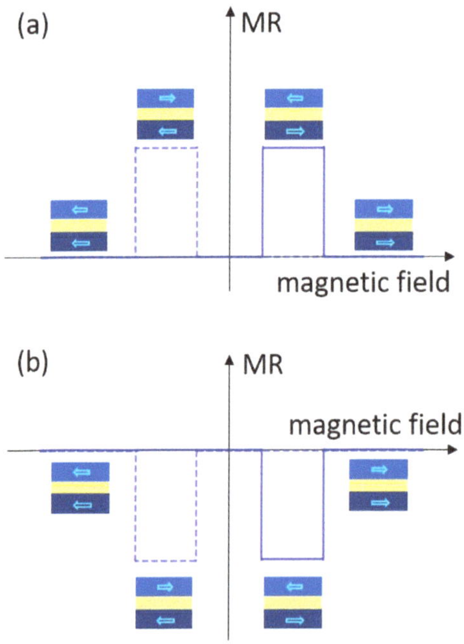

Figure 2. A schematic representation of the magnetoresistance hysteretic behavior with the magnetic field for (**a**) positive MR and (**b**) negative MR, respectively.

Table 1 presents most of the results encountered for magnetic junctions with an organic semiconductor spacer, i.e., organic spin valve junctions, which are composed of different materials, including small molecules that contain acenes, thiophenes, fullerenes, metal complexes and their derivatives, but also larger molecules and polymers. This table includes the chemical representation of each organic layer, the junction sequence mentioning the thickness of each system and the MR at various temperatures extracted from the corresponding references. The organic films of the junctions were deposited by physical vapor deposition techniques or wet chemical routes. The MR strongly depends on the nature of the used materials, the thickness of the organic spacer and the temperature.

Table 1. Organic spin valves. Abbreviations of molecule names are as follows: **PC71BM**—phenyl-C71-butyric-acid-metyl-ester; **TCNE**—tetracyanoethylene; **BCP**—buthocuproine; **TPD**—N,N-bis(3-methylphenyl)-N,N-diphenylbenzidine; **BTQBT**—bis(l,2,5-thiadiazolo)-P-quinobis(l,3-dithiole); **TIPS-pentacene**—6,13-bis(triisopropylsilylethynyl)-pentacene; **DOO-PPV**—poly(dioctyloxy)phenylenevinylene; **Alq$_3$**—tris(8-hydroxyquinolinato)aluminum; **Gaq$_3$**—tris-(8-hydroxyquinoline)gallium; **IC$_{12}$H$_2$(PO$_3$Et$_2$)$_2$**—diethyl(11-iodoundecyl)phosphonate; **Tb[Pc(C$_{12}$OPO$_3$Et$_2$)]$_2$**—terbium bis-phthalocyaninato-diethyl-phosphonate; **NitPO**—4-methylbenzylphosphonate)-4,4,5,5-tetramethylimidazoline-1-oxyl-3-oxide; **P3HT**—poly(3-hexylthiophene); **F$_4$TCNQ**—2,3,5,6-Tetrafluoro-7,7,8,8-tetracyanoquinodimethane; **P3MT**—poly(3-methylthiophene); **T$_6$**—sexithienyl; **CVB**—4,4′-bis99-(ethyl-3-carbazovinylene)-1,1′-bipheny; **C8-BTBT**—2,7-dioctyl [1]benzothieno[3,2-*b*][1]benzothiophene; **P(NDI2OD-T2)**—poly{[N,N′-bis(2-octyldodecyl)-1,4,5,8-naphthalenedicarboximide)-2,6-diyl]-alt-5,5′-(2,2′-bithio-phene)}; **F8BT**—poly (9,9-dioctylfluorene-co-benzothidiazole); **Py-Np**—4,4′,4″,4‴-(pyrene-1,3,6,8-tetrayl)tetraaniline-naphthalene-2,6-dicarbaldehyde; **TPP**—tetraphenyl porphyrin; **H$_2$Pc**—hydrogen-phthalocyanine; **CuPc**—copper-phthalocyanine; **FePc**—iron-phthalocyanine; **F$_{16}$CuPc**—fluorinated copper-phthalocyanine; and **BDMT**—1,4 benzenedimethanethiol.

Organic Material	System (Thickness)	MR/Temp. (K)	Ref.
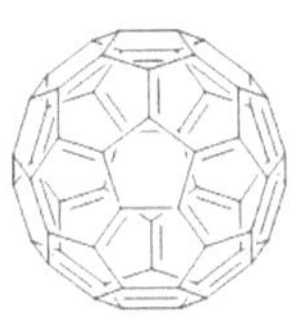 Fullerene	LSMO(50 nm)/**C$_{60}$**(120 nm)/Co(15 nm)	−13.3%/20 K	[13]
	LSMO(50 nm)/**C$_{60}$**(180 nm)/Co(15 nm)	~−7%/20 K	[13]
	LSMO(50 nm)/**C$_{60}$**(180 nm)/Co(15 nm)	~−4.5%/120 K	[13]
	LSMO(50 nm)/**C$_{60}$**(180 nm)/Co(15 nm)	~−1%/290 K	[13]
	Co(15 nm)/AlO$_x$(1 nm)/**C$_{60}$**(5 nm)/Py(20 nm)	9%/300 K	[14]
	Co(15 nm)/AlO$_x$(1 nm)/**C$_{60}$**(28 nm)/Py(20 nm)	5.5%/300 K	[14]
	Co(15 nm)/AlO$_x$(1 nm)/**C$_{60}$**(18 nm)/Py(20 nm)	~14%/80 K	[14]
	Fe$_3$O$_4$/AlO$_x$/**C$_{60}$**(80 nm)/Co(10 nm)	6.9%/150 K	[15]
	Fe$_3$O$_4$/AlO$_x$/**C$_{60}$**(80 nm)/Co(10 nm)	5.3%/300 K	[15]
	Fe$_3$O$_4$/AlO$_x$/**C$_{60}$**(110 nm)/Co(10 nm)	~2.2%/150 K	[15]
	Fe$_3$O$_4$/AlO$_x$/**C$_{60}$**(110 nm)/Co(10 nm)	~0.5%/300 K	[15]
	LSMO(50 nm)/**C$_{70}$**(120 nm)/Co(15 nm)	−8.9%/20 K	[13]
	LSMO(50 nm)/**C$_{70}$**(180 nm)/Co(15 nm)	~−6%/20 K	[13]
	LSMO(50 nm)/**C$_{70}$**(180 nm)/Co(15 nm)	~−4%/120 K	[13]
	LSMO(50 nm)/**C$_{70}$**(180 nm)/Co(15 nm)	~−0.8%/290 K	[13]
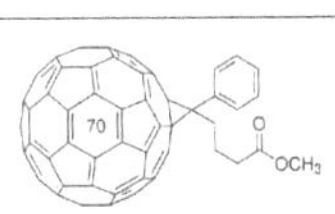 PC71BM	Co(20 nm)/LiF(1.4 nm)/**PC71BM**(45 nm)/Py(12 nm)	~6.2%/300 K	[16]
	Co(20 nm)/LiF(1.5 nm)/**PC71BM**(45 nm)/Py(12 nm)	~1.3%/300 K	[16]
	Co(20 nm)/LiF(1.6 nm)/**PC71BM**(45 nm)/Py(12 nm)	~−0.6%/300 K	[16]
	Co(20 nm)/LiF(1.8 nm)/**PC71BM**(45 nm)/Py(12 nm)	~−2.4%/300 K	[16]
	Co(20 nm)/LiF(2 nm)/**PC71BM**(45 nm)/Py(12 nm)	~−1.2%/300 K	[16]

Table 1. *Cont.*

Organic Material	System (Thickness)	MR/Temp. (K)	Ref.
	LSMO(15–20 nm)/**Alq3** (200 nm)/Al_2O_3 (2 nm)/Co(35 nm)	~−5%/20 K	[33]
	LSMO(15–20 nm)/**Alq3** (100 nm)/Al_2O_3 (2 nm)/Co(35 nm)	−11%/20 K	[33]
	LSMO(15–20 nm)/**Alq3** (100 nm)/Al_2O_3 (2 nm)/Co(35 nm)	−6%/100 K	[33]
	LSMO(15–20 nm)/**Alq3** (100 nm)/Al_2O_3 (2 nm)/Co(35 nm)	−0.15%/300 K	[33]
	LSMO/**Alq3** (23 nm)/Co-nanodots/Co(7 nm)	−1%/10 K	[34]
	LSMO/**Alq3** (67 nm)/Co-nanodots/Co(7 nm)	−7%/10 K	[34]
	LSMO/**Alq3** (93 nm)/Co-nanodots/Co(7 nm)	−300%/10 K	[34]
	LSMO/**Alq3** (135 nm)/Co-nanodots/Co(7 nm)	−13%/10 K	[34]
	LSMO/**Alq3** (93 nm)/Co(7 nm)	−35%/10 K	[34]
	LSMO/**Alq3** (135 nm)/Co(7 nm)	−4%10 K	[34]
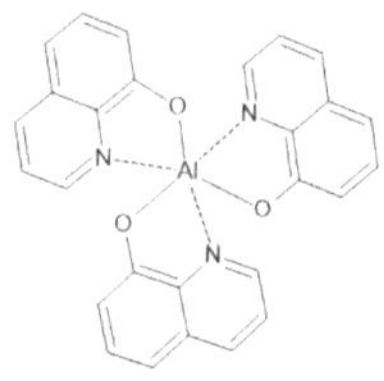	LSMO/**Alq3** (40 nm)/Co(20 nm)	~−18%/100 K	[35]
	LSMO/**Alq3** (40 nm)/Co(20 nm)	~−10%/150 K	[35]
	LSMO/**Alq3** (40 nm)/Co(20 nm)	−0.07%/300 K	[35]
	LPCMO(60 nm)/**Alq3**(60 nm)/Co(10 nm)	100–440% (1−7T)/10 K	[36]
Alq3	LPCMO(60 nm)/**Alq3**(60 nm)/Co(10 nm)	~200%(7T)/25 K	[36]
	LPCMO(60 nm)/**Alq3**(60 nm)/Co(10 nm)	~100%(7T)/75 K	[36]
	LPCMO(60 nm)/**Alq3**(60 nm)/Co(10 nm)	~1%(7T)/150 K	[36]
	LSMO/**Alq3**(60 nm)/Co(10 nm)	−37%/10 K	[36]
	SFMO(150 nm)/**Alq3**(45 nm)/Co(16 nm)	30%/10 K	[37]
	SFMO(150 nm)/**Alq3**(45 nm)/Co(16 nm)	25%/100 K	[37]
	SFMO(150 nm)/**Alq3**(45 nm)/Co(16 nm)	10%/200 K	[37]
	SFMO(150 nm)/**Alq3**(45 nm)/Co(16 nm)	2%/300 K	[37]
	Fe_3O_4(110 nm)/AlO_x(2 nm)/**Alq3**(2 nm)/Co(10 nm)	~12.5%/150 K	[38]
	Fe_3O_4(110 nm)/AlO_x(2 nm)/**Alq3**(2 nm)/Co(10 nm)	~6%/300 K	[38]
	Fe_3O_4(110 nm)/AlO_x(2 nm)/**Alq3**(10 nm)/Co(10 nm)	~4%/150 K	[38]
	Fe_3O_4(110 nm)/AlO_x(2 nm)/**Alq3**(10 nm)/Co(10 nm)	~2%/300 K	[38]
	Fe_3O_4(110 nm)/AlO_x(2 nm)/**Alq3**(20 nm)/Co(10 nm)	~1%/150 K	[38]
	Fe_3O_4(110 nm)/AlO_x(2 nm)/**Alq3**(20 nm)/Co(10 nm)	~0.5%/300 K	[38]
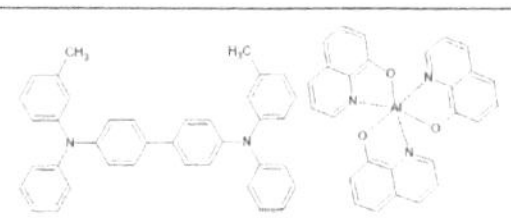 TPD/Alq3	FeCo(17 nm)/**TPD**(50 nm)/**Alq3**(200 nm)/LiF(1.9 nm)/NiFe(17 nm)	~0.3%/40 K	[39]

Table 1. *Cont.*

Organic Material	System (Thickness)	MR/Temp. (K)	Ref.
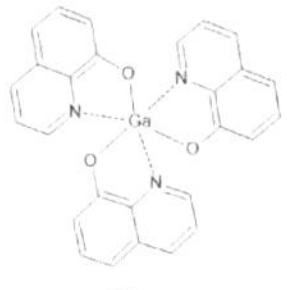 **Gaq3**	LSMO(20 nm)/**Gaq3** (10–15 nm)/AlO_x(2 nm)/Co(7 nm) electrically controlled-high-resistive state (oxygen migration)	−2%/100 K	[40]
	LSMO(20 nm)/**Gaq3** (10–15 nm)/AlO_x(2 nm)/Co(7 nm) electrically controlled-high-resistive state (oxygen migration)	−8%/100 K	[40]
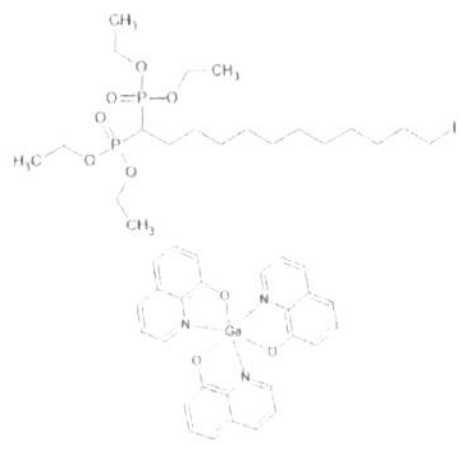 **$IC_{12}H_2(PO_3Et_2)_2$/ Gaq3**	LSMO/$IC_{12}H_2(PO_3Et_2)_2$(ML)/**Gaq3**(40 nm)/ AlO_x(2 nm)/Co	~−1.9%/200 K	[41]
	LSMO/$IC_{12}H_2(PO_3Et_2)_2$(ML)/**Gaq3**(40 nm)/ AlO_x(2 nm)/Co	~−9.8%/100 K	[41]
	LSMO/$IC_{12}H_2(PO_3Et_2)_2$(ML)/**Gaq3**(40 nm)/ AlO_x(2 nm)/Co	~−18%/10 K	[41]
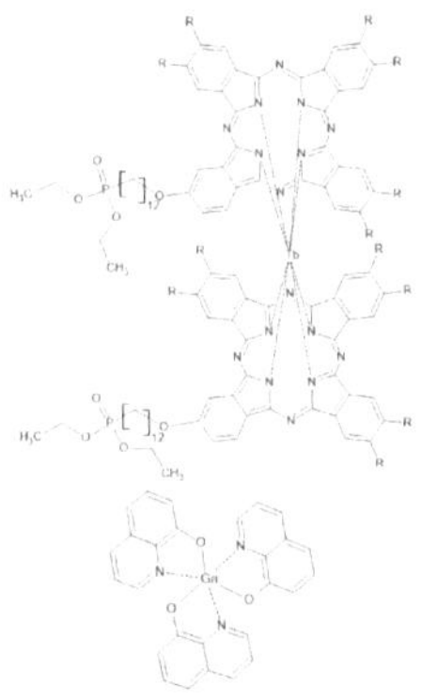 **Tb[Pc($C_{12}OPO_3Et_2$)]$_2$ /Gaq3**	LSMO/Tb[Pc($C_{12}OPO_3Et_2$)]$_2$(ML)/ **Gaq3**(40 nm)/AlO_x(2 nm)/Co	~−2%/200 K	[41]
	LSMO/Tb[Pc($C_{12}OPO_3Et_2$)]$_2$(ML)/ **Gaq3**(40 nm)/AlO_x(2 nm)/Co	~−9.3%/100 K	[41]
	LSMO/Tb[Pc($C_{12}OPO_3Et_2$)]$_2$(ML)/ **Gaq3**(40 nm)/AlO_x(2 nm)/Co	~−13%/100 K	[41]
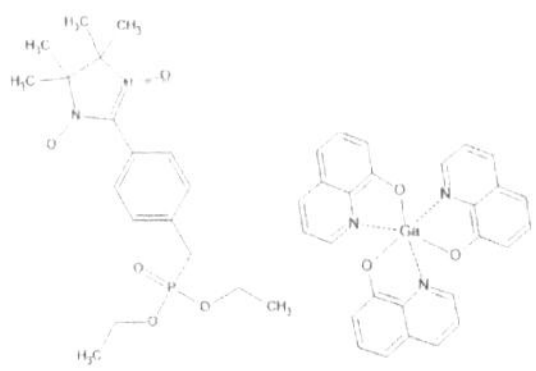 **NitPO/Gaq3**	LSMO(15 nm)/NitPO(ML)/**Gaq3**(150 nm)/ AlO_x(2 nm)/Co	−4.34%/100 K	[42]
	LSMO(15 nm)/NitPO(ML)/**Gaq3**(150 nm)/ AlO_x(2 nm)/Co	−7.1%/3 K	[42]

Table 1. *Cont.*

Organic Material	System (Thickness)	MR/Temp. (K)	Ref.
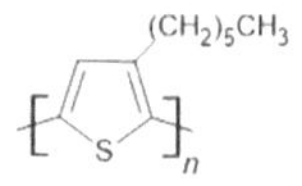 **P3HT**	$Fe_{50}Co_{50}$(20 nm)/**P3HT**(75 nm)/$Ni_{81}Fe_{19}$(20 nm)	0.1%/300 K	[43]
	$Fe_{50}Co_{50}$(40 nm)/**P3HT**(150 nm)/$Ni_{81}Fe_{19}$(20 nm)	−0.04%/300 K	[43]
	$Ni_{81}Fe_{19}$(12 nm)/**P3HT**(30 nm)/AlOx(1 nm)/Co(10 nm)	~1%/10 K	[44]
	Fe_3O_4(80 nm)/**P3HT**(12 nm)/AlO_x(1 nm)/Co(12 nm)	~2.3%/100 K	[45]
	Fe_3O_4(80 nm)/**P3HT**(12 nm)/AlO_x(1 nm)/Co(12 nm)	~0.75%/200 K	[45]
	Fe_3O_4(80 nm)/**P3HT**(25 nm)/AlO_x(1 nm)/Co(12 nm)	~1.75%/100 K	[45]
	LSMO/**P3HT**(100 nm)/Co(10 nm)	80%/5 K	[46]
	LSMO/**P3HT**(100 nm)/Co(10 nm)	1.5%/300 K	[46]
	LSMO(100 nm)/**P3HT**(80 nm)/AlOx(1 nm)/Co(10 nm)	−0.2%/2 K 15.6%/2 K	[47]
	LSMO/**P3HTannealed**(45 nm)/Co	~15%/20 K	[48]
	LSMO/**P3HTannealed**(72 nm)/Co	~11%/20 K	[48]
	LSMO/**P3HTannealed**(103 nm)/Co	~8%/20 K	[48]
	LSMO/**P3HTannealed**(175 nm)/Co	~4%/20 K	[48]
	LSMO/**P3HT**(45 nm)/Co	~12%/20 K	[48]
	LSMO/**P3HT**(72 nm)/Co	~6%/20 K	[48]
	LSMO/**P3HT**(103 nm)/Co	~3%/20 K	[48]
	LSMO/**P3HT**(175 nm)/Co	~2%/20 K	[48]
	LPCMO(60 nm)/**P3HT(30 nm)**/AlO_x/Co(10 nm)	93%/30 K	[49]
	LPCMO(60 nm)/**P3HT(30 nm)**/AlO_x/Co(10 nm)	~53%/50 K	[49]
	LPCMO(60 nm)/**P3HT(30 nm)**/AlO_x/Co(10 nm)	~30%/75 K	[49]
	LSMO(100 nm)/**P3HT(20 nm)**/AlO_x(1 nm)/Co(10 nm)	~7.5%/2 K	[50]
	LSMO(100 nm)/**P3HT(20 nm)**/AlO_x(1 nm)/Co(10 nm)	~4%/5 K	[50]
	LSMO(100 nm)/**P3HT(20 nm)**/AlO_x(1 nm)/Co(10 nm)	~3%/10 K	[50]
	LSMO(100 nm)/**P3HT(20 nm)**/AlO_x(1 nm)/Co(10 nm)	~0.1%/30 K	[50]
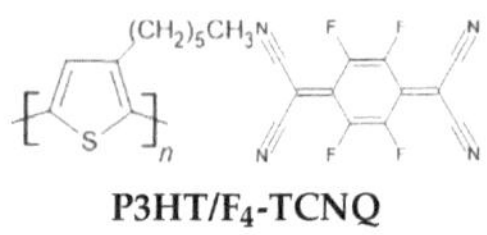 **P3HT/F_4-TCNQ**	LSMO(100 nm)/**P3HT**(40 nm)/F_4TCNQ dopant/Co(15 nm)	~19%/2 K	[51]
	LSMO(100 nm)/**P3HT**(40 nm)/F_4TCNQ dopant/Co(15 nm)	~16%/5 K	[51]
	LSMO(100 nm)/**P3HT**(40 nm)/F_4TCNQ dopant/Co(15 nm)	~14%/10 K	[51]
	LSMO(100 nm)/**P3HT**(40 nm)/F_4TCNQ dopant/Co(15 nm)	~13%/15 K	[51]
	LSMO(100 nm)/**P3HT**(40 nm)/F_4TCNQ dopant/Co(15 nm)	~11%/20 K	[51]
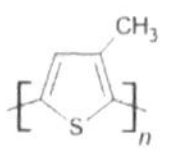 **P3MT**	LSMO/**P3MT**(15 nm)/Co(15 nm)	~−43%/20 K	[52]
	LSMO/**P3MT**(15 nm)/Co(15 nm)	~−0.3%/280 K	[52]
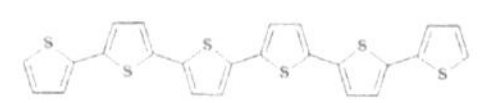 **T_6**	LSMO/**T_6**(70–140 nm)/LSMO	15–30%/300 K	[9]
	LSMO/**T_6**(200 nm)/LSMO	7–10%/300 K	[9]

Table 1. *Cont.*

Organic Material	System (Thickness)	MR/Temp. (K)	Ref.
	LSMO/**CVB**(100 nm)/Co	~−15%/14 K	[53]
CVB	LSMO/**CVB**(100 nm)/Co	~−3.5%/140 K	[53]
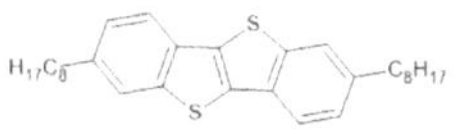	$Ni_{78}Fe_{22}$/**C8-BTBT**(2 nm)/$Ni_{78}Fe_{22}$	~−0.3%/300 K	[54]
	$Ni_{78}Fe_{22}$/**C8-BTBT**(4 nm)/$Ni_{78}Fe_{22}$	~0.5%/300 K	[54]
	Py/**C8-BTBT**(1ML)/Co	~−1%/10 K	[55]
C8-BTBT	Py/**C8-BTBT**(2ML)/Co	~−12%/10 K	[55]
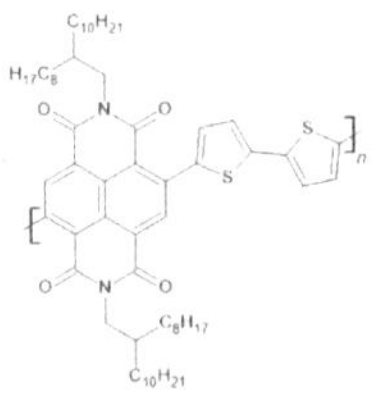	LSMO(100 nm)/**P(NDI2OD-T2)**(35 nm)/ AlO_x(1.5 nm)/Co(10 nm)	~30%/4.2 K	[56]
	LSMO(100 nm)/**P(NDI2OD-T2)**(35 nm)/ AlO_x(1.5 nm)/Co(10 nm)	~18%/50 K	[56]
	LSMO(100 nm)/**P(NDI2OD-T2)**(35 nm)/ AlO_x(1.5 nm)/Co(10 nm)	~5%/150 K	[56]
	LSMO(100 nm)/**P(NDI2OD-T2)**(35 nm)/ AlO_x(1.5 nm)/Co(10 nm)	~1%/250 K	[56]
P(NDI2OD-T2)	LSMO(100 nm)/**P(NDI2OD-T2)**(100 nm)/ AlO_x(1.5 nm)/Co(10 nm)	~10%/4.2 K	[56]
	LSMO(100 nm)/**P(NDI2OD-T2)**(100 nm)/ AlO_x(1.5 nm)/Co(10 nm)	~5%/50 K	[56]
	LSMO(100 nm)/**P(NDI2OD-T2)**(100 nm)/ AlO_x(1.5 nm)/Co(10 nm)	~2%/150 K	[56]
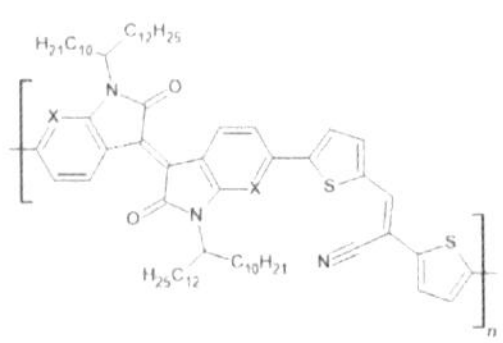	LSMO(50 nm)/**PIID-CNTVT-C1**(40 nm)/Py(10 nm)	~19%/50 K	[57]
	LSMO(50 nm)/**PIID-CNTVT-C1**(40 nm)/Py(10 nm)	~14%/100 K	[57]
	LSMO(50 nm)/**PIID-CNTVT-C1**(40 nm)/Py(10 nm)	~8%/150 K	[57]
	LSMO(50 nm)/**PIID-CNTVT-C1**(40 nm)/Py(10 nm)	~3%/200 K	[57]
	LSMO(50 nm)/**PAIID-CNTVT-C1**(40 nm)/Py(10 nm)	~25%/50 K	[57]
	LSMO(50 nm)/**PAIID-CNTVT-C1**(40 nm)/Py(10 nm)	~17%/100 K	[57]
PIID-CNTVT-C1: X=C **PAIID-CNTVT-C1: X=N**	LSMO(50 nm)/**PAIID-CNTVT-C1**(40 nm)/Py(10 nm)	~10%/150 K	[57]
	LSMO(50 nm)/**PAIID-CNTVT-C1**(40 nm)/Py(10 nm)	~5%/200 K	[57]
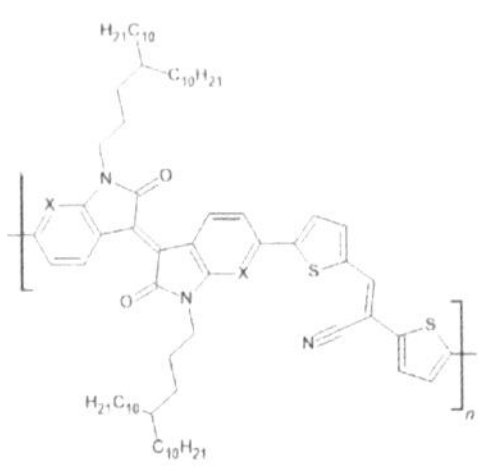	LSMO(50 nm)/**PIID-CNTVT-C3**(40 nm)/Py(10 nm)	~16%/50 K	[57]
	LSMO(50 nm)/**PIID-CNTVT-C3**(40 nm)/Py(10 nm)	~12%/100 K	[57]
	LSMO(50 nm)/**PIID-CNTVT-C3**(40 nm)/Py(10 nm)	~6%/150 K	[57]
	LSMO(50 nm)/**PIID-CNTVT-C3**(40 nm)/Py(10 nm)	~2%/200 K	[57]
	LSMO(50 nm)/**PAIID-CNTVT-C3**(40 nm)/Py(10 nm)	~23%/50 K	[57]
	LSMO(50 nm)/**PAIID-CNTVT-C3**(40 nm)/Py(10 nm)	~15%/100 K	[57]
PIID-CNTVT-C3: X=C-H **PAIID-CNTVT-C3: X=N**	LSMO(50 nm)/**PAIID-CNTVT-C3**(40 nm)/Py(10 nm)	~10%/150 K	[57]
	LSMO(50 nm)/**PAIID-CNTVT-C3**(40 nm)/Py(10 nm)	~5%/200 K	[57]

Table 1. *Cont.*

Organic Material	System (Thickness)	MR/Temp. (K)	Ref.
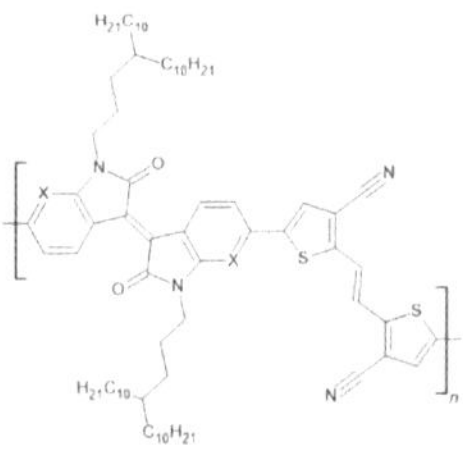 **PIID-TVTCN: X=C-H** **PAIID-TVTCN: X=N** **PFIID-TVTCN: X=C-F**	LSMO(30 nm)/**PIID-TVTCN**(25 nm)/Co(10 nm)	~−23%/50 K	[58]
	LSMO(30 nm)/**PIID-TVTCN**(25 nm)/Co(10 nm)	~−8%/200 K	[58]
	LSMO(30 nm)/**PAIID-TVTCN**(25 nm)/Co(10 nm)	~−10%/50 K	[58]
	LSMO(30 nm)/**PAIID-TVTCN**(25 nm)/Co(10 nm)	~−3%/200 K	[58]
	LSMO(30 nm)/**PFIID-TVTCN**(25 nm)/Co(10 nm)	~−16%/50 K	[58]
	LSMO(30 nm)/**PFIID-TVTCN**(25 nm)/Co(10 nm)	~−5%/200 K	[58]
	LSMO(30 nm)/**PIID-TVTCN**(25 nm)/ AlO_x(2 nm)/Co(10 nm)	~10%/50 K	[58]
	LSMO(30 nm)/**PIID-TVTCN**(25 nm)/ AlO_x(2 nm)/Co(10 nm)	~4%/200 K	[58]
	LSMO(30 nm)/**PAIID-TVTCN**(25 nm)/ AlO_x(2 nm)/Co(10 nm)	~17%/50 K	[58]
	LSMO(30 nm)/**PAIID-TVTCN**(25 nm)/ AlO_x(2 nm)/Co(10 nm)	~5%/200 K	[58]
	LSMO(30 nm)/**PFIID-TVTCN**(25 nm)/ AlO_x(2 nm)/Co(10 nm)	~10%/50 K	[58]
	LSMO(30 nm)/**PFIID-TVTCN**(25 nm)/ AlO_x(2 nm)/Co(10 nm)	~3%/200 K	[58]
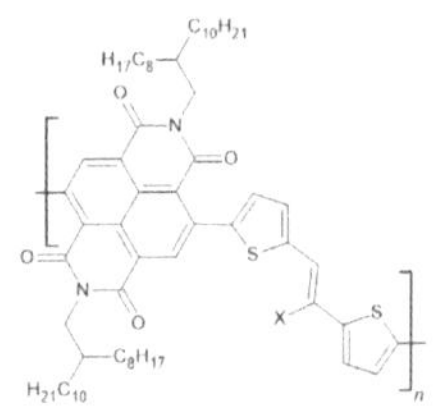 **PNVT-8: X=H** **PNVT-CN-8:X=CN**	NiFe(12 nm)/**PNVT-8**(15 nm)/Co(12 nm)	~0.35%/10 K	[59]
	NiFe(12 nm)/**PNVT-8**(15 nm)/Co(12 nm)	~0.18/150 K	[59]
	NiFe(12 nm)/**PNVT-8**(15 nm)/Co(12 nm)	~0.12/300 K	[59]
	NiFe(12 nm)/**PNVT-CN-8**(15 nm)/Co(12 nm)	~0.22/10 K	[59]
	NiFe(12 nm)/**PNVT-CN-8**(15 nm)/Co(12 nm)	~0.1%/150 K	[59]
	NiFe(12 nm)/**PNVT-CN-8**(15 nm)/Co(12 nm)	~0.08%/300 K	[59]
	NiFe(12 nm)/Au(3 nm)/**PNVT-CN-8**(25 nm)/Co(12 nm)	~−0.1%/100 K	[60]
	NiFe(12 nm)/Au(3 nm)/**PNVT-CN-8**(25 nm)/Co(12 nm)	~−0.08%/300 K	[60]
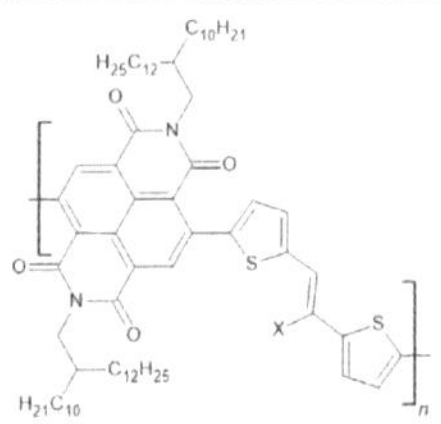 **PNVT-10: X=H** **PNVT-CN-10:X=CN**	NiFe(12 nm)/**PNVT-10**(15 nm)/Co(12 nm)	~0.19%/10 K	[59]
	NiFe(12 nm)/**PNVT-10**(15 nm)/Co(12 nm)	~0.07/150 K	[59]
	NiFe(12 nm)/**PNVT-10**(15 nm)/Co(12 nm)	~0.02/300 K	[59]
	NiFe(12 nm)/**PNVT-CN-10**(15 nm)/Co(12 nm)	~0.13%/10 K	[59]
	NiFe(12 nm)/**PNVT-CN-10**(15 nm)/Co(12 nm)	~0.05/150 K	[59]
	NiFe(12 nm)/**PNVT-CN-10**(15 nm)/Co(12 nm)	~0.02/200 K	[59]
	NiFe(12 nm)/Au(3 nm)/**PNVT-CN-10**(25 nm)/Co(12 nm)	~0.05%/10 K	[60]
	NiFe(12 nm)/Au(3 nm)/**PNVT-CN-10**(25 nm)/Co(12 nm)	~−0.03%/100 K	[60]
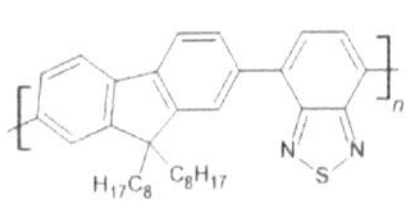 **F8BT**	LSMO(20 nm)/**F8BT**(45 nm)/MoO_x(3 nm)/Co(20 nm) magneto-electroluminescence	~9%/20 K ~2.4%/20 K	[61]
	LSMO(20 nm)/**F8BT**(45 nm)/MoO_x(3 nm)/Co(20 nm)	~7%/30 K	[61]
	LSMO(20 nm)/**F8BT**(45 nm)/MoO_x(3 nm)/Co(20 nm) magneto-electroluminescence	~3%/100 K ~0.4%/100 K	[61]
	LSMO(20 nm)/**F8BT**(45 nm)/MoO_x(3 nm)/Co(20 nm)	~1%/150 K	[61]

Table 1. *Cont.*

Organic Material	System (Thickness)	MR/Temp. (K)	Ref.
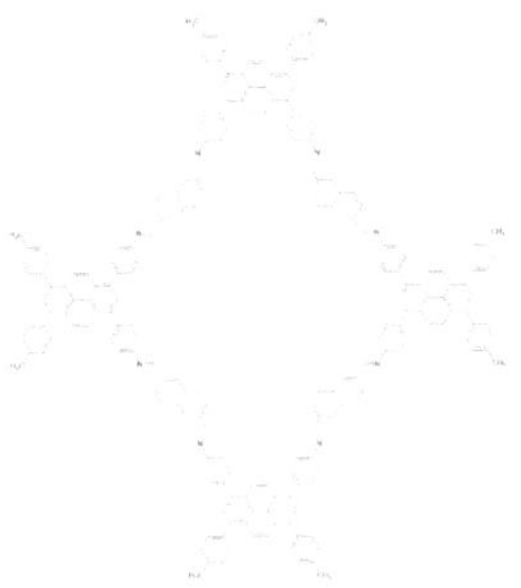 **Py-Np**	LSMO/**Py-Np**(90 nm)/Co(20 nm)	~−27%/30 K	[62]
	LSMO/**Py-Np**(90 nm)/Co(20 nm)	~−19%/100 K	[62]
	LSMO/**Py-Np**(90 nm)/Co(20 nm)	~−9%/150 K	[62]
	LSMO/**Py-Np**(90 nm)/Co(20 nm)	~−2%/200 K	[62]
 **TPP**	LSMO(50 nm)/**TPP**(20 nm)/Co(5 nm)	~−15%/80 K	[63]
	LSMO(50 nm)/**TPP**(20 nm)/Co(5 nm)	~−6%/80 K	[63]
	LSMO(50 nm)/**TPP**(20 nm)/Co(5 nm)	~−3%/80 K	[63]
 **H_2Pc**	Co(12 nm)/AlO_x(1.5 nm)/**H_2Pc**(90 nm)/Py spin-photovoltaic-resistive levels control with light intensity	~7%/300 K	[64]
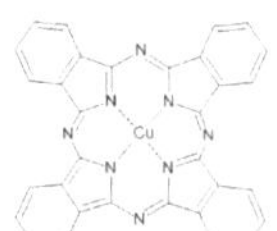 **CuPc**	Fe(25 nm)/**CuPc**(100 nm)/Co(5 nm)	6.4%/40 K	[65]
	Fe(25 nm)/**CuPc**(100 nm)/Co(5 nm)	3.2%/80 K	[65]
	Fe(25 nm)/**CuPc**(100 nm)/Co(5 nm)	1.8%/120 K	[65]
	LSMO/**CuPc**(110 nm)/Co	~−6%/10 K	[66]
	LSMO/**CuPc**(110 nm)/Co	~−4.5%/110 K	[66]
	LSMO/**CuPc**(110 nm)/Co	~−1%/300 K	[66]
	LSMO/**CuPc**(50 nm)/Co	~−20%/10 K	[66]
	LSMO/**CuPc**(160 nm)/Co	~−2.5%/10 K	[66]
	LSMO/**CuPc**(200 nm)/Co	~−0.5%/10 K	[66]
 **FePc**	Co/**FePc**(50 nm)/Co	4%/10 K	[67]
	Co/**FePc**(50 nm)/Co	1.2%/50 K	[67]
	Co/**FePc**(50 nm)/Co	0.5%/100 K	[67]

Table 1. *Cont.*

Organic Material	System (Thickness)	MR/Temp. (K)	Ref.
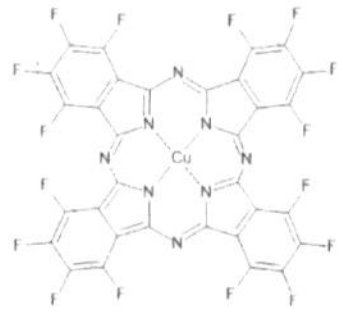	Co/AlOx/**F_{16}CuPc**(15 nm)/Py	~4%/300 K	[68]
F_{16}CuPc	Co/AlOx/**F_{16}CuPc**(90 nm)/Py Spin-photovoltaic (vacuum or air)	~4%/295 K −30/−25%/295 K	[68]
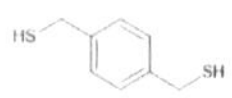	NiFe(15 nm)/FeCo(10 nm)/**BDMT**(1 nm)/AlOx(1 nm)FeCo(35 nm)	2.4%/20 K	[69]
BDMT	NiFe(15 nm)/FeCo(10 nm)/**BDMT**(1 nm)/AlOx(1 nm)/FeCo(35 nm)	1.9%/100 K	[69]
	NiFe(15 nm)/FeCo(10 nm)/**BDMT**(1 nm)/AlOx(1 nm)/FeCo(35 nm)	0.25%/300 K	[69]

The variation in the MR values of organic spin valves is related to several parameters, such as the following:

- The spin polarization of the ferromagnetic electrodes;
- Tunneling/Spin injection from the ferromagnetic electrode into the organic layer;
- Spin transport through the organic layer;
- Spin detection at the other electrode.

Therefore, the MR of the organic spin valves can also be expressed by adapting the Julliere formula as follows:

$$\mathrm{MR} = \frac{2P_1P_2e^{-d/\lambda_s}}{1+P_1P_2e^{-d/\lambda_s}}100(\%) \quad (2)$$

where d is usually the thickness of the spin transport medium and λ_s the spin diffusion length.

For efficient spin valves with high MR, the junctions preferably contain ferromagnetic electrodes with a high spin polarization. In particular, from oxides with the perovskite-based crystal structure, lanthanum strontium manganite with the general formula $La_{1-x}Sr_xMnO_3$, where x describes the doping level, known by its abbreviation LSMO, has a 100% spin polarization at low temperatures [70] and is one of the most used ferromagnetic electrodes (especially as the bottom electrode) in organic spin valves (see Table 1). As another La-comprising perovskite oxide material, $(La_{2/3}Pr_{1/3})_{5/8}Ca_{3/8}MnO_3$ (LPCMO) has a nearly maximum spin polarization (~97%) [71,72] and is starting to be employed in organic spin valves [36,49], leading to an incredibly large MR effect of up to 440% [36]. Moreover, it is worth to mention here the fact that LPCMO presents domains of electronic phase separation characterized by the coexistence of ferromagnetic metallic and antiferromagnetic insulating phases that can be magnetically controlled with preset magnetic fields, permitting the achievement of a nonvolatile tunable MR response in organic spin valves [36]. Starting from this result, one can imagine complex structured hybrid systems, like lateral arrays of tranches and nanowires [73–76], as templates for molecular assemblies that can provide an adjustable MR. The double perovskite Sr_2FeMoO_6 (SFMO) is also a potential material for organic spintronics with a high spin polarization (of 100% theoretically predicted [77] and ~85% experimentally detected [78]), used recently in organic-based junctions [37]. Additionally, magnetic oxides, such as Fe_3O_4 which has a high spin polarization of about -80% [79], are often used in organic spin valves [15,17,38,45]. Other materials with full spin polarization and a great potential in spintronics and multifunctional materials are Heusler compounds [80–82], from which Co_2MnSi has a 100% spin polarization [83] and can be employed in organic spin valves [25]. Magnetic transition metals Fe (44%), Co (34%) [84] and/or alloys of these metals, including FeCo (50%) [85], NiFe (45%) [86], $Ni_{80}Fe_{20}$ (~30%) [86,87] also called permalloy (Py) and close composition alloys such as $Ni_{81}Fe_{19}$ and $Ni_{78}Fe_{22}$ have an elevated Curie temperature (>700 °C) and are often used (mostly as a top electrode) in organic junctions (see Table 1).

Depending on the thickness of the separating layer, this can serve as a spin-conserved tunneling barrier or a spin transport medium. In the case of magnetic junctions with an organic spacer, the concept of tunneling is controversial, due to hybridization effects at interfaces and to the fact that the MR response is very similar either for tunneling or spin injection. However, for very thin thicknesses of only a few nm of the organic spacer (see Table 1), the tunneling mechanism is considered [63,88,89]. For organic spin valves, there are different mechanisms regarding the spin injection. A prime factor to be considered not only for the spin injection but also for the transport and the spin collection or detection is the energy level alignment at the metal–organic interface. Thus, the position of the Fermi level (E_F) of the metal with the work function (φ) in rapport with the positions of the molecular orbitals of the organic material with ionization energy (I_p) and the electron affinity energy (E_a) aligned at the vacuum level (Ev) determines the type of charge carrier. An example is shown in Figure 3. If the E_F of the metal is closer to the lowest unoccupied molecular orbital (LUMO), the main charge carriers are electrons (Figure 3a). If the E_F of the metal is closer to the highest occupied molecular orbital (HOMO), the main charge carriers are holes (Figure 3b). Moreover, one has to consider that because the spin polarization of the electronic states at different energy levels of the ferromagnetic electrode may be different, the spin injection and detection are likely different for different types of carriers and different energy alignments at the metal–organic interface. Additionally, alignment mismatches at interfaces may induce additional spin injection or collection problems (see, for example, Figure 3a).

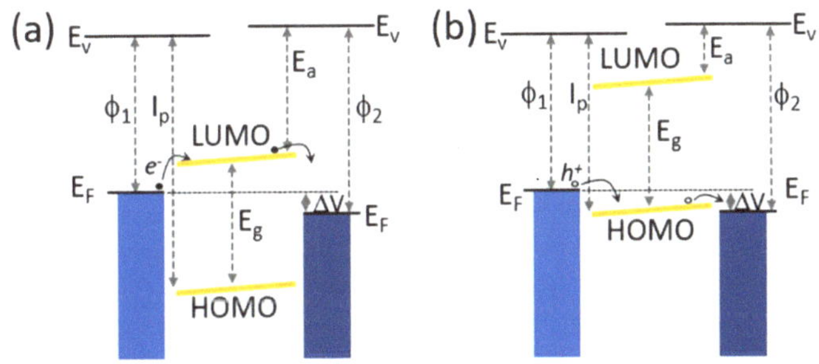

Figure 3. A schematic representation of the charge carrier transport consisting of (**a**) electrons (e^-) and (**b**) holes (h^+) in metal–organic–metal junctions for an applied bias voltage ΔV, in function of the energy level alignment, where E_v is the vacuum level, E_F is the Fermi level, φ_1 and φ_2 are the work functions of the two metals, LUMO and HOMO are the lowest unoccupied and the highest occupied molecular orbitals with an E_g energy gap between them and I_p and E_a are the ionization energy and the electron affinity energy, respectively.

Regarding the spin injection, every so often, a *tunnel spin injection* barrier consisting of a thin insulating film, such as AlO_x which is the most used (see Table 1), MoO_x [61], LAO ($LaAlO_3$) [20] or other films, like, for example, LiF [16] or organic layers [39,41,42,51], is inserted in between the ferromagnetic electrodes and the organic spin transport medium. These layers are used to optimize and tune the spin injection through processes such as solving the energy level mismatch at interfaces or tailoring the bend bending. Moreover, the inserted interfacial layer can improve the interface qualities, like, for example, in vertical junctions, it can reduce the penetration of the top ferromagnetic metallic electrode, or an organic spinterface [90–93] can ensure a less rough interface or better anchoring and tune the interface resistance [94–97].

Recently, several innovative spin injection methods based on studies on inorganic materials were proposed. From these, *ferromagnetic resonance spin pumping injection* is a pure spin injection process conducted by applying an external microwave field at its resonance frequency that generates a spin precession within the ferromagnetic layer, leading to an excess of spin angular momentum. When the precession frequency fits the microwave field, pure spin currents are injected into the organic layer through exchange interactions at the metal–organic interface [98–100], which lead to an induced electric field based on the inverse spin Hall effect, excluding the need for a bias voltage and resolving, thus, energy level alignment mismatches at interfaces.

Hot-electron spin injection or *ballistic spin injection* can be implemented in a three-terminal junction that contains three metal electrodes: an emitter, a base and a collector. From these layers, either the emitter or, in most cases, because a higher spin-polarized current is generated, the base should be ferromagnetic. The emitter and the base are separated by a tunnel barrier, while between the base and the collector is the spin transport medium, i.e., the organic layer. By applying a bias voltage between the emitter and the base, hot carrier electrons (or holes) are generated into the base that exhibit a ballistic spin filter effect based on different mean free paths of the majority and minority spin in the ferromagnet leading to a very high spin-polarized current that can be injected above the LUMO level for electrons (or below the HOMO level for holes) and measured at the collector if the applied voltage exceeds the injection barrier, i.e., the Schottky barrier, at the metal–organic interface [101–103]. In addition, hot electrons can be generated optically, using a photon photoemission process by applying an ultrashort laser pulse that generates hot electrons into the ferromagnet, from which a fraction can be ballistically injected at the metal–organic interface into the LUMO of the organic layer. Afterward, by applying a second laser pulse, the spin-polarized electrons remaining in the organic material can be excited into the vacuum and then photoemitted and detected [104]. Regarding spin-related optical effects, it is worth mentioning the coupling with luminescent and photovoltaic process that can lead to an enhancement in the electroluminescence intensity by the spin-polarized carriers [61] and to the manipulation of the spin-related magnetic response of the junction by the intensity of an irradiating light [64,68].

Another strategy to generate spin polarization in organic materials includes the *chirality-induced spin selectivity injection*. This can be reached when electrons with a spin orientation like the chirality of their transport medium are selected in favor of others that do not match this criterium [105–108]. The approach can lead to very high spin polarizations, also more than 80% [109].

Moreover, *charge transfer spin polarization* induced by an asymmetric population of molecular orbitals can be generated at the molecular level [110–112].

The transport of spin-polarized carriers through the organic layer is realized mainly through a hopping mechanism; however, for molecular crystals and polymers, a band transport is possible. Exchange coupling modes may be considered as an additional contribution for a very high concentration of transport carriers in doped or impurity band organic films. The spin diffusion thorough the organic layer for the hopping mode is defined principally by two factors, the spin diffusion length (λ_s) and the spin relaxation time (τ), as follows:

$$\lambda_s = \sqrt{\frac{k_B T}{q}\mu\tau} \tag{3}$$

where k_B is the Boltzmann constant, T is the temperature, q is the carrier charge and μ is the mobility. Therefore, a long spin diffusion length can be reached for a long spin relaxation time and high mobility. The spin relaxation time is mainly influenced by the spin orbit coupling and the hyperfine interactions. The spin orbit coupling characterizes the interaction between the spin and the orbital angular momentum of the charge carrier, which contributes to the spin–charge conversion and the spin relaxation. Usually, the spin–orbit coupling is proportional to the fourth power of the atomic number. Thus, in organic materials that contain light-weight elements, such as C, H, N and O, the strength of the spin–orbit coupling is weak leading to an extremely long spin relaxation time. The hyperfine interaction, reported to be strong only for localized carriers, characterizes the exchange interactions between the spin of the charge carriers and of the nuclei (which are the half-integer nuclear spins of the atoms) and causes spin relaxations, as demonstrated, for instance, for deuterium/hydrogen substitution is DOO-PPV [30] where λ_s was increased significantly by deuteration. The mobility of the charge carriers is influenced by the molecular structure, carrier type (the mobility of electrons is higher than the one of holes), morphology of the films, packing modes, aggregations, disorder, defects, impurities and temperature. Crystals, polymers and the highly ordered stacked molecular aggregation of π-conjugated systems have a high mobility and weak spin scattering effects leading to long spin transport distances. The impact of various parameters on the spin transport and the related value of the MR can be deduced by a comparison of various organic junctions mentioned in Table 1.

The mobility for organic materials is the factor that strongly decreases λ_s as the spin relaxation time is high compared to inorganic materials [113].

2.2. Ferroelectric Junctions and Organic Ferroelectrics

Ferroelectric materials are dielectrics that possess a spontaneous electrical polarization whose orientation can be switched by an external electric field, with a critical value denoted the coercive field. Ferroelectrics are also pyroelectrics, piezoelectrics and present second harmonic generation properties. In ferroelectric junctions that consist of a ferroelectric film sandwiched between metallic electrodes, the current across the thick ferroelectric film (ferroelectric diode [114,115]) or the thin ferroelectric barrier (ferroelectric tunnel junction [5–8]) can be modulated by the orientation of the electrical polarization of the ferroelectric film.

The electroresistance modulation associated with the electric field-induced polarization reversal is in principle related to electrostatic effects at ferroelectric–metal interfaces. At the interface, polarization-generated charges that have different polarities in function of the orientation of the polarization will have, accordingly, an attraction or repulsion effect on the transport charge carriers. This effect arises over a short distance δ in the metal within which the electrons screen the polarization charges depending on the density of states at the Fermi level, giving rise to different δ for different metals. The incomplete screening modifies the depolarization field [7] and the electrostatic potential at each ferroelectric–metal interface (Figure 4) seen by the transport electrons [8,116]. Thus, the electronic potential profile is asymmetric and is modulated by the orientation of the polarization (Figure 5), which is controlled with an external electric field. In Figure 5, for simplicity, it is considered that the metals have a similar work function ($\varphi_1 \approx \varphi_2$) marked as φ, while $\delta_1 > \delta_2$. The interface electrostatic effects are stronger as the thickness of the ferroelectric is smaller; thus, the electroresistance modulation by the orientation of the polarization is larger for ferroelectric tunnel junctions than for ferroelectric diodes. Moreover, the electroresistance values can be modulated to several resistive or also called memristive levels by applying progressive voltage pulses that partially switch the polar ferroelectric domains, which can reach a very high of a few orders of magnitude on–off ratio for the opposite orientation on the electrical polarization [117,118]. The electroresistance modulation can also be implemented in metal –organic ferroelectric–metal junctions [119–125]. Moreover, an electroresistance percentage (ER) can be deduced in a similar way as for magnetic spin valves described by the following:

$$\mathrm{ER} = \frac{R_{up} - R_{down}}{R_{down}} 100\ (\%) \tag{4}$$

where R_{up} and R_{down} denote the corresponding resistance for opposite orientations of the electrical polarizations of the ferroelectric field.

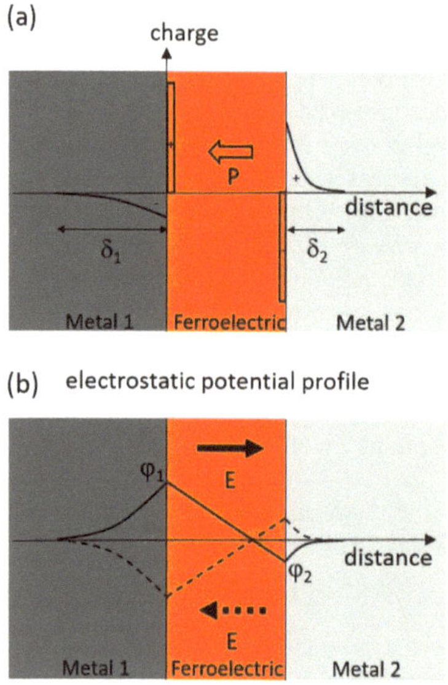

Figure 4. A schematic representation of the charge distribution (**a**) and the electrostatic potential profile (**b**) in metal–ferroelectric–metal junctions for a polarization P and a depolarizing electric field E, where δ_1 and δ_2 are the screening distances in each metal electrode and ϕ_1 and ϕ_2 the corresponding electrostatic potentials. The dashed lines in panel (**b**) are related to the reversed orientation of the polarization P.

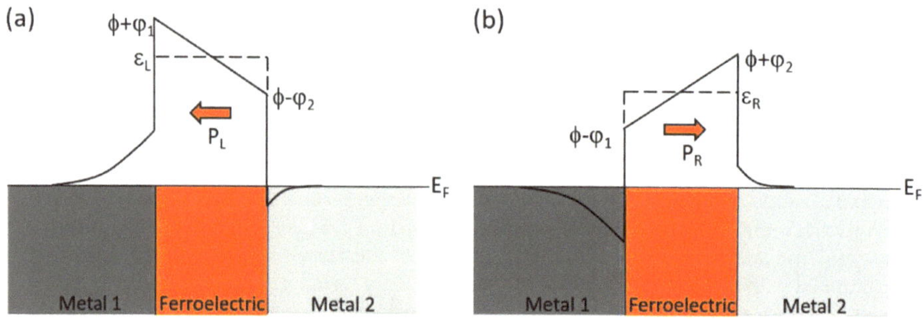

Figure 5. A schematic representation of the asymmetric electronic potential with an average height ε_L and ε_R, respectively, in a metal–ferroelectric–metal junction for different orientations of the polarizations, P_L (**a**) and P_R (**b**), respectively, assuming that $\varphi_1 \approx \varphi_2 = \varphi$ and $\delta_1 > \delta_2$ (resulting $\phi_1 > \phi_2$).

Ferroelectricity appears in films or crystals of organic materials by following a few conditions, such as the following:

- the presence of electrical dipoles;
- the arrangement of the dipoles in non-centrosymmetric structures to avoid canceling each other through dipole–dipole interactions;
- the compliance of properties with electrical polarization switching transitions that should have a sufficiently low energy-switching barrier affordable by the organic system.

The switching transitions can be associated with disorder–order transitions (which consist of rearrangements of individual molecules or of polar components) or displacive transitions (which consist of molecular displacement or of charge or ion displacement) [126–128].

Nevertheless, complex organic materials and metal–organic structures may include combinations of these processes. Organic ferroelectrics, depending on the composition, can be classified in the following: **1.** single molecular components; **2.** polymers and oligomers; **3.** charge transfer complexes, or metal–organic complexes and hydrogen-bonded co-crystals. Table 2 contains a few examples for each category. In order to be included in ferroelectric junctions, ferroelectric films can be deposited by thermal evaporation, e.g., organic molecular beam epitaxy, similarly to the organic spin valves for small single-component systems, or wet chemical methods that are mostly used [119–125], including spin coating, drop casting, Langmuir–Blodget, etc., for polymers and other macromolecular structures.

Table 2. Organic ferroelectrics. Abbreviations of molecule names are as follows: **TCAA**—trich loroacetamide; **TEMPO** or **tanane**—tetramethyl-2,2,6,6-piperidinyl-oxyle; **CDA**—cyclohexan-1,1′-diacetic acid; **TCHM**—tricyclohexylmethanol; **DNP**—1,6-bis(2,4-dinitrophenoxy)-2,4-hexadiyne; **CBDC**—cyclobutene-1,2-dicarboxylic acid; **PhMDA**—2-phenylmalondialdehyde; **3-HPLN**—3-hydroxyphenalenone; **MBI**—2-methylbenzimidazole; **DC-MBI**—5,6-dichloro-2-methylbenzimidazole; **1P**—2-(p-tolyl)-1H-phenanthro[9,10-d]imidazole; **SF-PFA**—N-salicylidene-2,3,4,5,6-pentafluoroaniline; **FF**—diphenylalanine; **PVDF**—poly(vinylidene fluoride); **P(VDF-TrFE)**—poly(vinylidene fluoride—trifluoroethylene); **P(VDF-HFP)**—poly(vinylidene fluoride-hexafluoropropylene); **TTF/CA**—tetra thiafulvalene/chloranil; **TTF/BA**—tetrathiafulvalene/bromanil; **Phz/H_2ca**—phenazine/2,5-dicloro-3,6-dihydroxy-p-benzoquinones; **Phz/H_2ba**—phenazine/2,5-dib-romo-3,6-dihydroxy-p-benzoquinones; **DMBP/H_2ia**—5,5′-dimethyl-2,2′-bipyridine/2,5-diiodo-3,6-dihydroxy-p-benzoquinones; **Hdabco/ReO_4**—1,4-diazabicyclo[2.2.2]octane perrhenate; **TGS**—triglycine sulfate; **TSCC**—tris-sarcosine calcium chloride; and **Rochelle salt**—potassium sodium tartrate tetrahydrate.

1.

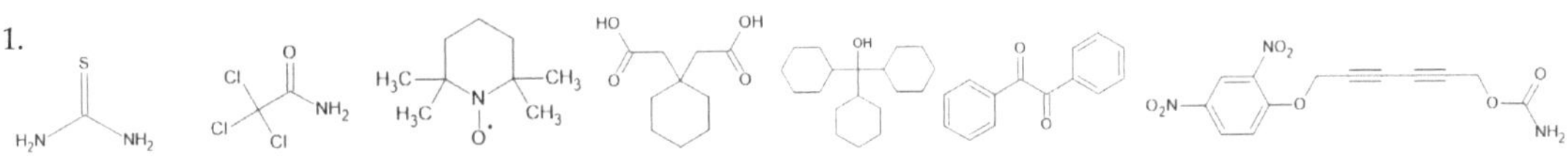

thiourea [129] **TCAA** [130] **TEMPO** [131] **CDA** [132] **TCHM** [133] **Benzil** [134] **DNP** [135]

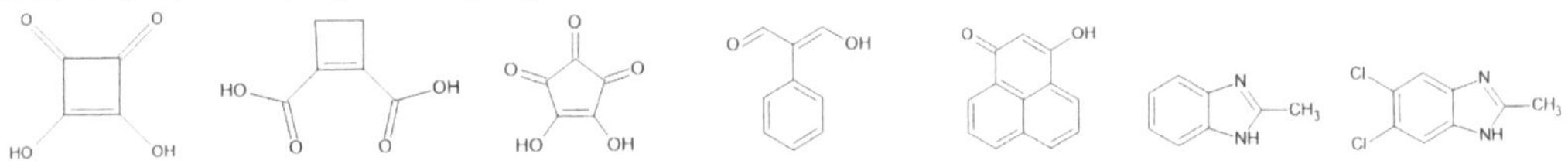

squaric acid [136] **CBDC** [137] **croconic acid** [138] **PhMDA** [137] **3-HPLN** [137] **MBI** [139] **DC-MBI** [139]

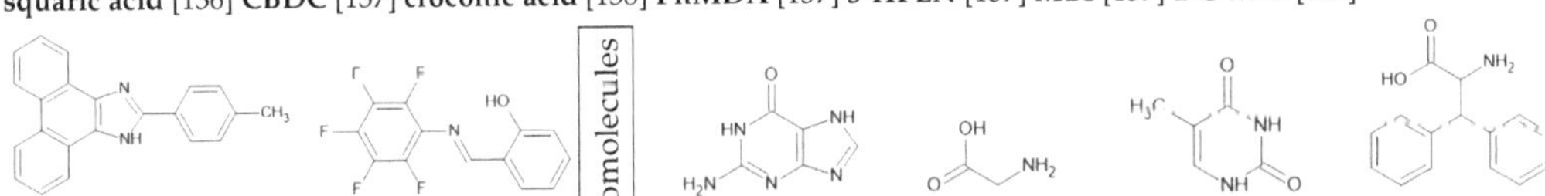

biomolecules

1P [140] **SF-PFA** [141] **guanine** [142] **glycine** [143] **thymine** [144] **FF** [145]

2.

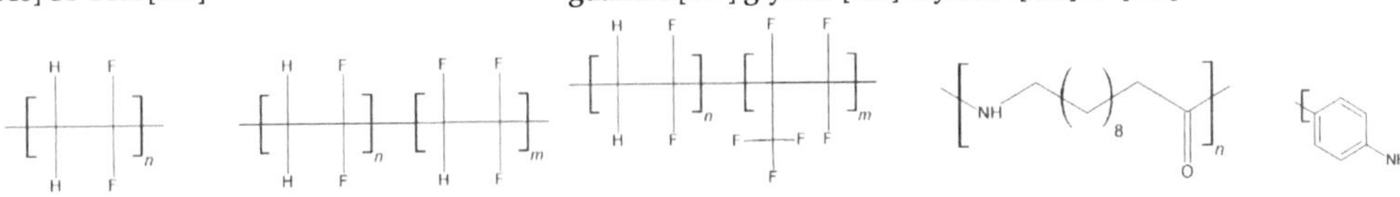

PVDF [146,147] **P(VDF-TrFE)** [148] **P(VDF-HFP)** [149] **nylon** 11 [150] 6,10,11,12-12 [151] **PANI** [152]

3.

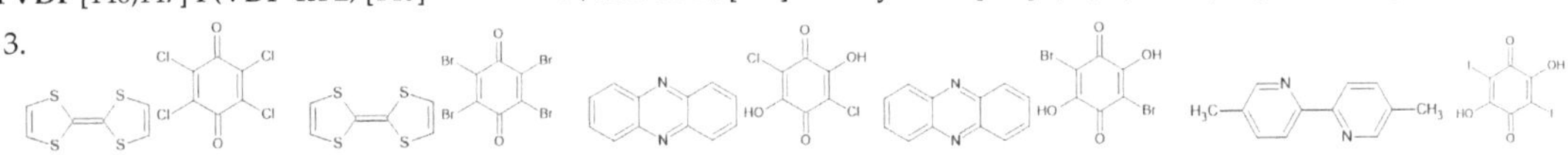

TTF/CA [153,154] **TTF/BA** [154] **Phz/H_2ca** [155] **Phz/H_2ba** [155] **DMBP/H_2ia** [156]

Table 2. *Cont.*

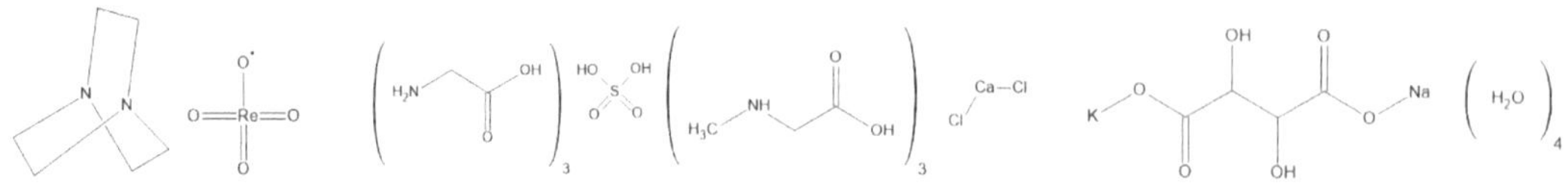

Hdabco/ReO$_4$ [157] **TGS** [158,159] **TSCC** [160] **Rochelle salt** [161]

The electrical polarization switching mechanism of organic ferroelectrics can be related to their structure and composition. The majority of single-component systems, which are mentioned in part 1 of Table 2, form crystallographic structures through hydrogen bonds such as N-H···O, O-H···N, O-H···O and N-H···N. These intermolecular interactions allow for both the disorder–order and the displacive transitions in the polarization orientation processes. Thus, electrical polarization switching involves molecular rotations (such as, for example, for thiourea [129], CDA [132], glycine [143], thymine [144]), the reorientation of functional groups (such as, for example, for TCAA [130], TEMPO [131], TCHM [133], benzil [134], DNP [135]) and the intermolecular displacement of H$^+$ within hydrogen bonds (such as, for example, within O-H···O for squaric acid [136], CBDC [137], croconic acid [138], PhMDA [137] and 3-HPLN [137]; within N-H···N for MBI [139], DC-MBI [139], 1P [140] and guanine [142] and within N-H···O for FF [145]). The latest example can be photo-activated. Another photo-induced switching is produced by a H$^+$ intramolecular displacement within the O-H···N bonds for SA-PFA molecules [141]. The polymeric systems that are included in part 2 of Table 2 all switch their electrical polarization by the reorientation of functional groups and rotations of polymeric chains [146–152]. The bicomponent systems from part 3 of Table 2 change their polarization state by electronic transfer for TTF/CA [153,154] and TTF/BA [154] and by H$^+$ transfer within O-H···N bonds for Phz/H$_2$ca [155], Phz/H$_2$ba [155] and DMBP/H$_2$ia [156]. The polarization changes for the systems in the last row of Table 2 are all produced by molecular reorientations and order–disorder transitions [157–161].

3. Organic Multiferroic Junctions

OMFJs can be obtained by combining the previously described devices, for instance, introducing an organic ferroelectric film as the separating medium in an organic spin valve. The main characteristic of OMFJs is the possibility to electrically control the spin polarization at the ferromagnet–organic–ferroelectric interfaces by switching the electrical polarization of the ferroelectric and thus manipulating the MR and ER effects.

These coupled processes were first demonstrated for inorganic multiferroic tunnel junctions [2,3,162]. Experimentally, recent studies on OMFJs using a few organic ferroelectric materials, such as PVDF [163–165], P(VDF-TrFE) [166,167] and croconic acid [168], have confirmed their memristive behavior manipulated with electric and magnetic stimuli. The rich variety of possibilities (see Table 2) of including other ferroelectric organic and biomolecular materials in OMFJs opens further routes for tuning the magnetoelectric and electronic/spintronic properties of these systems. As for any other organic spin valves and ferroelectric junctions, the MR and ER effects depend on other parameters and conditions. Moreover, the ferroelectric control of the spin polarization can induce the reversal of the magnetoresistance hysteresis (see Figure 2), i.e., the spin rectification effect, by switching the electrical polarization.

The MR effect of OMFJs, as for any organic spin valve, strongly depends on temperature and the thickness of the organic film. Expected consequences of these factors can be observed on the examples represented in Figure 6a–d. Moreover, the ER effect is also affected by these parameters (Figure 6d). Furthermore, the interfaces play a crucial role in defining the MR, ER and the coupled effects. For LSMO/PVDF/Co systems presented in Figure 6c,e, the PVDF/Co interface is different as the Co layer was deposited at low and room temperature, respectively. The MR for a supposedly rougher interface, with diffusion channels that also affect the thickness of the spin transport medium, is negative and does not show spin rectification for any electrical polarization orientation of the ferroelectric film (Figure 6e). For a similar system, by intercalating a thin MgO layer, the MR becomes positive (Figure 6e,f).

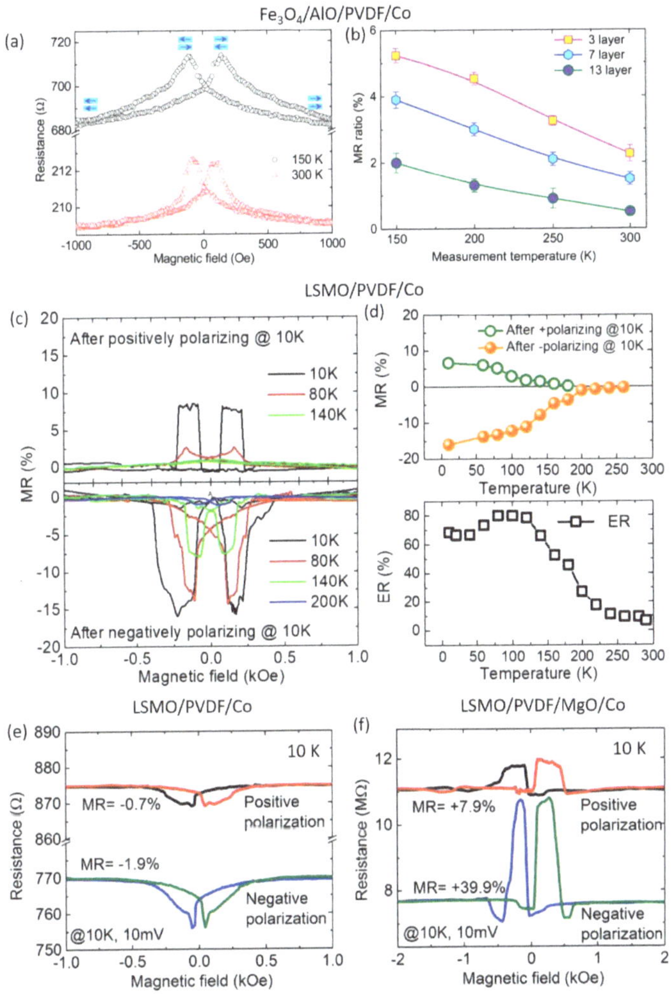

Figure 6. MR and ER effects influenced by various factors such as interface, thickness and temperature on OMFJs containing PVDF in (**a**,**b**) Fe_3O_4/AlO/PVDF/Co (adapted with permission from Ref. [164]); (**c**,**d**) LSMO/PVDF/Co (adapted with permission from Ref. [165]); (**e**,**f**) LSMO/PVDF/Co and LSMO/PVDF/MgO/Co junctions (adapted with permission from Ref. [163]).

A similar behavior of the MR, ER and spin rectification effect with the temperature and the thickness is encountered in the LSMO/P(VDF-TrFE)/Co system (Figure 7). As expected, the MR and ER decrease by increasing the temperature and the thickness (Figure 7b,d,f,g). Different tendencies at certain temperature ranges may be related to a different spin transport mechanism [164]. Moreover, the thickness is important for determining the sign of the MR and the spin rectification effect (Figure 7a–f). In addition, the memristive properties of the ferroelectric spacer and the ferromagnetic–ferroelectric interface coupling induce mainly an increase in the MR and ER with the magnitude of the external electric field applied successively to the OMFJ (Figure 7i).

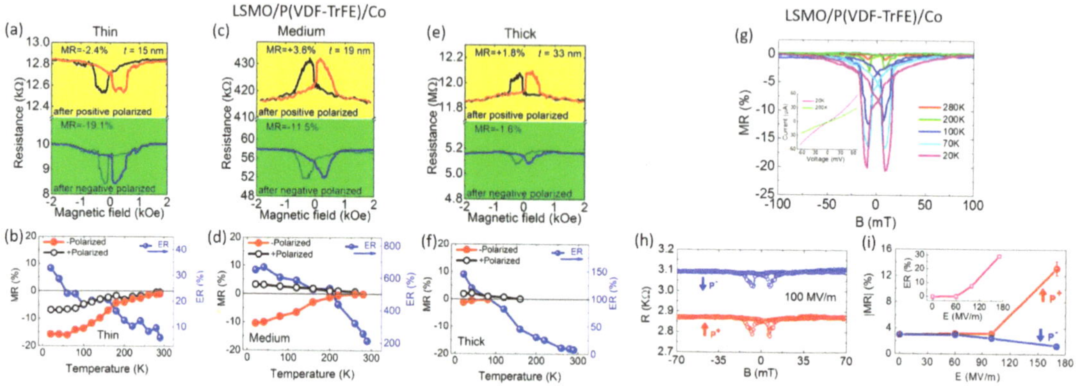

Figure 7. MR and ER effects influenced by thickness, temperature and the applied electric field on OMFJs containing P(VDF-TrFE) films in LSMO/P(VDF-TrFE)/Co systems. (**a**–**f**) panels adapted with permission from Ref. [167]; (**g**–**i**) panels adapted with permission from Ref. [166].

The role of the electric field and measurement values of the bias voltage in the MR was carefully investigated for the LSMO/croconic acid/SiO_2/Co system [168]. Positive and negative electrical pulses are used to switch the polarization of the ferroelectric film to on and off states. But, additionally, a negative applied voltage with a magnitude above a certain threshold for measuring the MR induces the spin rectification effect (Figure 8). The ferroelectric–metallic interface effects of the OMFJs are used to fine-tune the MR.

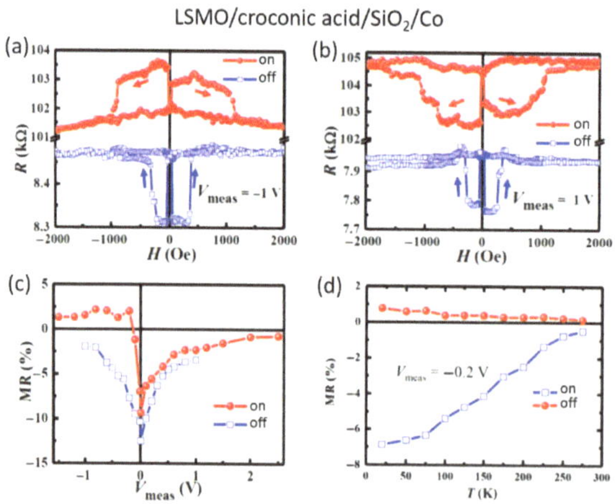

Figure 8. MR and ER effects of a LSMO/croconic acid/SiO_2/Co OMFJ influenced by the polarity and magnitude of the measurement voltage (**a**–**c**) and temperature (**d**). The figure was adapted with permission from Ref. [168].

4. Conclusions and Outlook

The recent realization of OMFJs by combining the knowledge on organic spin valves and organic ferroelectric junctions shows encouraging results on exploiting the magnetoelectric coupling, electrostatic and energy alignment conditions at the ferroelectric–ferromagnetic interfaces to tune the MR and ER. Depending on these factors and the related variable parameters like, for example, the materials, the thickness, an intercalated oxide layer at the organic–metal interface and the temperature, the memristive states of OMFJs can be manipulated by triggering the external magnetic and electric stimuli. The memristive properties of switching and retaining the electrical resistance using the spin degree of freedom of electrons can principally be used as nonvolatile memory elements to store and process information with less power consumption and can serve as models of synaptic and neuronal operations. The promising results of OMFJs [163–168] open routes for new research contributions exploring various materials and conditions to optimize their properties. For instance, a large variety of molecular and biomolecular ferroelectrics [129–161] can be used as the spin transport medium in OMFJs. Extensively used magnetic films may be replaced by other magnetic compounds with high polarization in spin [25,80–83], new magnetic constituents such as 2D van der Waals materials [169–175], or 2D networks of organic and metal–organic magnets on surfaces [110–112,176–183], metal–organic frameworks [20,21], heterostructures [110,111,184–186] and organic spinterfaces [13,49,51,90–92,163,187] as electrodes to improve the spin injection and/or the functionality of OMFJs. A few recent successful examples already include some of these new alternative types of electrodes for photovoltaic [188], capacitive [189] and spin-crossover [190] devices. Further studies may be important to also explore the potential of OMFJs for flexible electronics, wearable electronics, transient electronics, bioelectronics and perchance coupling their properties with light-related phenomena in magneto-optics, optoelectronics or photovoltaics contributing to the advancement of nanostructures and thin films [191].

Funding: This work was funded by the Romanian Ministry of Research, Innovation and Digitalization through the projects PN-III-P2-2.1-PED-2021-0378 (contract nr. 575PED/2022) and the Core Program PC2-PN2308020.

Institutional Review Board Statement: Not applicable.

Informed Consent Statement: Not applicable.

Data Availability Statement: Data are contained within the article.

Conflicts of Interest: The author declares no conflicts of interest.

References

1. Gajek, M.; Bibes, M.; Fusil, S.; Bouzehouane, K.; Fontcuberta, J.; Barthélémy, A.; Fert, A. Tunnel junctions with multiferroic barriers. *Nat. Mater.* **2007**, *6*, 296–302. [CrossRef] [PubMed]
2. Garcia, V.; Bibes, M.; Bocher, L.; Valencia, S.; Kronast, F.; Crassous, A.; Moya, X.; Enouz-Vedrenne, S.; Gloter, A.; Imhoff, D.; et al. Ferroelectric control of spin polarization. *Science* **2010**, *327*, 1106–1110. [CrossRef] [PubMed]
3. Pantel, D.; Goetze, S.; Hesse, D.; Alexe, M. Reversible electrical switching of spin polarization in multiferroic tunnel junctions. *Nat. Mater.* **2012**, *11*, 289–293. [CrossRef] [PubMed]
4. Jullière, M. Tunneling between ferromagnetic films. *Phys. Lett. A* **1975**, *54*, 225–226. [CrossRef]
5. Maekawa, S.; Gafvert, U. Electron tunneling between ferromagnetic films. *IEEE Trans. Magn.* **1982**, *18*, 707–708. [CrossRef]
6. Zhuravlev, M.Y.; Sbirianov, R.F.; Jaswal, S.S.; Tsymbal, E.Y. Giant electroresistance in ferroelectric tunnel junctions. *Phys. Rev. Lett.* **2005**, *94*, 246802. [CrossRef]
7. Kohlstedt, H.; Pertsev, N.A.; Rodriguez-Contreras, J.; Waser, R. Theoretical current-voltage characteristics of ferroelectric tunnel junctions. *Phys. Rev. B* **2005**, *72*, 125341. [CrossRef]
8. Tsymbal, E.Y.; Kohlstedt, H. Tunneling across a ferroelectric. *Science* **2006**, *313*, 181–183. [CrossRef] [PubMed]
9. Dediu, V.; Murgia, M.; Matacotta, F.C.; Taliani, C.; Barbanera, S. Room temperature spin polarized injection in organic semiconductor. *Solid State Commun.* **2002**, *122*, 181–184. [CrossRef]
10. Pramanik, S.; Stefanita, C.G.; Patibandla, S.; Bandyopadhyay, S.; Garre, K.; Harth, N.; Cahay, M. Observation of extremely long spin relaxation times in an organic nanowire spin valve. *Nat. Nanotechnol.* **2007**, *2*, 216–219. [CrossRef]
11. Fu, K.K.; Wang, Z.; Dai, J.; Carter, M.; Hu, L. Transient electronics: Materials and devices. *Chem. Mater.* **2016**, *28*, 3527–3539. [CrossRef]
12. Cosseddu, P.; Caironi, M. (Eds.) *Organic Flexible Electronics: Fundamentals, Devices, and Applications*; Woodhead Publishing Series in Electronic and Optical Materials; Woodhead Publishing: Cambridge, UK, 2020. [CrossRef]
13. Liang, S.; Geng, R.; Yang, B.; Zhao, W.; Subedi, R.C.; Li, X.; Han, X.; Nguyen, T.D. Curvature-enhanced spin-orbit coupling and spinterface effect in fullerene-based spin valves. *Sci. Rep.* **2016**, *6*, 19461. [CrossRef]

14. Gobbi, M.; Golmar, F.; Llopis, R.; Casanova, F.; Hueso, L.E. Room-temperature spin transport in C_{60}-based spin valves. *Adv. Mater.* **2011**, *23*, 1609–1613. [CrossRef]
15. Zhang, X.; Mizukami, S.; Kubota, T.; Ma, Q.; Oogane, M.; Naganuma, H.; Ando, Y.; Miyazaki, T. Observation of a large spin-dependent transport length in organic spin valves at room temperature. *Nat. Commun.* **2013**, *4*, 1392. [CrossRef] [PubMed]
16. Hu, S.; Liu, W.; Guo, L.; Zhang, R.; Gu, X.; Meng, K.; Qin, Y.; Guo, A.; Yang, T.; Zhang, C.; et al. Continuous room-temperature spin-injection modulation achieved by spin-filtering competition in molecular spin valves. *Adv. Mater.* **2023**, *35*, 2300055. [CrossRef]
17. Zhang, X.; Ma, Q.; Suzuk, K.; Sugihara, A.; Qin, G.; Miyazaki, T.; Mizukami, S. Magnetoresistance effect in rubrene-based spin valves at room temperature. *ACS Appl. Mater. Interfaces* **2015**, *7*, 4685–4692. [CrossRef]
18. Shim, J.H.; Raman, K.V.; Park, Y.J.; Santos, T.S.; Miao, G.X.; Satpati, B.; Moodera, J.S. Large spin diffusion length in an amorphous organic semiconductor. *Phys. Rev. Lett.* **2008**, *100*, 226603. [CrossRef] [PubMed]
19. Lin, R.; Wang, F.; Rybicki, J.; Wohlgenannt, M.; Hutchinson, K.A. Distinguishing between tunneling and injection regimes of ferromagnet/organic semiconductor/ferromagnet junctions. *Phys. Rev. B* **2010**, *81*, 195214. [CrossRef]
20. Yoo, J.-W.; Jang, H.W.; Prigodin, V.N.; Kao, C.; Eom, C.B.; Epstein, A.J. Tunneling vs. giant magnetoresistance in organic spin valve. *Synth. Met.* **2010**, *160*, 216–222. [CrossRef]
21. Li, B.; Kao, C.-Y.; Lu, Y.; Yoo, J.-W.; Prigodin, V.N.; Epstein, A.J. Room-temperature organic-based spin polarizer. *Appl. Phys. Lett.* **2011**, *99*, 153503. [CrossRef]
22. Li, B.; Kao, C.-Y.; Yoo, J.-W.; Prigodin, V.N.; Epstein, A.J. Magnetoresistance in an all-organic-based spin valve. *Adv. Mater.* **2011**, *23*, 3382–3386. [CrossRef]
23. Sun, X.; Gobbi, M.; Bedoya-Pinto, A.; Txoperena, O.; Federico Golmar, F.; Llopis, R.; Andrey Chuvilin, A.; Casanova, F.; Hueso, L.E. Room-temperature air-stable spin transport in bathocuproine-based spin valves. *Nat. Commun.* **2013**, *4*, 2794. [CrossRef]
24. Sun, X.; Bedoya-Pinto, A.; Llopis, R.; Casanova, F.; Hueso, L.E. Flexible semi-transparent organic spin valve based on bathocuproine. *Appl. Phys. Lett.* **2014**, *105*, 083302. [CrossRef]
25. Kawasugi, Y.; Ujino, T.; Tada, H. Room-temperature magnetoresistance in organic spin-valves based on a Co_2MnSi Heusler alloy. *Org. Electron.* **2013**, *14*, 3186–3189. [CrossRef]
26. Ikegami, T.; Kawayama, I.; Tonouchi, M.; Nakao, S.; Yamashita, Y.; Tada, H. Planar-type spin valves based on low-molecular-weight organic materials with $La_{0.67}Sr_{0.33}MnO_3$ electrodes. *Appl. Phys. Lett.* **2008**, *92*, 153304. [CrossRef]
27. Jiang, S.W.; Wang, P.; Jiang, S.C.; Chen, B.B.; Wang, M.; Jiang, Z.S.; Wu, D. Fabrication of lateral organic spin valves based on $La_{0.7}Sr_{0.3}MnO_3$ electrodes. *Spin* **2014**, *4*, 1440008. [CrossRef]
28. Mooser, S.; Cooper, J.F.K.; Banger, K.K.; Wunderlich, J.; Sirringhaus, H. Spin injection and transport in a solution-processed organic semiconductor at room temperature. *Phys. Rev. B* **2012**, *85*, 235202. [CrossRef]
29. Wang, Y.; Yao, J.; Ding, S.; Guo, S.; Cui, D.; Wang, X.; Yang, S.; Zhang, L.; Tian, X.; Wu, D.; et al. Spin injection and transport in single-crystalline organic spin valves based on TIPS-pentacene. *Sci. China Mater.* **2021**, *64*, 2795–2804. [CrossRef]
30. Nguyen, T.D.; Hukic-Markosian, G.; Wang, F.; Wojcik, L.; Li, X.-G.; Ehrenfreund, E.; Vardeny, Z.V. Isotope effect in spin response of π-conjugated polymer films and devices. *Nat. Mater.* **2010**, *9*, 345–352. [CrossRef]
31. Xiong, Z.H.; Wu, D.; Vardeny, Z.V.; Shi, J. Giant magnetoresistance in organic spin-valves. *Nature* **2004**, *427*, 821–824. [CrossRef]
32. Santos, T.S.; Lee, J.S.; Migdal, P.; Lekshmi, I.C.; Satpati, B.; Moodera, J.S. Room-temperature tunnel magnetoresistance and spin-polarized tunneling through an organic semiconductor barrier. *Phys. Rev. Lett.* **2007**, *98*, 16601. [CrossRef]
33. Dediu, V.; Hueso, L.E.; Bergenti, I.; Riminucci, A.; Borgatti, F.; Graziosi, P.; Newby, C.; Casoli, F.; De Jong, M.P.; Taliani, C.; et al. Room-temperature spintronic effects in Alq_3-based hybrid devices. *Phys. Rev. B* **2008**, *78*, 115203. [CrossRef]
34. Sun, D.; Yin, L.; Sun, C.; Guo, H.; Gai, Z.; Zhang, X.-G.; Ward, T.Z.; Cheng, Z.; Shen, J. Giant magnetoresistance in organic spin valves. *Phys. Rev. Lett.* **2010**, *104*, 236602. [CrossRef] [PubMed]
35. Wang, S.; Shi, Y.J.; Lin, L.; Chen, B.B.; Yue, F.J.; Du, J.; Ding, H.F.; Zhang, F.M.; Wu, D. Room-temperature spin valve effects in $La_{0.67}Sr_{0.33}MnO_3/Alq_3/Co$ devices. *Synth. Met.* **2011**, *161*, 1738–1741. [CrossRef]
36. Yang, W.; Shi, Q.; Miao, T.; Li, Q.; Cai, P.; Liu, H.; Lin, H.; Bai, Y.; Zhu, Y.; Yu, Y.; et al. Achieving large and nonvolatile tunable magnetoresistance in organic spin valves using electronic phase separated manganites. *Nat. Commun.* **2019**, *10*, 3877. [CrossRef] [PubMed]
37. Angervo, I.; Saloaro, M.; Palonen, H.; Huhtinen, H.; Paturi, P.; Mäkelä, T.; Majumdar, S. Giant magnetoresistance response in Sr2FeMoO6 based organic spin valves. *Appl. Surf. Sci.* **2022**, *589*, 152854. [CrossRef]
38. Zhang, X.; Mizukami, S.; Ma, Q.; Kubota, T.; Oogane, M.; Naganuma, H.; Ando, Y.; Miyazaki, T. Spin-dependent transport behavior in C_{60} and Alq_3 based spin valves with a magnetite electrode (invited). *J. Appl. Phys.* **2014**, *115*, 172608. [CrossRef]
39. Drew, A.J.; Hoppler, J.; Schulz, L.; Pratt, F.L.; Desai, P.; Shakya, P.; Kreouzis, T.; Gillin, W.P.; Suter, A.; Morley, N.A.; et al. Direct measurement of the electronic spin diffusion length in a fully functional organic spin valve by low-energy muon spin rotation. *Nat. Mater.* **2009**, *8*, 109–114. [CrossRef] [PubMed]
40. Bergenti, I.; Borgatti, F.; Calbucci, M.; Riminucci, A.; Cecchini, R.; Graziosi, P.; MacLaren, D.A.; Giglia, A.; Rueff, J.P.; Céolin, D.; et al. Oxygen impurities link bistability and magnetoresistance in organic spin valves. *ACS Appl. Mater. Interfaces* **2018**, *10*, 8132–8140. [CrossRef]
41. Cucinotta, G.; Poggini, L.; Pedrini, A.; Bertani, F.; Cristiani, N.; Torelli, M.; Graziosi, P.; Cimatti, I.; Cortigiani, B.; Otero, E.; et al. Tuning of a vertical spin valve with a monolayer of single molecule magnets. *Adv. Funct. Mater.* **2017**, *27*, 1703600. [CrossRef]

42. Poggini, L.; Cucinotta, G.; Pradipto, A.-M.; Scarrozza, M.; Barone, P.; Caneschi, A.; Graziosi, P.; Calbucci, M.; Cecchini, R.; Dediu, A.A.; et al. An Organic Spin Valve Embedding a Self-Assembled Monolayer of Organic Radicals. *Adv. Mater. Interfaces* **2016**, *3*, 1500855. [CrossRef]
43. Morley, N.A.; Rao, A.; Dhandapani, D.; Gibbs, M.R.J.; Grell, M.; Richardson, T. Room temperature organic spintronics. *J. Appl. Phys.* **2008**, *103*, 07F306. [CrossRef]
44. Ding, S.; Tian, Y.; Dong, H.; Zhu, D.; Hu, W. Anisotropic magnetoresistance in NiFe-based polymer spin valves. *ACS Appl. Mater. Interfaces* **2019**, *11*, 11654–11659. [CrossRef]
45. Ding, S.; Tian, Y.; Liu, X.; Zou, Y.; Dong, H.; Mi, W.; Hu, W. Unveiling the role of Fe_3O_4 in polymer spin valve near Verwey transition. *Nano Res.* **2021**, *14*, 304–310. [CrossRef]
46. Majumdar, S.; Laiho, R.; Laukkanen, P.; Väyrynen, I.J.; Majumdar, H.S.; Österbacka, R. Application of regioregular polythiophene in spintronic devices: Effect of interface. *Appl. Phys. Lett.* **2006**, *89*, 122114. [CrossRef]
47. Ding, S.; Tian, Y.; Li, Y.; Mi, W.; Dong, H.; Zhang, X.; Hu, W.; Zhu, D. Inverse magnetoresistance in polymer spin valves. *ACS Appl. Mater. Interfaces* **2017**, *9*, 15644–15651. [CrossRef]
48. Geng, R.; Subedi, R.C.; Luong, H.M.; Pham, M.T.; Huang, W.; Li, X.; Hong, K.; Shao, M.; Xiao, K.; Hornak, L.A.; et al. Effect of charge localization on the effective hyperfine interaction in organic semiconducting polymers. *Phys. Rev. Lett.* **2018**, *120*, 086602. [CrossRef]
49. Zhang, C.; Ding, S.; Tian, Y.; Wang, J.; Chen, Y.; Zhao, T.; Hu, F.; Hu, W.; Shen, B. The in situ optimization of spinterface in polymer spin valve by electronic phase separated oxides. *Small* **2023**, *19*, 2303375. [CrossRef]
50. Ding, S.; Tian, Y.; Wang, H.; Zhou, Z.; Mi, W.; Ni, Z.; Zou, Y.; Dong, H.; Gao, H.; Zhu, D.; et al. Reliable spin valves of conjugated polymer based on mechanically transferrable top electrodes. *ACS Nano* **2018**, *12*, 12657–12664. [CrossRef]
51. Yu, D.; Ding, S.; Li, J.; Mi, W.; Tian, Y.; Hu, W. Molecular spinterface in F4TCNQ-doped polymer spin valves. *J. Mater. Chem. C* **2022**, *10*, 2608–2615. [CrossRef]
52. Geng, R.; Roy, A.; Zhao, W.; Subedi, R.C.; Li, X.; Locklin, J.; Nguyen, T.D. Engineering of spin injection and spin transport in organic spin valves using π-conjugated polymer brushes. *Adv. Funct. Mater.* **2016**, *26*, 3999–4006. [CrossRef]
53. Wang, F.J.; Yang, C.G.; Valy Vardeny, Z.; Li, X.G. Spin response in organic spin valves based on $La_{2/3}Sr_{1/3}MnO_3$ electrodes. *Phys. Rev. B* **2007**, *75*, 245324. [CrossRef]
54. Matsuzaka, M.; Sasaki, Y.; Hayashi, K.; Misawa, T.; Komine, T.; Akutagawa, T.; Fujioka, M.; Nishii, J.; Kaiju, H. Room-temperature magnetoresistance in $Ni_{78}Fe_{22}/C_8$-BTBT/$Ni_{78}Fe_{22}$ nanojunctions fabricated from magnetic thin-film edges using a novel technique. *Nanoscale Adv.* **2022**, *4*, 4739. [CrossRef]
55. Luo, Z.; Song, X.; Liu, X.; Lu, X.; Yao, Y.; Zeng, J.; Li, Y.; He, D.; Zhao, H.; Gao, L.; et al. Revealing the key role of molecular packing on interface spin polarization at two-dimensional limit in spintronic devices. *Sci. Adv.* **2023**, *9*, eade9126. [CrossRef]
56. Li, F.; Li, T.; Chen, F.; Zhang, F. Excellent spin transport in spin valves based on the conjugated polymer with high carrier mobility. *Sci. Rep.* **2015**, *5*, 9355. [CrossRef]
57. Li, D.; Wang, X.; Lin, Z.; Zheng, Y.; Jiang, Q.; Zheng, N.; Zhang, W.; Kui-juan Jin, K.-j.; Yu, G. Tuning charge carrier and spin transport properties via structural modification of polymer semiconductors. *ACS Appl. Mater. Interfaces* **2019**, *11*, 30089–30097. [CrossRef]
58. Li, D.; Zheng, Y.; Yang, M.; Wei, C.; Liu, X.; Zheng, N.; Zhang, W.; Kuijuan Jin, K.; Yu, G. Molecular and interfacial adjustment of magnetoresistance in organic spin valves using isoindigo-based polymers. *ACS Mater. Lett.* **2022**, *4*, 1065–1073. [CrossRef]
59. Zheng, N.; Lin, Z.; Zheng, Y.; Li, D.; Yang, J.; Zhang, W.; Wang, L.; Yu, G. Room-temperature stable organic spin valves using solution-processed ambipolar naphthalenediimide-based conjugated polymers. *Org. Electron.* **2020**, *81*, 105684. [CrossRef]
60. Zheng, N.; Wang, X.; Zheng, Y.; Li, D.; Lin, Z.; Zhang, W.; Jin, K.-j.; Yu, G. Negative Magnetoresistance Behavior in Polymer Spin Valves Based on Donor-Acceptor Conjugated Molecules. *Adv. Mater. Interfaces* **2020**, *7*, 2000868. [CrossRef]
61. Prieto-Ruiz, J.P.; Gómez Miralles, S.; Prima-García, H.; López-Muñoz, A.; Riminucci, A.; Graziosi, P.; Aeschlimann, M.; Cinchetti, M.; Dediu, V.A.; Coronado, E. Enhancing light emission in interface engineered spin-OLEDs through spin-polarized injection at high voltages. *Adv. Mater.* **2019**, *31*, 1806817. [CrossRef]
62. Liu, X.; Li, H.; Zhang, W.; Yang, Z.; Li, D.; Liu, M.; Jin, K.; Wang, L.; Yu, G. Magnetoresistance in organic spin valves based on acid-exfoliated 2D covalent organic frameworks thin films. *Angew. Chem. Int. Ed.* **2023**, *62*, e202308921. [CrossRef]
63. Xu, W.; Szulczewski, G.J.; LeClair, P.; Navarrete, I.; Schad, R.; Miao, G.; Guo, H.; Gupta, A. Tunneling magnetoresistance observed in LSMO/organic molecule/Co junctions. *Appl. Phys. Lett.* **2007**, *90*, 072506. [CrossRef]
64. Bairagi, K.; Garcia Romero, D.; Calavalle, F.; Catalano, S.; Zuccatti, E.; Llopis, R.; Casanova, F.; Hueso, L.E. Room-temperature operation of a p-type molecular spin photovoltaic device on a transparent substrate. *Adv. Mater.* **2020**, *32*, 1906908. [CrossRef]
65. Liu, Y.; Lee, T.; Katz, H.E.; Reich, D.H. Effects of carrier mobility and morphology in organic semiconductor spin valves. *J. Appl. Phys.* **2009**, *105*, 07C708. [CrossRef]
66. Jiang, S.W.; Wang, P.; Chen, B.B.; Zhou, Y.; Ding, H.F.; Wu, D. Tuning carrier mobility without spin transport degrading in copper-phthalocyanine. *Appl. Phys. Lett.* **2015**, *107*, 042407. [CrossRef]
67. Tong, J.; Ruan, L.; Yao, X.; Qin, G.; Zhang, X. Defect states dependence of spin transport in iron phthalocyanine spin valves. *Phys. Rev. B* **2019**, *99*, 054406. [CrossRef]

68. Sun, X.; Bedoya-Pinto, A.; Mao, Z.; Gobbi, M.; Yan, W.; Guo, Y.; Atxabal, A.; Llopis, R.; Yu, G.; Liu, Y.; et al. Active morphology control for concomitant long distance spin transport and photoresponse in a single organic device. *Adv. Mater.* **2016**, *28*, 2609–2615. [CrossRef]
69. Hong, J.Y.; Chang, S.-H.; Ou Yang, K.H.; Yeh, P.C.; Shiu, H.W.; Chen, C.-H.; Chiang, W.C.; Lin, M.T. A multifunctional molecular spintronic platform with magnetoresistive and memristive responses via a self-assembled monolayer. *J. Appl. Phys.* **2019**, *125*, 142905. [CrossRef]
70. Bowen, M.; Bibes, M.; Barthélémy, A.; Contour, J.-P.; Anane, A.; Lemaître, Y.; Fert, A. Nearly total spin polarization in $La_{2/3}Sr_{1/3}MnO_3$ from tunneling experiments. *Appl. Phys. Lett.* **2003**, *82*, 233–235. [CrossRef]
71. Uehara, M.; Mori, S.; Chen, C.; Cheong, S.W. Percolative phase separation underlies colossal magnetoresistance in mixed-valent manganites. *Nature* **1999**, *399*, 560–563. [CrossRef]
72. Cao, G.; Zhang, J.; Xu, Y.; Wang, S.; Yu, J.; Cao, S.; Jing, C.; Shen, X. Action of strong coupling on steplike magnetization and transport properties in phase-separated manganite. *Appl. Phys. Lett.* **2005**, *87*, 232501. [CrossRef]
73. Stanciu, A.E.; Schinteie, G.; Kuncser, A.C.; Locovei, C.; Trupina, L.; Iacob, N.; Leca, A.; Borca, B.; Kuncser, V. Magnetic properties of nanosized Fe and FeCo systems on trenched Mo templates. *Coatings* **2022**, *12*, 1366. [CrossRef]
74. Borca, B.; Fruchart, O.; David, P.; Rousseau, A.; Meyer, C. Kinetic self-organization of trenched templates for the fabrication of versatile ferromagnetic nanowires. *Appl. Phys. Lett.* **2007**, *90*, 142507. [CrossRef]
75. Borca, B.; Fruchart, O.; Cheynis, F.; Hasegawa, M.; Meyer, C. Growth and magnetism of self-organized arrays of Fe(110) wires formed by deposition on kinetically grooved W(110). *Surf. Sci.* **2007**, *601*, 4358–4361. [CrossRef]
76. Borca, B.; Fruchart, O.; Kritsikis, E.; Cheynis, F.; Rousseau, A.; David, P.; Meyer, C.; Toussaint, J.-C. Tunable magnetic properties of arrays of Fe(110) nanowires grown on kinetically grooved W(110) self-organized templates. *J. Magn. Magn. Mater.* **2010**, *322*, 257–264. [CrossRef]
77. Kobayashi, K.I.; Kimura, T.; Sawada, H.; Terakura, K.; Takura, Y. Room-temperature magnetoresistance in an oxide material with an ordered double-perovskite structure. *Nature* **1998**, *395*, 677–680. [CrossRef]
78. Ray, S.; Kumar, A.; Sarma, D.D.; Cimino, R.; Turchini, S.; Zennaro, S.; Zema, N. Electronic and magnetic structures of Sr_2FeMoO_6. *Phys. Rev. Lett.* **2001**, *87*, 097204. [CrossRef]
79. Dedkov, Y.S.; Rüdiger, U.; Güntherodt, G. Evidence for the half-metallic ferromagnetic state of Fe3O4 by spin-resolved photoelectron spectroscopy. *Phys. Rev. B* **2002**, *65*, 064417. [CrossRef]
80. Graf, T.; Felser, C.; Parkin, S.S.P. Simple rules for the understanding of Heusler compounds. *Prog. Solid State Chem.* **2011**, *39*, 1–50. [CrossRef]
81. Tolea, F.; Tolea, M.; Sofronie, M.; Valeanu, M. Distribution of plates' sizes tell the thermal history in a simulated martensitic-like phase transition. *Solid State Commun.* **2015**, *213–214*, 37–41. [CrossRef]
82. Sofronie, M.; Tolea, F.; Tolea, M.; Popescu, B.; Valeanu, M. Magnetic and magnetostrictive properties of the ternary Fe67.5Pd30.5Ga2 ferromagnetic shape memory ribbons. *J. Phys. Chem. Solids* **2020**, *142*, 109446. [CrossRef]
83. Jourdan, M.; Minár, J.; Braun, J.; Kronenberg, A.; Chadov, S.; Balke, B.; Gloskovskii, A.; Kolbe, M.; Elmers, H.J.; Schönhense, G.; et al. Direct observation of half-metallicity in the Heusler compound Co_2MnSi. *Nat. Commun.* **2014**, *5*, 3974. [CrossRef]
84. Tedrow, P.M.; Meservey, R. Spin polarization of electrons tunneling from films of Fe, Co, Ni, and Gd. *Phys. Rev. B* **1973**, *7*, 318. [CrossRef]
85. Monsma, D.J.; Parkin, S.S.P. Spin polarization of tunneling current from ferromagnet/Al_2O_3 interfaces using copper-doped aluminum superconducting films. *Appl. Phys. Lett.* **2000**, *77*, 720–722. [CrossRef]
86. Paraskevopoulos, D.; Meservey, R.; Tedrow, P.M. Spin polarization of electrons tunneling from 3D ferromagnetic metals and alloys. *Phys. Rev. B* **1977**, *16*, 4907. [CrossRef]
87. Soulen, R.J., Jr.; Byers, J.M.; Osofsky, M.S.; Nadgorny, B.; Ambrose, T.; Cheng, S.F.; Broussard, P.R.; Tanaka, C.T.; Nowak, J.; Moodera, J.S.; et al. Measuring the spin polarization of a metal with a superconducting point contact. *Science* **1998**, *282*, 85–88. [CrossRef]
88. Barraud, C.; Seneor, P.; Mattana, R.; Fusil, S.; Bouzehouane, K.; Deranlot, C.; Graziosi, P.; Hueso, L.; Bergenti, I.; Dediu, V.; et al. Unravelling the role of the interface for spin injection into organic semiconductors. *Nat. Phys* **2010**, *6*, 615–620. [CrossRef]
89. Jiang, J.S.; Pearson, J.E.; Bader, S.D. Absence of spin transport in the organic semiconductor Alq_3. *Phys. Rev. B* **2008**, *77*, 035303. [CrossRef]
90. Metzelaars, M.; Schleicher, S.; Hattori, T.; Borca, B.; Matthes, F.; Sanz, S.; Bürgler, D.E.; Rawson, J.; Schneider, C.M.; Kögerler, P. Cyclophane with eclipsed pyrene units enables construction of spin interfaces with chemical accuracy. *Chem. Sci.* **2021**, *12*, 8430–8437. [CrossRef]
91. Schleicher, S.; Borca, B.; Rawson, J.; Matthes, F.; Bürgler, D.E.; Kögerler, P.; Schneider, C.M. Ultra-high vacuum deposition of pyrene molecules on metal surfaces. *Phys. Status Solidi B* **2018**, *255*, 1800235. [CrossRef]
92. Atodiresei, N.; Brede, J.; Lazić, P.; Caciuc, V.; Hoffmann, G.; Wiesendanger, R.; Blügel, S. Design of the local spin polarization at the organic-ferromagnetic interface. *Phys. Rev. Lett.* **2010**, *105*, 066601. [CrossRef] [PubMed]
93. Raman, K.V.; Kamerbeek, A.M.; Mukherjee, A.; Atodiresei, N.; Sen, T.K.; Lazić, P.; Caciuc, V.; Michel, R.; Stalke, D.; Mandal, S.K.; et al. Interface-engineered templates for molecular spin memory devices. *Nature* **2013**, *493*, 509–513. [CrossRef]
94. Lennartz, M.C.; Caciuc, V.; Atodiresei, N.; Karthäuser, S.; Blügel, S. Electronic mapping of molecular orbitals at the molecule-metal interface. *Phys. Rev. Lett.* **2010**, *105*, 066801. [CrossRef] [PubMed]

95. Michnowicz, T.; Borca, B.; Pétuya, R.; Schendel, V.; Pristl, M.; Pentegov, I.; Kraft, U.; Klauk, H.; Wahl, P.; Mutombo, P.; et al. Controlling single molecule conductance by a locally induced chemical reaction on individual thiophene units. *Angew. Chem. Int. Ed.* **2020**, *59*, 6207–6212. [CrossRef] [PubMed]
96. Borca, B.; Michnowicz, T.; Aguilar-Galindo, F.; Pétuya, R.; Pristl, M.; Schendel, V.; Pentegov, I.; Kraft, U.; Klauk, H.; Wahl, P.; et al. Chiral and catalytic effects of site-specific molecular adsorption. *J. Phys. Chem. Lett.* **2023**, *14*, 2072–2077. [CrossRef] [PubMed]
97. Su, T.A.; Neupane, M.; Steigerwald, M.L.; Venkataraman, L.; Nuckolls, C. Chemical principles of single-molecule electronics. *Nat. Rev. Mater.* **2016**, *1*, 16002. [CrossRef]
98. Ando, K.; Watanabe, S.; Mooser, S.; Saitoh, E.; Sirringhaus, H. Solution-processed organic spin–charge converter. *Nat. Mater.* **2013**, *12*, 622–627. [CrossRef] [PubMed]
99. Jiang, S.W.; Liu, S.; Wang, P.; Luan, Z.Z.; Tao, X.D.; Ding, H.F.; Wu, D. Exchange-dominated pure spin current transport in Alq_3 molecules. *Phys. Rev. Lett.* **2015**, *115*, 086601. [CrossRef] [PubMed]
100. Li, Z.; Li, T.; Qi, D.-C.; Tong, W.; Xu, L.; Zhu, J.; Zhang, Z.; Xu, H.; Zhang, W.; Guo, Y.; et al. Quantitative study of spin relaxation in rubrene thin films by inverse spin Hall effect. *Appl. Phys. Lett.* **2019**, *115*, 053301. [CrossRef]
101. Jiang, J.S.; Pearson, J.E.; Bader, S.D. Direct determination of energy level alignment and charge transport at metal−Alq_3 interfaces via ballistic-electron-emission spectroscopy. *Phys. Rev. Lett.* **2011**, *106*, 156807. [CrossRef]
102. Gobbi, M.; Bedoya-Pinto, A.; Golmar, F.; Llopis, R.; Casanova, F.; Hueso, L.E. C_{60}-based hot-electron magnetic tunnel transistor. *Appl. Phys. Lett.* **2012**, *101*, 102404. [CrossRef]
103. Gobbi, M.; Pietrobon, L.; Atxabal, A.; Bedoya-Pinto, A.; Sun, X.; Golmar, F.; Llopis, R.F.; Casanova, F.; Hueso, L.E. Determination of energy level alignment at metal/molecule interfaces by in-device electrical spectroscopy. *Nat. Commun.* **2014**, *5*, 4161. [CrossRef] [PubMed]
104. Cinchetti, M.; Heimer, K.; Wüstenberg, J.P.; Andreyev, O.; Bauer, M.; Lach, S.; Ziegler, C.; Gao, Y.; Aeschlimann, M. Determination of spin injection and transport in a ferromagnet/organic semiconductor heterojunction by two-photon photoemission. *Nat. Mater.* **2009**, *8*, 115–119. [CrossRef] [PubMed]
105. Ray, K.; Ananthavel, S.P.; Waldeck, D.H.; Naaman, R. Asymmetric scattering of polarized electrons by organized organic films of chiral molecules. *Science* **1999**, *283*, 814–816. [CrossRef] [PubMed]
106. Safari, M.R.; Matthes, F.; Schneider, C.M.; Ernst, K.-H.; Bürgler, D.E. Spin-selective electron transport through single chiral molecules. *Small* **2023**, 2308233. [CrossRef] [PubMed]
107. Safari, M.R.; Matthes, F.; Caciuc, V.; Atodiresei, N.; Schneider, C.M.; Ernst, K.-H.; Bürgler, D.E. Enantioselective adsorption on magnetic surfaces. *Adv. Mater.* **2024**, *36*, 2308666. [CrossRef]
108. Safari, M.R.; Matthes, F.; Ernst, K.-H.; Bürgler, D.E.; Schneider, C.M. Deposition of chiral heptahelicene molecules on ferromagnetic Co and Fe thin-film substrates. *Nanomaterials* **2022**, *12*, 3281. [CrossRef] [PubMed]
109. Jia, L.; Wang, C.; Zhang, Y.; Yang, L.; Yan, Y. Efficient spin selectivity in self-sssembled superhelical conducting polymer microfibers. *ACS Nano* **2020**, *14*, 6607–6615. [CrossRef] [PubMed]
110. Garnica, M.; Stradi, D.; Barja, S.; Calleja, F.; Díaz, C.; Alcamí, M.; Martín, N.; Vázquez de Parga, A.L.; Martín, F.; Miranda, R. Long-range magnetic order in a purely organic 2D layer adsorbed on epitaxial graphene. *Nat. Phys.* **2013**, *9*, 368–374. [CrossRef]
111. Garnica, M.; Calleja, F.; Vázquez de Parga, A.L.; Miranda, R. Mapping spin distributions in electron acceptor molecules adsorbed on nanostructured graphene by the Kondo effect. *Surf. Sci.* **2014**, *630*, 356–360. [CrossRef]
112. Mugarza, A.; Krull, C.; Robles, R.; Stepanow, S.; Ceballos, G.; Gambardella, P. Spin coupling and relaxation inside molecule–metal contacts. *Nat. Commun.* **2011**, *2*, 490. [CrossRef] [PubMed]
113. Jang, H.J.; Richter, C.A. Organic spin-valves and beyond: Spin injection and transport in organic semiconductors and the effect of interfacial engineering. *Adv. Mater.* **2017**, *29*, 1602739. [CrossRef]
114. Jiang, A.Q.; Wang, C.; Jin, K.J.; Liu, X.B.; Scott, J.F.; Hwang, C.S.; Tang, T.A.; Lu, H.B.; Yang, G.Z. A resistive memory in semiconducting BiFeO3 thin-film capacitors. *Adv. Mater.* **2011**, *23*, 1277–1281. [CrossRef] [PubMed]
115. Blom, P.W.M.; Wolf, R.M.; Cillessen, J.F.M.; Krijn, M.P.C.M. Ferroelectric Schottky diode. *Phys. Rev. Lett.* **1994**, *73*, 2107. [CrossRef] [PubMed]
116. Garcia, V.; Bibes, M. Ferroelectric tunnel junctions for information storage and processing. *Nat. Commun.* **2014**, *5*, 4289. [CrossRef]
117. Yamada, H.; Garcia, V.; Fusil, S.; Boyn, S.; Marinova, M.; Gloter, A.; Xavier, S.; Grollier, J.; Jacquet, E.; Carrétéro, C.; et al. Giant electroresistance of super-tetragonal $BiFeO_3$-based ferroelectric tunnel junctions. *ACS Nano* **2013**, *7*, 5385–5390. [CrossRef] [PubMed]
118. Fang, H.; Wang, J.; Nie, F.; Zhang, N.; Yu, T.; Zhao, L.; Shi, C.; Zhang, P.; He, B.; Lü, W.; et al. Giant electroresistance in ferroelectric tunnel junctions via high-throughput designs: Toward high-performance neuromorphic computing. *ACS Appl. Mater. Interfaces* **2024**, *16*, 1015–1024. [CrossRef]
119. Tian, B.B.; Wang, J.L.; Fusil, S.; Liu, Y.; Zhao, X.L.; Sun, S.; Shen, H.; Lin, T.; Sun, J.L.; Duan, C.G.; et al. Tunnel electroresistance through organic ferroelectrics. *Nat. Commun.* **2016**, *7*, 11502. [CrossRef]
120. Majumdar, S.; Chen, B.; Qin, Q.H.; Majumdar, H.S.; van Dijken, S. Electrode dependence of tunneling electroresistance and switching stability in organic ferroelectric P(VDF-TrFE)-based tunnel junctions. *Adv. Funct. Mater.* **2018**, *28*, 1703273. [CrossRef]
121. Majumdar, S.; Tan, H.; Qin, Q.H.; van Dijken, S. Energy-efficient organic ferroelectric tunnel junction memristors for neuromorphic computing. *Adv. Electron. Mater.* **2019**, *5*, 1800795. [CrossRef]

122. He, S.; Guo, M.; Dan, Z.; Lan, S.; Ren, W.; Zhou, L.; Wang, Y.; Liang, Y.; Zheng, Y.; Pan, J.; et al. Large-area atomic-smooth polyvinylidene fluoride Langmuir-Blodgett film exhibiting significantly improved ferroelectric and piezoelectric responses. *Sci. Bull.* **2021**, *66*, 1080–1090. [CrossRef]
123. Tu, L.; Yuan, S.; Xu, J.; Yang, K.; Wang, P.; Cui, X.; Zhang, X.; Wang, J.; Zhan, Y.-Q.; Zheng, L.-R. A wide-range operating synaptic device based on organic ferroelectricity with low energy consumption. *RSC Adv.* **2018**, *8*, 26549. [CrossRef]
124. Kim, K.L.; Lee, W.; Hwang, S.K.; Joo, S.H.; Cho, S.M.; Song, G.; Cho, S.H.; Jeong, B.; Hwang, I.; Ahn, J.-H.; et al. Epitaxial growth of thin ferroelectric polymer films on graphene layer for fully transparent and flexible nonvolatile memory. *Nano Lett.* **2016**, *16*, 334–340. [CrossRef]
125. Cheng, L.; Sun, H.; Xu, J.; Yu, C.; Xiao, H.; Wang, R.; Xu, L.; Zeng, Z.; Liang, S. Emulation of synaptic behavior by organic ferroelectric tunnel junctions. *Phys. Lett. A* **2021**, *392*, 127138. [CrossRef]
126. Horiuchi, S.; Tokura, Y. Organic ferroelectrics. *Nat. Mater* **2008**, *7*, 357–366. [CrossRef]
127. Tayi, A.S.; Kaeser, A.; Matsumoto, M.; Aida, T.; Stupp, S.I. Supramolecular ferroelectrics. *Nat. Chem.* **2015**, *7*, 281–294. [CrossRef]
128. Zhang, W.; Xiong, R.-G. Ferroelectric metal–organic frameworks. *Chem. Rev.* **2012**, *112*, 1163–1195. [CrossRef]
129. Goldsmith, G.J.; White, J.G. Ferroelectric behavior of thiourea. *J. Chem. Phys.* **1959**, *31*, 1175–1187. [CrossRef]
130. Kamishima, Y.; Akishige, Y.; Hashimoto, M. Ferroelectricity activity on organic crystal trichloroacetamide. *J. Phys. Soc. Jpn.* **1991**, *60*, 2147–2150. [CrossRef]
131. Bordeaux, D.; Bornarel, J.; Capiomont, A.; Lajzerowicz-Bonneteau, J. New ferroelastic-ferroelectric compound: Tanane. *Phys. Rev. Lett.* **1973**, *31*, 314–317. [CrossRef]
132. Kroupa, J.; Vaněk, P.; Krupkova, R.; Zikmund, Z. Dielectric and optical properties of weak ferroelectric cyclohexan-1,1′-diacetic acid. *Ferroelectrics* **1997**, *202*, 229–234. [CrossRef]
133. Szklarz, P.; Bator, G. Pyroelectric properties of tricyclohexylmethanol (TCHM) single crystal. *J. Phys. Chem. Solids* **2005**, *66*, 121–125. [CrossRef]
134. Terauchi, H.; Kojima, T.; Sakaue, K.; Tajiri, F.; Maeda, H. Structural phase transition of benzil. *J. Chem. Phys.* **1982**, *76*, 612–615. [CrossRef]
135. Gruner-Bauer, P.; Dormann, E. The ferroelectric low-temperature phase of single crystals of the substituted diacetylene 1,6-bis(2,4-dinitrophenoxy)-2,4-hexadiyne (DNP). *J. Phys. Condens. Matter.* **1992**, *4*, 5599–5609. [CrossRef]
136. Feder, J. Two-dimensional ferroelectricity. *Ferroelectrics* **1976**, *12*, 71–84. [CrossRef]
137. Horiuchi, S.; Kumai, R.; Tokura, Y. Hydrogen-bonding molecular chains for high-temperature ferroelectricity. *Adv. Mater.* **2011**, *23*, 2098–2103. [CrossRef]
138. Horiuchi, S.; Tokunaga, Y.; Giovannetti, G.; Picozzi, S.; Itoh, H.; Shimano, R.; Kumai, R.; Tokura, Y. Above-room-temperature ferroelectricity in a single-component molecular crystal. *Nature* **2010**, *463*, 789–792. [CrossRef]
139. Horiuchi, S.; Kagawa, F.; Hatahara, K.; Kobayashi, K.; Kumai, R.; Murakami, Y.; Tokura, Y. Above-room-temperature ferroelectricity and antiferroelectricity in benzimidazoles. *Nat. Commun.* **2012**, *3*, 1308. [CrossRef]
140. Dutta, S.; Vikas; Yadav, A.; Boomishankar, R.; Bala, A.; Kumar, V.; Chakraborty, T.; Elizabeth, S.; Munshi, P. Record-high thermal stability achieved in a novel single-component all-organic ferroelectric crystal exhibiting polymorphism. *Chem. Commun.* **2019**, *55*, 9610–9613. [CrossRef]
141. Liao, W.-Q.; Deng, B.-B.; Wang, Z.-X.; Cheng, T.-T.; Hu, Y.-T.; Cheng, S.-P.; Xiong, R.-G. Optically induced ferroelectric polarization switching in a molecular ferroelectric with reversible photoisomerization. *Adv. Sci.* **2021**, *8*, 2102614. [CrossRef]
142. Socol, M.; Trupina, L.; Galca, A.-C.; Chirila, C.; Stan, G.E.; Vlaicu, A.-M.; Stanciu, A.E.; Boni, A.G.; Botea, M.; Stanculescu, A.; et al. Electro-active properties of nanostructured films of cytosine and guanine nucleobases. *Nanotechnology* **2021**, *32*, 415702. [CrossRef]
143. Zelenovskii, P.S.; Vasileva, D.S.; Vasilev, S.G.; Kopyl, S.; Kholkin, A. Ferroelectricity in glycine: A mini-review. *Front. Mater.* **2022**, *9*, 918890. [CrossRef]
144. Bdikin, I.; Heredia, A.; Neumayer, S.M.; Bystrov, V.S.; Gracio, J.; Rodriguez, B.J.; Kholkin, A.L. Local piezoresponse and polarization switching in nucleobase thymine microcrystals. *J. Appl. Phys.* **2015**, *118*, 072007. [CrossRef]
145. Gan, Z.; Wu, X.; Zhu, X.; Shen, J. Light-induced ferroelectricity in bioinspired self-assembled diphenylalanine nanotubes/microtubes. *Angew. Chem. Int. Ed.* **2013**, *52*, 2055–2059. [CrossRef] [PubMed]
146. Lovinger, A.J. Ferroelectric polymers. *Science* **1983**, *220*, 1115–1121. [CrossRef] [PubMed]
147. Noda, K.; Ishida, K.; Kubono, A.; Horiuchi, T.; Yamada, H.; Matsushige, K. Remanent polarization of evaporated films of vinylidene fluoride oligomers. *J. Appl. Phys.* **2003**, *93*, 2866–2870. [CrossRef]
148. Furukawa, T. Ferroelectric properties of vinylidene fluoride copolymers. *Phase Transit.* **1989**, *18*, 143–211. [CrossRef]
149. He, X.; Yao, K.; Gan, B.K. Phase transition and properties of a ferroelectric poly(vinylidene fluoride-hexafluoropropylene) copolymer. *J. Appl. Phys.* **2005**, *97*, 084101. [CrossRef]
150. Katz, D.; Gelfandbein, V. Ferroelectric behaviour of α-nylon 11. *J. Phys. D Appl. Phys.* **1982**, *15*, L115. [CrossRef]
151. Yanakaa, A.; Sakai, W.; Kinashi, K.; Tsutsum, N. Ferroelectric performance of nylons 6–12, 10–12, 11–12, and 12–12. *RSC Adv.* **2020**, *10*, 15740–15750. [CrossRef]
152. Bahgat, A.A.; Sayyah, S.M.; Abd-Elsalam, H.M. Study of ferroelectricity in polyaniline. *Int. J. Polym. Mater.* **2003**, *52*, 499–515. [CrossRef]

153. Okamoto, H.; Mitani, T.; Tokura, Y.; Koshihara, S.; Komatsu, T.; Iwasa, Y.; Koda, T.; Saito, G. Anomalous dielectric response in tetrathiafulvalene-p-chloranil as observed in temperature- and pressure-induced neutral-to-ionic phase transition. *Phys. Rev. B* **1991**, *43*, 8224–8232. [CrossRef] [PubMed]
154. Horiuchi, S.; Okimoto, Y.; Kumai, R.; Tokura, Y. Anomalous valence fluctuation near a ferroelectric transition in an organic charge-transfer complex. *J. Phys. Soc. Jpn* **2000**, *69*, 1302–1305. [CrossRef]
155. Horiuchi, S.; Ishii, F.; Kumai, R.; Okimoto, Y.; Tachibana, H.; Nagaosa, N.; Tokura, Y. Ferroelectricity near room temperature in co-crystals of nonpolar organic molecules. *Nat. Mater.* **2005**, *4*, 163–166. [CrossRef] [PubMed]
156. Horiuchi, S.; Kumai, R.; Tokura, Y. A supramolecular ferroelectric realized by collective proton transfer. *Angew. Chem. Int. Ed.* **2007**, *46*, 3497–3501. [CrossRef]
157. Szafrański, M.; Katrusiak, A.; McIntyre, G.J. Ferroelectric order of parallel bistable hydrogen bonds. *Phys. Rev. Lett.* **2002**, *89*, 215507. [CrossRef]
158. Hoshino, S.; Mitsui, T.; Jona, F.; Pepinsky, R. Dielectric and thermal study of tri-glycine sulfate and tri-glycine fluoberyllate. *Phys. Rev.* **1957**, *107*, 1255–1258. [CrossRef]
159. Khanduri, P.C.; Upadhyay, T.C. Study of ferroelectric phase transition in triglycine fluoberyllate crystal. *Ferroelectrics* **2024**, *618*, 375–384. [CrossRef]
160. Levstik, A.; Filipič, C.; Blinc, R. Critical exponents of ferroelectric TSCC determined from dielectric measurements. *Solid State Commun.* **1976**, *18*, 1231–1234. [CrossRef]
161. Valasek, J. Piezo-electric and allied phenomena in Rochelle salt. *Phys. Rev.* **1921**, *17*, 475–481. [CrossRef]
162. Yang, Y.; Xi, Z.; Dong, Y.; Zheng, C.; Hu, H.; Li, X.; Jiang, Z.; Lu, W.-C.; Wu, D.; Wen, Z. Spin-filtering ferroelectric tunnel junctions as multiferroic synapses for neuromorphic computing. *ACS Appl. Mater. Interfaces* **2020**, *12*, 56300–56309. [CrossRef] [PubMed]
163. Liang, S.; Yang, H.; Yang, H.; Tao, B.; Djeffal, A.; Chshiev, M.; Huang, W.; Li, X.; Ferri, A.; Desfeux, R.; et al. Ferroelectric control of organic/ferromagnetic spinterface. *Adv. Mater.* **2016**, *28*, 10204–10210. [CrossRef] [PubMed]
164. Zhang, X.; Tong, J.; Zhu, H.; Wang, Z.; Zhou, L.; Wang, S.; Miyashita, T.; Mitsuishi, M.; Qin, G. Room temperature magnetoresistance effects in ferroelectric poly(vinylidene fluoride) spin valves. *J. Mater. Chem. C* **2017**, *5*, 5055. [CrossRef]
165. Xiao, C.; Sun, H.; Cheng, L.; Devaux, X.; Ferri, A.; Huang, W.; Desfeux, R.; Li, X.-G.; Migot, S.; Chshiev, M.; et al. Temperature dependence of transport mechanisms in organic multiferroic tunnel junctions. *J. Phys. D Appl. Phys.* **2020**, *53*, 325301. [CrossRef]
166. Subedi, R.C.; Geng, R.; Luong, H.M.; Huang, W.; Li, X.; Hornak, L.A.; Nguyen, T.D. Large magnetoelectric effect in organic ferroelectric copolymer-based multiferroic tunnel junctions. *Appl. Phys. Lett.* **2017**, *110*, 053302. [CrossRef]
167. Liang, S.; Yu, Z.; Devaux, X.; Ferri, A.; Huang, W.; Yang, H.; Desfeux, R.; Li, X.; Migot, S.; Chaudhuri, D.; et al. Quenching of spin polarization switching in organic multiferroic tunnel junctions by ferroelectric "ailing-channel" in organic barrier. *ACS Appl. Mater. Interfaces* **2018**, *10*, 30614–30622. [CrossRef]
168. Yin, Y.; Jiang, X.; Koten, M.A.; Shield, J.E.; Chen, X.; Yun, Y.; N'Diaye, A.T.; Hong, X.; Xu, X. Spin rectification and electrically controlled spin transport in molecular-ferroelectrics-based spin valves. *Phys. Rev. Appl.* **2020**, *13*, 064011. [CrossRef]
169. Hu, Z.; Zhang, L.; Chakraborty, A.; D'Olimpio, G.; Fujii, J.; Ge, A.; Zhou, Y.; Liu, C.; Agarwal, A.; Vobornik, I.; et al. Terahertz nonlinear Hall rectifiers based on spin-polarized topological electronic states in 1T-$CoTe_2$. *Adv. Mater.* **2023**, *35*, 2209557. [CrossRef]
170. Jugovac, M.; Cojocariu, I.; Sánchez-Barriga, J.; Gargiani, P.; Valvidares, M.; Feyer, V.; Blügel, S.; Bihlmayer, G.; Perna, P. Inducing single spin-polarized flat bands in monolayer graphene. *Adv. Mater.* **2023**, *35*, 2301441. [CrossRef]
171. Chakraborty, A.; Fujii, J.; Kuo, C.-N.; Lue, C.S.; Politano, A.; Vobornik, I.; Agarwal, A. Observation of highly anisotropic bulk dispersion and spin-polarized topological surface states in $CoTe_2$. *Phys. Rev. B* **2023**, *107*, 085406. [CrossRef]
172. Muñiz Cano, B.; Ferreiros, Y.; Pantaleón, P.A.; Dai, J.; Tallarida, M.; Figueroa, A.I.; Marinova, V.; García-Díez, K.; Mugarza, A.; Valenzuela, S.O.; et al. Experimental demonstration of a magnetically induced warping transition in a topological insulator mediated by rare-earth surface dopants. *Nano Lett.* **2023**, *23*, 6249–6258. [CrossRef]
173. Vobornik, I.; Sarkar, A.B.; Zhang, L.; Boukhvalov, D.W.; Ghosh, B.; Piliai, L.; Kuo, C.-N.; Mondal, D.; Fujii, J.; Lue, C.S.; et al. Kitkaite NiTeSe, an ambient-stable layered Dirac semimetal with low-energy type-II Fermions with application capabilities in spintronics and optoelectronics. *Adv. Funct. Mater.* **2021**, *31*, 2106101. [CrossRef]
174. Jugovac, M.; Cojocariu, I.; Brondin, C.A.; Crotti, A.; Petrović, M.; Bonetti, S.; Locatelli, A.; Menteş, T.O. Coupling borophene to graphene in air-stable heterostructures. *Adv. Electron. Mater.* **2023**, *9*, 2300136. [CrossRef]
175. Fujii, J.; Ghosh, B.; Vobornik, I.; Sarkar, A.B.; Mondal, D.; Kuo, C.-N.; Bocquet, F.C.; Zhang, L.; Boukhvalov, D.W.; Lue, C.S.; et al. Mitrofanovite Pt_3Te_4: A topological metal with termination-dependent surface band structure and strong spin polarization. *ACS Nano* **2021**, *15*, 14786–14793. [CrossRef] [PubMed]
176. Decker, R.; Schlickum, U.; Klappenberger, F.; Zoppellaro, G.; Klyatskaya, S.; Ruben, M.; Barth, J.V.; Brune, H. Using metal-organic templates to steer the growth of Fe and Co nanoclusters. *Appl. Phys. Lett.* **2008**, *93*, 243102. [CrossRef]
177. Barja, S.; Stradi, D.; Borca, B.; Garnica, M.; Díaz, C.; Rodriguez-García, J.M.; Alcamí, M.; Vázquez de Parga, A.L.; Martín, F.; Miranda, R. Ordered arrays of metal–organic magnets at surfaces. *J. Phys. Condens. Matter.* **2013**, *25*, 484007. [CrossRef]
178. Martín-Fuentes, C.; Parreiras, S.O.; Urgel, J.I.; Rubio-Giménez, V.; Muñiz Cano, B.; Moreno, D.; Lauwaet, K.; Valvidares, M.; Valbuena, M.A.; Gargiani, P.; et al. On-surface design of a 2D cobalt-organic network preserving large orbital magnetic moment. *J. Am. Chem. Soc.* **2022**, *144*, 16034–16041. [CrossRef] [PubMed]

179. Schlickum, U.; Klappenberger, F.; Decker, R.; Zoppellaro, G.; Klyatskaya, S.; Ruben, M.; Kern, K.; Brune, H.; Barth, J.V. Surface-confined metal−organic nanostructures from Co-directed assembly of linear terphenyl-dicarbonitrile linkers on Ag(111). *J. Phys. Chem. C* **2010**, *114*, 15602–15606. [CrossRef]
180. Cojocariu, I.; Carlotto, S.; Jugovac, M.; Floreano, L.; Casarin, M.; Feyer, V.; Schneider, C.M. Distortion-driven spin switching in electron-doped metal porphyrins. *J. Mater. Chem. C* **2022**, *10*, 9748–9757. [CrossRef]
181. Stradi, D.; Borca, B.; Barja, S.; Garnica, M.; Díaz, C.; Rodríguez-García, J.M.; Alcamí, M.; Vázquez de Parga, A.L.; Miranda, R.; Martín, F. Understanding the self-assembly of TCNQ on Cu(111): A combined study based on scanning tunnelling microscopy experiments and density functional theory simulations. *RSC Adv.* **2016**, *6*, 15071–15079. [CrossRef]
182. Vijayaraghavan, S.; Auwärter, W.; Ecija, D.; Seufert, K.; Rusponi, S.; Houwaart, T.; Sautet, P.; Bocquet, M.-L.; Thakur, P.; Stepanow, S.; et al. Restoring the Co magnetic moments at interfacial Co-porphyrin arrays by site-selective uptake of iron. *ACS Nano* **2015**, *9*, 3605–3616. [CrossRef]
183. Ballav, N.; Wäckerlin, C.; Siewert, D.; Oppeneer, P.M.; Jung, T.A. Emergence of on-surface magnetochemistry. *J. Phys. Chem. Lett.* **2013**, *4*, 2303–2311. [CrossRef]
184. Cuxart, M.G.; Valbuena, M.A.; Robles, R.; Moreno, C.; Bonell, F.; Sauthier, G.; Imaz, I.; Xu, H.; Nistor, C.; Barla, A.; et al. Molecular approach for engineering interfacial interactions in magnetic/topological insulator heterostructures. *ACS Nano* **2020**, *14*, 6285–6294. [CrossRef]
185. Bystrov, V.; Paramonova, E.; Meng, X.; Shen, H.; Wang, J.; Lin, T.; Fridkin, V. Ferroelectric thin films and composites based on polyvinylidene fluoride and graphene layers: Molecular dynamics study. *Coatings* **2024**, *14*, 356. [CrossRef]
186. Bychkov, I.; Belim, S.; Maltsev, I.; Shavrov, V. Phase transition and magnetoelectric effect in 2D ferromagnetic films on a ferroelectric substrate. *Coatings* **2021**, *11*, 1325. [CrossRef]
187. Chylarecka, D.; Kim, T.K.; Tarafder, K.; Müller, K.; Gödel, K.; Czekaj, I.; Wäckerlin, C.; Cinchetti, M.; Ali, M.E.; Piamonteze, C.; et al. Indirect magnetic coupling of manganese porphyrin to a ferromagnetic cobalt substrate. *J. Phys. Chem. C* **2011**, *115*, 1295–1301. [CrossRef]
188. Aryal, U.K.; Ahmadpour, M.; Turkovic, V.; Rubahn, H.-G.; Di Carlo, A.; Madsen, M. 2D materials for organic and perovskite photovoltaics. *Nano Energy* **2022**, *94*, 106833. [CrossRef]
189. Forouzandeh, P.; Pillai, S.C. Two-dimensional (2D) electrode materials for supercapacitors. *Mater. Today Proc.* **2021**, *41*, 498–505. [CrossRef]
190. Hao, G.; Mosey, A.; Jiang, X.; Yost, A.J.; Sapkota, K.R.; Wang, T.; Zhang, X.; Zhang, J.; N'Diaye, A.T.; Cheng, R.; et al. Nonvolatile voltage controlled molecular spin state switching. *Appl. Phys. Lett.* **2019**, *114*, 032901. [CrossRef]
191. Borca, B.; Bartha, C. Advances of nanoparticles and thin films. *Coatings* **2022**, *12*, 1138. [CrossRef]

MDPI AG
Grosspeteranlage 5
4052 Basel
Switzerland
Tel.: +41 61 683 77 34

Coatings Editorial Office
E-mail: coatings@mdpi.com
www.mdpi.com/journal/coatings